Messaging and Queuing
Using the MQI

Messaging and Queuing Using the MQI

Concepts & Analysis, Design & Development

Burnie Blakeley

Harry Harris

Rhys Lewis

McGraw-Hill, Inc.

New York   San Francisco   Washington, D.C.   Auckland   Bogotá
Caracas   Lisbon   London   Madrid   Mexico City   Milan
Montreal   New Delhi   San Juan   Singapore
Sydney   Tokyo   Toronto

Library of Congress Cataloging-in-Publication Data

Blakeley, Burnie.
    Messaging and queuing using the MQI : concepts & analysis, design
& development / Burnie Blakeley, Harry Harris, Rhys Lewis.
        p.   cm.
    Includes bibliographical references and index.
    ISBN 0-07-005730-3
    1. Application software.   2. MQI.   I. Harris, Harry.   II. Lewis,
Rhys.   III. Title.
QA76.76.A65B56   1995
005.7′1—dc20   95-10241
                                                                              CIP

    3 4 5 6 7 8 9 0   DOC/DOC 9 0 0 9 8 7 6

ISBN 0-07-005730-3

*The sponsoring editor for this book was Jerry Papke, and the
production supervisor was Pamela A. Pelton. This book was set in
Century Schoolbook by North Market Street Graphics.*

Printed and bound by R. R. Donnelley & Sons Company.

McGraw-Hill books are available at special quantity discounts to use as
premiums and sales promotions, or for use in corporate training programs. For more information, please write to the Director of Special
Sales, McGraw-Hill, Inc., 11 West 19th Street, New York, NY 10011. Or
contact your local bookstore.

To the MQSeries team,
for their immensely hard work
and dedication in creating this
new messaging and queuing technology
and making it available around the world.

Trademarks

Appletalk and Macintosh are trademarks or registered trademarks of Apple Computer Corporation.

DEC, DECnet, VAX, and VMS are trademarks or registered trademarks of Digital Equipment Corporation.

Encina is a trademark or registered trademark of Transarc Corporation.

ezBRIDGE and Transact are trademarks of Systems Strategies Incorporated.

Guardian and Nonstop are trademarks of Tandem Computers Incorporated.

IBM, SNA, AIX, RISC System/6000, Application System/400, AS/400, CICS, CICS/ESA, CICS/VSE, IMS, IMS/ESA, MQSeries, MVS/ESA, Operating System/2, OS/2, OS/400, Personal System/2, PS/2, System/370, System/390, and System 88 are trademarks or registered trademarks of International Business Machines Corporation in the USA and/or other countries.

OSF, Motif, OSF/1, DCE, and DME are trademarks or registered trademarks of the Open Software Foundation.

TCP/IP is a trademark or registered trademark of Defense Advanced Research Projects Administration.

UNIX and Tuxedo are trademarks or registered trademarks of AT&T Bell Laboratories.

Windows is a trademark of Microsoft Corporation.

X/Open is a trademark or registered trademark of X/Open Corporation.

Contents

Preface

Reading and Using This Book

This book serves as both a tutorial and a reference for messaging and queuing computer software. It presents both technical and business perspectives on the subject, as well as details about programming and products.

Messaging and queuing computer software is simple-to-use *middleware* that supports not only client/server computing but, indeed, all forms of distributed computing in a very efficient and effective manner.

Both applications-oriented and systems-oriented persons will find this book useful. Application developers (analysts, designers, programmers, business planners, and administrators), systems developers (platform writers, distributed services providers, communication programmers, network designers, and performance analysts), and others (teachers, systems integrators, independent software vendors, and students) will all find this book to be complete, accurate, easy to understand, and valuable. It is the first book of its kind in the computer industry.

A Quick Skimming

To maximize the value of this book to you, the reader, we recommend that you first skim the table of contents and then skim the chapters and the appendixes. You will notice that the material is divided into parts to match the subtitles of the book. Abundant illustrations complement the text to make reading easier and understanding quicker, especially for the more visually oriented reader.

Front-to-Back Detail Reading

The chapters of this book are arranged in a logical progression for easy digestion of information, from the front cover to the back cover. Topics are introduced and expounded upon, one by one.

The first part of the book introduces the concepts of messaging (a short term used henceforth throughout this book to signify combined messaging and queuing). The second part of the book deals with understanding messaging principles. The third and fourth parts deal with using messaging products. Thus, the book gives you not only messaging concepts and principles for application planning and development, but also information on how you personally can immediately begin using the message queue interface (MQI) to access the many messaging products already available.

Later Reference

This book can also serve as a reference book if you begin by searching the extensive index for the topic of interest. You can then read about the selected topic of interest on each page that is referenced. This may lead you to additional reading material identified in the bibliography, provided as App. E.

Acknowledgments

We especially thank Mr. Tony Cripps of IBM UK Laboratories, Limited, in Hursley, Hampshire, England, for his many contributions to the material within this book and for his encouragement during its preparation.

We acknowledge and thank Dr. Barron Housel and Mr. John Hind, coauthors along with Burnie Blakeley of the original Message Queue Interface (MQI) specification of 1989.

We owe much to Mr. Rob Drew and Mr. Dick Dievendorff, both of whom were major contributors to the development of the IBM MQSeries product family (and to this book).

We appreciate the special help given to us by Mr. Steve Smith, Ms. Loretta Mirelli, Ms. Debbie Boyd, and others in collecting reference materials for Part 3 of the book. Ms. Gray Heffner did a splendid job creating most of the graphics in Part 3.

We value the MQI and MQSeries product material that Mr. Curt Bury and others contributed to this book. It is always useful to see information about available messaging and queuing products and how they can be used with applications and other software products. Dr. Paul Williams helped tremendously with his many new insights into message-driven processing (MDP) application design and development.

Most of all, we express our appreciation to the many members of the MQSeries team who planned, created, explained, and advertised messaging and queuing and to whom this book is dedicated.

And, of course, we thank our wives and families who patiently endured the inconvenience of our absences while we prepared this book.

Burnie Blakeley
Harry Harris
Rhys Lewis
1995

Welcome to *Messaging and Queuing* and the *Message Queue Interface*

By permission of Johnny Hart and Creators Syndicate, Inc.

Messaging and Queuing
Using the MQI

The Concepts

This part of the book introduces messaging and queuing concepts, after first giving a very brief history of and rationalization for messaging.

Messaging (and queuing) is not new; it is, in fact, a heavily used, tried-and-tested technique that is several decades old! The messaging technique has never been undiscovered, ignored, or unused; it has, instead, existed in an extremely wide variety of forms with no single, consolidated form naturally emerging until recently (the late 1980s). That is, we in the computer industry have been troubled not by the nonexistence of any forms of messaging but rather by the existence of too many forms. Unfortunately, many of these forms of messaging have been buried within operating systems or hidden within application software and not generally available to ordinary application programmers—but now uniform and easy messaging and queuing is available to the masses.

This part of the book describes:

- *The beginnings of messaging*
- *The purposes of messaging*
- *Messaging concepts*
- *Queuing concepts*

If you are already a little bit familiar with messaging, you may wish to either skim or skip the first two chapters.

Whenever he wanted directions he sent a message, or note with neither heading nor signature, to which she was obliged to reply in the same off-hand style.

THOMAS HARDY
Far From the Madding Crowd

The Beginnings of Messaging

Messaging had its beginnings (in the computer industry) many decades ago. This chapter describes the early beginnings of messaging.

1.1 Early Messaging (The 1960s)

Messaging and queuing is not new to the computer industry. In fact, you could consider that messaging in one form or another has been with computing since the start of commercial operations.

1.1.1 Batch operations

In the early 1960s, the punched card could be considered the message and the tray of cards could be considered the queue of messages. These trays of cards were loaded into a card reader, read into the system, and usually operated on by a batch program that was previously loaded from the program library into the processor, as shown in Fig. 1.1. The processor executed the batch program, which read and/or wrote data on a storage device before producing a printed report.

The cards were the source of input data for the program. So we had discrete elements of work, controlled by the arrival of an external set of information (i.e., a tray of cards). These were the elements of a message-based system. However, at that time this was all the systems were doing. The system could be considered a single-user system with the batch processor having control of all the required resources. This meant that there was no need to synchronize changes; security was relatively simple and, when the changes were applied, there was no danger of them being removed.

What we were primarily concerned with here was the input and output of information to and from a system. The method used to store it in

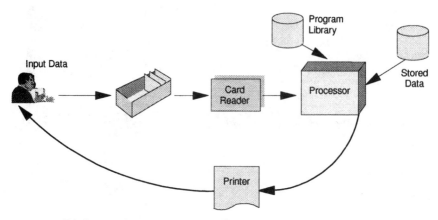

Figure 1.1 Batch operations.

the system did not concern us. With a card/batch system, the information, as already mentioned, was held on punched cards. These cards had the information punched into them remotely from the computer system and could contain either programs or data. Normally the programs were kept separate from the data, but it was possible to have the deck of cards contain both the program and the data.

The general form of operation was that the input data was prepared remotely from the processing system and processed in a batched manner that handled all the input at one time. The input data was submitted as a deck of cards with a header containing control language that defined the programs to be executed, where intermediate storage was, and where the output was to be directed. This led to a structure that was contained in the messaging systems: a *header,* which contained process information, followed by data.

This system was good for handling large amounts of data and for allowing the data, once it had been processed, to be accessed by other users. The drawback was that data was prepared remotely from the system; therefore, it had to be validated during a separate step. Also, it meant that other users could not have rapid access to the updated data.

1.1.2 Clerical input

Many messages are the result of clerically creating large amounts of input.

1.1.2.1 Record-oriented messages.   What we have with a batch operation is a system that is ideally suited to handle a large amount of data that are prepared remotely. Messages are batched into very large queues of message records for periodic processing. However, the keypunch machine that was generally used to prepare the cards had several drawbacks. The data had to be verified by a separate step, away from the system. Also, decks of cards were vulnerable to damage by mishandling, and they were bulky. This led to methods of improving the way the data was prepared and submitted to the systems.

Key-to-tape improved on the cardpunch by allowing the operator to input the data directly onto a ½-in magnetic tape that would be read by the system. This improved the speed of data input and allowed the verification and correction of the data to be done in the same step, by holding the data in a small amount of memory before writing it to tape. It also improved on the durability of the data; tapes were more reliable and took up less space. However, it still did not give the end user, the person preparing the data, direct access to the system. Nor did it allow more than one operator access to the tape at any one time.

This multiple access came with the *key-to-disk* and *key-to-diskette* workstations. Here the data was keyed onto a disk or removable diskette; the information was validated by a process in the station and then transferred to tape for access by the system.

So, during the 1960s, we had large enterprise systems based on batch operations with large numbers of data operators separated from the system. Some of these batch operations are still playing an important role in many enterprises.

1.1.2.2 Screen-oriented messages. During the 1960s, the *video display unit* (VDU) was developed. See Fig. 1.2. Here the data was held in volatile storage in the device and was displayed on a screen before being sent to the system for processing. This removed the need for an intermediate transfer medium. It also allowed the user to easily see and correct mistakes at the same station before sending the data to the system for processing.

The VDU dramatically changed the way systems were used by eventually giving the end user direct access to the programs needed to act on the input data. This provided far better access to data, but put strains on the system that existing processing could not handle, and so we moved into the world of *transaction processing*.

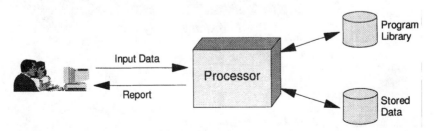

Figure 1.2 Video display unit (VDU) operations.

1.2 Growing Awareness (The 1970s and 1980s)

With the growth of communications networks, larger processing power, and direct access storage devices (DASD), there was an explosive growth of applications made available to computer users. These were largely based on the *transaction monitor*. The most significant examples of these were IMS™ and CICS™. A transaction monitor controls a sequence of messages that constitute a transaction. For many years, these systems dominated enterprise computing and they still play a major role in the systems of today.

1.2.1 Information Management System (IMS)

IMS is one of the dominant data communication and database systems in use today. Its roots go back to an inventory tracking system used for the U.S. moon landing and it was used in the aerospace industry in general. It provides the ability to update online indexed data files.

In the early days this was considered very advanced because, up until then, access to files had been via local or remote card reader/punches and batch jobs.

IMS uses the *queued transaction* approach. In this model, a message arrives containing a transaction code, the name of the process to be executed, and the data that the process is to use. These messages are processed by IMS and the data is placed on a *queue*. Each of these input queues represents a service, and all messages on a queue are similar. These messages are then processed asynchronously. It is the queue that provides the interface between the application and the services provided by the IMS subsystem, as shown in Fig. 1.3.

A transaction processing system is a style of computing that supports interactive applications in which requests submitted by terminal users are processed as soon as they are received. Results are returned

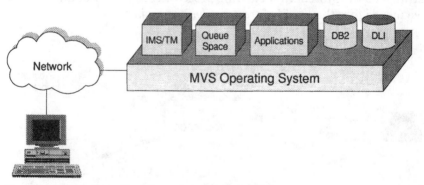

Figure 1.3 IBM Information Management System (IMS).

to the requestor in a relatively short period of time. A transaction processing system supervises the sharing of resources for processing multiple transactions at the same time. A transaction system can be considered as a system that provides ACID properties that are available to applications. ACID properties can be defined as:

Atomicity	The changes in state caused by a transaction either "all happen" or "none happen." These include changes to resources under the control of another resource manager but affected by the transaction.
Consistency	A transaction produces the same changes in state for each of its instances; that is, the changes are predictable and repeatable.
Isolation	Transactions may execute concurrently but the results must appear as if the execution is completely serial. There must be no overlapping effects of each execution.
Durability	When a transaction has completed successfully and all the changes are committed, they must remain and survive subsequent failures.

IMS has two major parts:

IMS / TM *(transaction management)*	This was known as IMS/DC (data communications) until around 1990. This is the transaction monitor of the system. It carries out such services as scheduling, authorization, presentation services, and operating functions.
IMS / DB (Database)	Better known as DL/1, this is the database system supporting the hierarchical data model. Applications running under IMS usually access DL/1 but they can access other resource managers such as DB2 (database 2, a relational database manager).

IMS runs on the multiple virtual storage (MVS) operating system, and it takes the approach of using the operating systems services rather than replacing these services. This means that it uses the operating systems processes, its scheduling and protection, file systems, program management, etc. IMS provides the transaction services. It is these processes that have been expanded on and developed to allow IMS to grow with the demands of the users of these large systems. It has developed its performance and throughput to a stage where a typical system will handle multiple MVS systems, sharing terabytes of data from thousands of terminals. All this is done with recoverability, security, and acceptable performance.

Further information about the messaging and queuing nature of IMS can be found in Martin and Chapman (1987) (see Bibliography).

1.2.2   Customer Information Control System (CICS)

CICS is the most popular and prevalent transaction processing system in the computing industry. Unlike IMS, which uses the operating system functions, CICS makes minimal demands on the operating system and implements the functions within its own address space. This means, in MVS, that CICS and its applications run in their own address space. This has the advantage that CICS can provide operating system-type functions with much less overhead. It also means that the CICS application environment is easily ported to other operating systems. So we have CICS, not only on MVS but also on the virtual storage extended (VSE) and virtual machine (VM) operating systems on the ES/390 range as well as CICS/OS2, CICS/6000, and CICS/400. It is now being made available on many non-IBM platforms.

CICS is a transaction environment and so provides the ACID properties required. However, unlike IMS, it does not use the queued transaction approach. Instead, it gives the application direct access to the services it provides. The application executes in the CICS address space, sharing this space with the CICS system code and controlling access via the EXEC CICS interface, as shown in Fig. 1.4. Within this address space, queuing facilities are provided by a *transient data* (TD) control and *temporary storage* (TS) control. TD provides a generalized queuing facility. Data can be stored for subsequent internal (within CICS) or external processing.

Selected data, specified in the application, can be routed to or from predefined destinations. These queues are either:

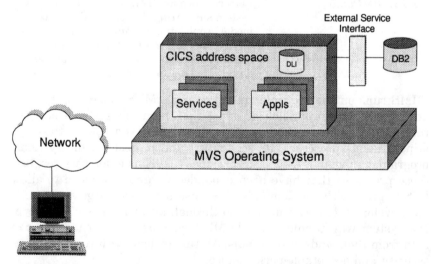

Figure 1.4   IBM Customer Information Control System (CICS).

Intrapartition   Associated with a facility within the CICS region

Extrapartition   Associated with a facility outside the CICS region

With intrapartition queues there is a facility, *automatic transaction initiation,* which allows for the automatic starting of CICS transactions. The transactions are initiated when a nonzero trigger level is accomplished.

Temporary storage provides the application with the ability to store data in temporary storage queues. These queues can be held either in main storage or as auxiliary on direct access storage devices.

Development of CICS has continued. Interfaces to other system resource managers have been produced and there has been more exploitation of operating system services. Initially, there was access to IMS DL/1. This has been extended to cover DB2, the security managers, and the cross-memory services.

Further information about the messaging and queuing nature of CICS can be found in Lamb (1993) Wipfler (1989) (see Bibliography).

1.2.3   Telecommunications Access Method (TCAM)

TCAM is an MVS subsystem that allows communication with *logical units* (LU) within a teleprocessing network. So why has it been included in a section that covers the introduction of messaging and queuing? This is because it provides the system with a method of handling messages and, in order to effectively process the in-balance of input and output performance, a queuing mechanism.

A TCAM system is a subsystem that is controlled by a single *message control program* (MCP) and consists of collections of LUs and applications. The LUs are in the terminals that are attached to the network. The TCAM subsystem is installed and operates under the control of an MVS operating system. TCAM is no longer a supported program product, as this function of controlling LUs has now been taken over by another program, Virtual Telecommunications Access Method (VTAM).

In a TCAM system, messages are transferred between LUs, which are part of the system, as shown in Fig. 1.5. TCAM schedules the transfer of messages and stores them on *destination queues.* A destination queue is a queue on which TCAM places messages that are bound for a particular LU, application program, the MCP, or a service program. The order in which the messages are sent to their destination depends on several variables:

- The relative order in which they were received at the destination queues

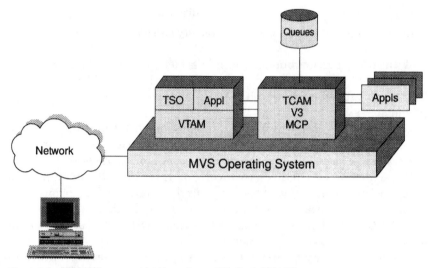

Figure 1.5   IBM Telecommunications Access Method (TCAM).

- What priorities were assigned to the messages
- The use of HOLD or LOCK macros within the MCP

These queues are held either in main storage or on disk or a combination of both. The choice between disk or main storage is made dependent on the storage available and the speed of access required. The messages consist of *user data* and *control data*. This control data is generated by both the application and TCAM.

As mentioned earlier, these messages are available not only to LUs but also to application programs. The application has access to these messages by a GET or READ operation. After the message has been sent to the application and has been processed, the response is placed on the queue by a PUT or WRITE operation.

It is these queued operations that have allowed applications to be developed using this communication technique. Many people consider TCAM to be the precursor to the messaging and queuing products that are available today.

Further information about the messaging and queuing nature of TCAM can be found in IBM GC30-3235 in the Bibliography.

1.2.4   Improving networks

During the 1970s and 1980s, the majority of applications were executed in a single computer with terminals attached, via local controllers and then by networks to the central processing unit (CPU) that executed the program. Networks were generally unreliable, so protocols were produced that could detect and inform the application of a network error. The correction and handling of the error was, in most cases, left up to the application. This added considerable overhead to the application and led to a hierarchical system structure with a strong central point of control that was capable of handling a large number of the errors in the network. It was in this environment that IBM's Systems Network Architecture (SNA) was developed. It was this architecture, with its defined layers, functions, and protocols, that enabled the growth of the workstation-based network. It later developed to allow for program-to-program communications.

Most communications were done within an enterprise, and there was little interenterprise communication at the system level. However, as the range of applications and the access devices became more sophisticated, demands were made on the network suppliers to improve speed and reliability. Also, the differences between WANs (wide area networks) and LANs (local area networks) were becoming apparent. Different sets of protocols and architectures were developed to handle the different circumstances within WANs and within LANs.

1.2.5   Local data centers

Early in the 1970s the price/performance of systems continued to improve, online storage continued to increase, and systems architecture improved. With the introduction of transactional subsystems, the dependency on accurate data became more acute. This led to the establishment of satellite (distributed) data centers, as shown in Fig. 1.6. Data was collected and processed locally. Then, during an off-shift period, the data was collected and shipped in bulk via tape to the main computer site. The primary driver for this was the lack of reliable, high-capacity, low-cost networks.

As these sites developed, specialized systems were developed to handle data entry, inquiry, and reporting functions. These systems addressed the needs of fast, cheap, and responsive data capture. They also offered a degree of resilience and unmanned operations that increased the apparent availability of the overall system. However, because of their nature, they were very specialized; thus, any change in the operations generally necessitated a change in hardware and operating systems.

1.2.6   Self-service devices

Generally at this stage of the computer's development, only trained DP staff or users knowledgeable in the application inputted information into the system. Computing devices such as early banking terminals

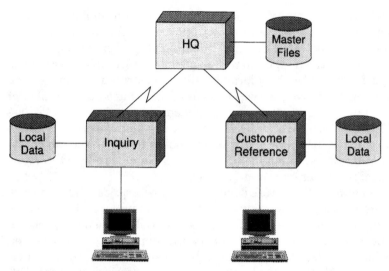

Figure 1.6 Local data centers.

were known as service-provided devices since the teller provided the service to the customer of the bank. But with the emergence of automatic teller machines, point-of-sale terminals, and the start of home banking with Videotext, the general, untrained public now had access to the applications and computers. Customers now had access to banking and other self-service devices. Thus, the quest for more user-friendly systems started. This increased access likewise increased the need for security, and introduced the use of the personal identification number (PIN). Also, there was a need for greater access to the system, 7 days a week, 24 hours a day. This allowed little time to apply maintenance and recover data.

1.2.6.1 Personal computers (PCs).   Into this scene of growing communications, more use of computers by the general public, and the rapid collection of data, appeared the personal computer.

Up until now, computer systems had been large and expensive, and were generally the domain of large industry. But now the use and control of the power of the medium-size computers were made generally available. Everybody could collect and share their own data. Each person could now be considered to be an enterprise with individual computing needs. Many companies took advantage of the cheap processing power to develop departmental applications. Many others, including the general public, brought in application packages that performed a specific function. This meant that the local data centers had now moved out of the main information technology (IT) or central computer operations function and were being distributed into the hundreds of small PCs.

1.2.7 Client-server computing

While many PCs were being used to execute local applications, another very good use for them was their working in conjunction with a mainframe and its local data center to provide an intelligent front end to the applications.

We now arrive at the age of client-server computing. Here, the access and handling of the end-user requirements can be carried out by the workstation, which supplies the necessary end-user friendliness while the mainframe provides the data storage and access. This is a good example of a specific client-server application and, indeed, the mainframe could be both a file server connected via a LAN or an enterprise processor connected via a WAN. The smaller talks to the larger computer.

However, in certain cases, the client-server relationship is not so clearly defined. For example, an application executing in the host wants access to a printer that is attached to a PC on a LAN. Here the host application is the client and the PC application is the server. This is the reverse of the earlier scenario. What is apparent is that there is a general requirement to place the function where the business requires it. This necessitates peer-to-peer application communications across a variety of operating systems and networking protocols.

We have seen that, initially, there was one application executing in a system, but today's environment generally requires that applications communicate with each other. This facility has been available in mainframes for some time. The difference now is that it is required at the individual workstation level. The client-server solution in all situations requires the type of management and administration that is needed for any distributed systems. These aspects are covered in Chap 9. It is at this level that messaging and queuing was recognized as a possible solution to this general problem.

1.2.8 Program-to-program messaging

The transaction monitors of the 1970s and 1980s provided mechanisms that allowed programs to communicate directly with each other, but these relied on the monitor controlling both programs. These communications were achieved by monitor-defined communications protocols and information formats contained in the messages being passed. This worked well in the large mainframes of these times because it was normal to have these monitors in most of the processors. However, when program-to-program communications is required in the emerging programmable workstation environment of today, the existence of a transaction monitor is an inhibitor. It therefore became necessary to define methods of program-to-program communication that cover the different environments and allow the exploitation of the facilities of the operating systems. These models will be examined in the next chapter. However, each of these models requires the passing of messages of different types and sizes.

What we saw in the 1980s was a growing need for programs to communicate with other programs without the intervention of a human operator. This requires a different set of control and recoverability characteristics in the application environment, because the application can no longer rely on the actions of a human operator to handle exceptional circumstances. This type of environment is provided by the transaction monitors and that is why we are seeing the need for these monitors in the workstations of the '90s.

1.2.9 Messaging and queuing applications

While transaction processing was becoming the cornerstone of enterprise applications, many enterprises realized that some form of queuing was a valuable asset to the application program. As a result, subsystems based on a queuing principle developed. These RYO (roll-your-own) application subsystems became, sometimes unknowingly, the bases for many large systems. This means that many enterprises are currently running and using systems and functions that are described in this book. You may, in fact, already be familiar with one or more messaging systems using the principles described in this book.

An example of a messaging and queuing-based application system is the Distributed Application Environment (DAE) subsystem. DAE provides an environment that supplies a range of facilities to enable the development of application and communication with heterogeneous manufacturing plant floor devices. The DAE environment comprises resources in a distributed network. The environments can be as a single node or a complex distributed network. The applications request services of the system through an application programming interface (API). The location of these services is outside the application. The applications are loosely coupled to the servers through a messaging interface. A range of services is provided by DAE:

Base system services: These control the environment and provide operating system interdependent services.

Communications services: These allow applications to communicate using a connectionless communication model.

Data services: These provide for efficient organization of data and access in a multiuser environment.

User services: These provide consistent presentation services.

Device services: These provide a logical view of devices.

Part of base system services is a message-queuing transportation mechanism that provides a priority-based disk queuing facility, giving either local or remote access that is transparent to the application.

1.3 The 1990s

The 1990s are full of challenges, as shown in Fig. 1.7. Many networking protocols, multiple vendors, and many unlike applications make intercommunication difficult.

1.3.1 Messaging and queuing

While there have been queuing subsystems available for some time, IMS is the prime example of one. We have seen that there is a requirement to make these facilities more generally available across multiple operating environments. Several applications are already available that use messaging and queuing as part of their infrastructure, but the facilities are not generally accessible by application programmers.

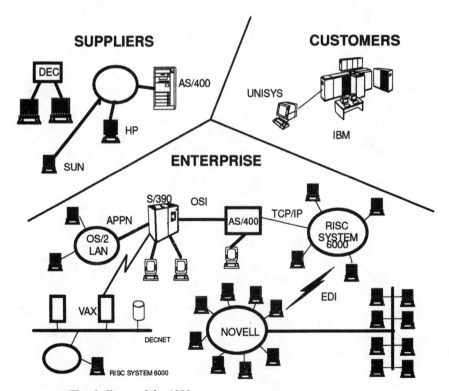

Figure 1.7 The challenge of the 1990s.

1.3.2 Open networking

The enterprise of today has evolved to contain several local data centers, many department applications, hardware from different manufacturers, and often a mixture of operating systems. Each of these applications and data centers has information that is required for the efficient running of the enterprise. But how do you access them?

One solution would be to merge onto one operating system base and, over time, migrate applications and data to that system. However, this is expensive and could take a long time. Another approach would be to supply a mechanism that would allow the existing applications to communicate with each other and with new applications, regardless of the operating systems, to share information. This is the concept of *open networking*. It does not require a single set of protocols or operating systems to be available in the enterprise. It only requires that the application communication mechanism can execute over any communication protocol and operate on any operating system.

Open networking is discussed in detail in Chap. 7.

The Purposes of Messaging

Messaging is an asynchronous (time-independent) communication technique in contrast with synchronous (time-dependent) communication techniques that abound in the computer industry today and which require constant availability of communicating parties and all resources between them.

Messaging serves many diverse purposes very efficiently. This chapter explains the purposes of messaging, from early on until now.

2.1 Introduction

Throughout history there has been an endless variety of methods of passing messages. But the purpose of messaging has always been the same: the reliable passing of information from one location to another. This applies whether the carrier is a human runner or a high-tech satellite communications link. In this chapter, we will look at machine-assisted communications and messaging.

We have seen that messaging has been with the computer industry from the earliest days and that its implementation and usage has changed radically over time. The business applications that gave rise to messaging were those that involved a large number of low-function information pairs, such as airline reservation or bank inquiry systems. The primary purpose of messaging was to provide a speedy exchange of information with expected low-integrity delivery and human-retry recovery. This has now matured into a complex, reliable network in which complex business-dependent information is passed regularly between processors with no need for human recovery. Connection-oriented solutions that do not use messaging have caused a "connection explosion" which may result in lengthy restart and recovery times, which are unacceptable in today's business environment. This, in turn, has led to the development of connectionless messaging, which will be explored further in the coming chapters.

Let's take a quick look at what we mean by (asynchronous and connectionless) messaging and some of the communication paradigms.

2.2 Person-to-Person Messaging

Messages were exchanged between persons long before they were exchanged between computer programs. Messaging between persons is at human speeds (seconds, minutes, hours, days); messaging between computer programs is at computer and communication network speeds (subseconds and seconds).

2.2.1 Telephone

The telephone necessarily brought about an amazing change in the ability of humans to communicate, as illustrated in Fig. 2.1, because now people were no longer bound by physical contact and, as interconnected networks grew, they were not even restricted by country boundaries, physical barriers or even distance. However, the telephone did constitute a structured method of communication. Because it was not possible to see the party with whom you were communicating, there was a need to establish a protocol and to have both people and the communication line between them constantly available. Were the two parties speaking the same language? Who had the right to speak first? How long did you wait before asking another question or making a comment so as not to interrupt the other party's response? Another requirement was that each of the parties was available to establish communications. Thus, protocols were established and understood naturally between humans because we had been communicating for generations. These types of protocols have likewise been established for computers and are also now taken for granted.

With the telephone, we established an effective global communications method with established protocols. The only problem was, what happened if nobody answered the phone?

2.2.2 Answering machine

Here technology provides the answer to the question of how to capture the telephone message and allow the receiving user to handle the message asynchronously (that is, later and not in step with the caller). The answering machine allows two parties to communicate effectively, even though they are not physically connected. It also introduces an element of time independence. Messages can be gathered together and handled when the owner of the answering machine wants to do so.

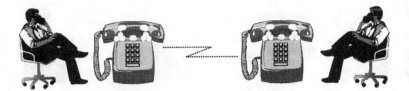

Successful call depends upon:

Constant communication line availability

Constant people availability

Common, defined language

Figure 2.1 Telephone messaging.

This introduces into human communications the concepts of messaging and queuing.

The answering machine has the ability to offer a welcoming message. This allows a degree of two-way communication without having both parties available.

Both the phone and the answering machine rely on vocal communications; however, the printed word coupled with an appropriate delivery mechanism is just as effective a method of asynchronous communication. Letters, books, and the postal system were established before the telephone. The difference made by telephones is the time it takes to move a message from one user to another. The telephone does it at a much higher speed than the mail system ever could. This speed of communication had never been achieved for the printed word until, once again, technology provided the facility: The facsimile reproduction or fax, as shown in the top of Fig. 2.2.

2.2.3 Fax

The fax machine integrates the written word with the telephone network. The message can be clearly identified as the page or pages of the fax being sent. However, unlike the telephone, this is not a conversational type of communication; it is strictly unidirectional. The receipt of a fax does not automatically generate a message answer. There is an acknowledgment that the message has been received, but that constitutes the total conversation.

The added value that a fax provides, as opposed to the telephone, is the ability to transmit written material at a low cost, with the message being recoverable. If, for some reason, the line breaks during transmission, the page can be resent, nothing is lost, and the meaning of the

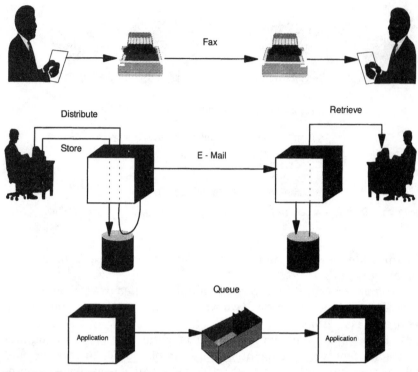

Figure 2.2   Facsimile (fax) messaging.

total message is preserved. The fax brings a degree of transactional processing to the telephone network.

None of the systems mentioned so far directly involves a computer system, and the communication has always been person to person. This means that the sender can rely on the receiver to apply some form of interpretation or error correction if necessary, always assuming that the initial protocols have been established.

2.2.4   Office systems

The telephone system and the fax machine provide many of the elements required for the office of today. The computerized office system, with electronic mail, as shown in the middle of Fig. 2.2, adds to these the flexibility of:

- Long-term storage, both permanent and changeable
- Directory and routing capabilities
- The possibility of integrating all the enterprise functions into a common workstation

But the essential characteristics remain the same as ever. It is still a person who is sending and receiving the messages.

There are many office systems currently available on the market. They all provide the basic functions that allow a user to create, alter, transmit, and receive messages. The total facilities provided by these systems may include:

- The immediate or delayed transmission of messages
- The creation and maintenance of large complex documents
- The management, both centrally and locally, of directories
- Diary and calendar functions

Because they operate person to person, most of these systems do not provide the same recovery-or-security features that are provided by a transaction system. Like with the fax machine, there is a degree of recovery in that the source of the message is preserved, but, if there is an interruption in the unit of work, the system depends on the end user to take immediate action to recover to the point of failure. Also, there is a limited degree of security as, once again, the end user is expected to become involved.

We mentioned before that standards had to be introduced to allow meaningful communications to take place. This applies to a greater degree when systems communicate, as with queue-based messaging shown in the bottom of Fig. 2.2. There are a large number of standards currently published on this subject and it is not the intention of this book to cover them. However, they are important to remember if meaningful data interchange is to occur between systems.

2.3 Program-to-Program Messaging

So far we have been considering person-to-person communication and the use of technology to assist with it. This type of communication requires that there be a person on each end who can apply intelligence to decode and recover any messages sent. This means that the enforcement of protocols, while still required, can be applied more flexibly. At the basic level, computer systems do not have this level of intelligence; therefore, we have to apply and enforce strict protocols if meaningful and assured communications are to be achieved with them. Person-to-person and program-to-program messaging, as identified in Fig. 2.2, are contrasted in more detail in App. D.

The remainder of this book will cover the topics concerned with the high-level functions provided by the messaging and queuing environment. Before we look in detail at this, it would be worthwhile to consider three of the main program-to-program communication models that are found in transactional environments. These are:

CPI-C   Common Programming Interface for Communications

RPC   Remote Procedure Call

MQI   Message Queue Interface

These three communication models are discussed in detail in Chap. 8.

Each of these models has its advocates, and there are certain cases where the use of one model has clear advantages over the use of another. Also, there are functional overlaps, and the capability supplied with the products utilizing these models makes these overlaps greater. We will attempt a nonjudgmental look at the models.

However, before we look at the models, we should be clear on a number of topics. We are examining program-to-program communication. The end user is not a person interacting with the system via a screen or service point, but it is a program reacting to input or creating it. So, in client-server terms, the client is requesting a "service" from the server; the client initiates and terminates the request. In peer-to-peer terms, each partner has equal standing to initiate a request.

Furthermore, the terms *synchronous* and *asynchronous* have been used in many ways. Here we will use "synchronous" to mean that the requesting program must have connectivity to its partner for the whole of the request and cannot continue with its operations until the partner has responded. "Asynchronous" implies that the requestor can continue its operation while the partner is processing the request.

2.3.1 CPI-C

CPI-C is a conversational, synchronous mode of communication. One of the pair of programs initiates the conversation and subsequently controls the direction of the flow of information, as shown in Fig. 2.3. The primary program can then SEND data while the secondary can RECEIVE it. This secondary program must BID to send information to the primary. Each program must keep track of the state of the conversation so that it can recover in case of an error. During the conversation, the partner program must be available all the time. If, for any reason, the connection breaks, it is up to the requestor to reestablish and recover the conversation. This adds logic to the application to handle the situation. While it is possible to have a peer-to-peer relationship, the state of who is the initiator of the conversation is fixed at the start of the program and is maintained. It is possible to change the relationship, but this must be done outside the conversation. This means that CPI-C supports both client-server and peer-to-peer environments.

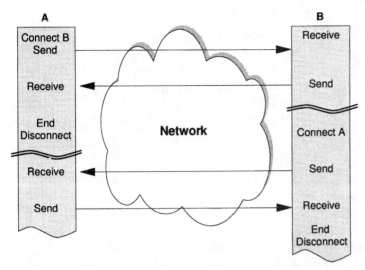

Figure 2.3 Connection-based, synchronous communication with CPI-C.

As we said earlier, this is a synchronous model in that a connection needs to be maintained and that, without a supporting environment, the program cannot execute until the request has been satisfied. However, the transaction environments that support CPI-C do allow for a degree of overlap and asynchronous nature. This can be achieved in a CICS environment with the use of *transient data queues.*

CPI-C has, until recently, been supported only with SNA protocols. However, it is now supported with both over TCP/IP and SNA protocols.

Because it is necessary for the application program to handle error conditions and recover from them, the interface (API), is very comprehensive. To some degree, this is made simpler by the supporting environments.

2.3.2 RPC

Remote Procedure Call, or RPC, is also a conversational, synchronous model, as shown in Fig. 2.4. With RPC, a calling program passes its request to a server for handling. The calling is done via a stub that performs the communication to the remote system that provides the function. This is a fixed relationship between caller and callee, and it is very difficult to produce the peer-to-peer functions. As for CPI-C, communication errors must be handled by the application program and the execution of the requestor is blocked until the request is satisfied.

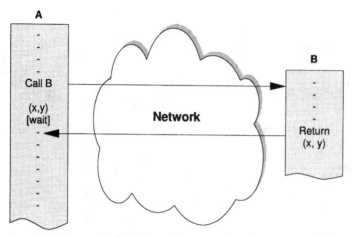

Figure 2.4   Connection-based, synchronous communication with RPC.

2.3.3 MQI

The Message Queue Interface, or MQI, provides an asynchronous method of communication between programs, as shown in Fig. 2.5. An application uses the medium of a queue to communicate with its partner program. These queues can be either local or remote to the application. If an application needs to communicate with another given application, it simply PUTs a message to the queue associated with the required application. The application doing the PUT is then free to continue processing. It is not aware that any communications are taking place. The communications and error recovery are all handled by the *queue manager*.

This style of communications is independent of the communication protocols used to transmit the information. Because the application

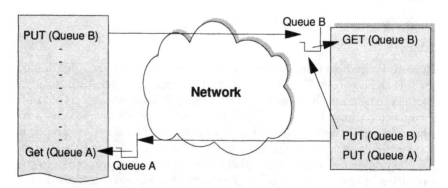

Figure 2.5   Queue-based, asynchronous communication with MQI.

program can be unaware that communications are taking place, any standard communications protocol could be used. This may be SNA, TCP/IP, a LAN protocol, or even some unique features provided by operating systems.

The partner application need not be available when the requesting application executes the PUT. Indeed, the same application can simply communicate with multiple applications by performing PUT operations to different queues.

The CPI-C, RPC, and MQI choices are discussed in detail in Chap. 8.

Messaging Concepts

Messaging and queuing concepts are quite simple. Messaging concepts are discussed in this chapter; queuing concepts are discussed in the next chapter. Messaging can be accomplished with or without queuing (see Sec. 3.2.3), but this book concentrates on accomplishing messaging with the assistance of queuing.*

Messaging is a NO-WAIT communication choice. Messaging partners do not wait on each other. Messaging is also referred to as asynchronous communication (that is, communication that is not synchronized between partners). Messaging is an *availability adaptation* technique used for the transfer of information between two entities (programs, people, devices, systems, etc.) without regard for the immediate availability and accessibility of either.

This chapter explains messaging concepts from two perspectives: first generally (as between persons) and then in more detailed fashion (as between programs). Messaging parties generally operate independently; they can communicate in different or partially overlapping time periods and are not forced to communicate only within a single time period.

3.1 General Concepts of Messaging

The general concepts of messaging are easily illustrated by an analogy that contrasts synchronous and asynchronous communication (that is, synchronized, necessarily in-step communication and nonsynchronized, probably out-of-step communication, respectively). For purposes of simplicity, let us assume that we are faced with 20 questions that

---

* Throughout this book, the term *messaging* is used to mean messaging and queuing.

need 20 answers. Let us further assume that the questions are not related to each other and that they can be asked and answered in any sequence by any qualified person. We are then free to choose a method for getting the 20 questions answered, either in a synchronous (non-messaging) or asynchronous (messaging) way.

3.1.1 Twenty Questions and Synchronous Communication

One possible method for getting our 20 answers is to connect ourselves up to a qualified person who can undoubtedly answer all of the questions. Then we can serially ask question 1, get answer 1; ask question 2, get answer 2; and so forth in a synchronous (serialized during a time period) manner, as shown in Fig. 3.1. The hand (at the left of the diagram) holding the card with a question mark (the question) belongs to the question maker. The little owl with the mortarboard hat (at the right of the diagram) is the qualified answer maker. The owl is providing the answer by returning the card with the exclamation mark (the answer).

This is, in fact, the natural manner in which we get answers to questions in a face-to-face conversation if we have all the questions and the other person has only answers and no questions. We ask the questions sequentially and get the answers alternating with the questions.

Synchronous communication has the distinct advantage of having dedicated parties intensely communicating for some period of time with no outside influence. It also has the distinct disadvantage of having the communicating parties tied up and out of circulation during their synchronous time period.

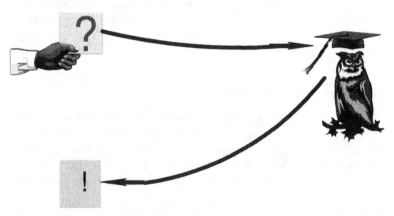

Figure 3.1 Twenty questions and synchronous communication.

A simple example of synchronous communication is to be found in the old-fashioned school classroom environment where the teacher and a single student can often be engaged in a discussion to the exclusion of the remainder of the students (except that they are listening). Students usually take turns synchronously communicating with the teacher.

In our 20 questions example, one question maker and one answer maker each always use up all of their time during a particular synchronous time period either waiting for the other or communicating with the other. There is no time left over to do other things, because they take turns since they are dedicated to each other.

The total time required for the entire process to occur is the aggregate of all time required to ask the questions, plus all time required to answer the questions, plus the time required to change gears between communicating parties. This is an entirely serial process with no time savings. If questions take an average of 10 seconds (to create and ask) and answers take an average of 20 seconds (to ponder and answer) and all others times are negligible, then the entire serial process takes (20 questions × 10 seconds/question) plus (20 answers × 20 seconds/ answer), which is (200 + 400 seconds) or 600 seconds or 10 minutes.

Synchronous communication means dedicated communication and serialized operations. The answer maker (owl) is necessarily available. Surely there is a more efficient way to get our 20 questions answered. But, we must be practical. There are some cases where the 20 questions must be asked and answered in strict sequence, especially if the answer to a previous question might influence the answer to the current question or might actually determine the selection of the remaining questions. The connection choice implies the synchronous choice.

3.1.2 Twenty questions and asynchronous communication

Another possible method for getting our 20 answers is to abandon the synchronous approach and use an asynchronous (without regard for time) approach by asking our questions without regard for the presence and availability of the answer maker (the owl in the previous diagram) and without regard for the period of time during which the answers will be created and made available. We can place our questions (our messages) somewhere where they can be found by the answer maker and then come back later for the answers, as shown in Fig. 3.2, where the question is placed "in a box" to be found later (or immediately) by the answer maker. The answer maker (owl) might have been busy, absent, or perhaps even available when the message was stored. The question maker does not depend upon the answer maker's immediate availability.

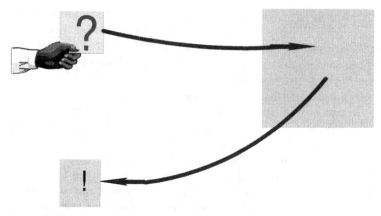

Figure 3.2 Twenty questions and asynchronous communication.

This simple asynchronous method of communication is known generally as *message passing*. Most of us have used basic message passing all of our lives to leave messages for other persons or to get messages from other persons in lieu of talking to them face to face (or phone to phone) in a synchronous way. Deep inside software operating systems, exchanging signals and other information by message passing has been a way of life for decades. Message passing (messaging) works efficiently, predictably, and constantly. The basic principle of messaging (asynchronous communication) is the absence of waiting, which promotes parallelism and concurrent operations.

Basic message passing need not involve multiple messages; it may involve just a single message. In this case, messaging does not always need queuing (since we can discount the trivial case of a queue with just one message). Also, the messages that are put somewhere need not necessarily be organized or arranged in any particular way. A case in point is the family refrigerator door or cabinet counter top that is often splattered with messages.

The first step in message passing (asynchronous messaging) is to place a message somewhere where it can later be found, as shown in Fig. 3.3. The second step in message passing (asynchronous messaging) is to get a message from wherever it was earlier stored, as shown in Fig. 3.4. When you get a message, there may be more than one or there might be just one. If multiple messages are available, they may or may not be organized or grouped in any particular recognizable order.

Notice that with message passing, we never make direct contact with the answer maker (the owl in the diagram). The work gets done (the questions all get answers) even without this direct contact. We cannot be certain of the order in which the questions are actually answered by

Figure 3.3   Putting the message someplace.

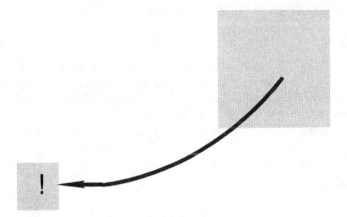

Figure 3.4   Getting the message from someplace.

the answer maker. Perhaps the answers are created serially, as with synchronous messaging.

Also notice that with message passing, we may find that there are no messages to retrieve even if we are looking in the right place. We might be early, before the answer maker created even one answer, be it the answer to the first question, the 13th question, or the last question. Even if there are one or more answers, they may not necessarily be in sequence (unless you specified such to the answer maker earlier). Out-of-sequence answers can be perfectly acceptable and wonderful. Everything in the world does not need to run sequentially and serially in order to function properly. Out-of-sequence messages save lots of time. Who wants to wait on answers 3 through 20 because the answer maker had to consult someone else about answer 2?

3.2   Messaging Concepts for Programs

The general concepts of messaging that we discussed in the previous section (illustrated with the 20 questions example) apply to messaging between programs. Most of the hundreds of thousands of computer programs used today have been built upon the (false) assumption that dependable communication can happen only in a synchronous manner. Fortunately, however, there are tens of thousands of computer programs that quietly operate (unadvertised) with asynchronous communication and perform splendidly with high message traffic volumes, unpredictable traffic patterns, and business-critical dependability.

Messaging starts with the composition of a message. Messages must be built, delivered, and interpreted before they have real value. Programs must be able to recognize what constitutes a message (its format) and what constitutes the information therein (its data content).

3.2.1 The message

A *message* is a string of bytes representing information. In general, some of the information (the first and sometimes the last) is *system-related* control information, while the middle is *application-related* user information, as shown in Fig. 3.5. The messaging and queuing software product needs the *header;* the messaging application program needs the *user data.*

3.2.2 Asynchronous (messaging) communication

Messaging programs are asynchronous programs. Asynchronous (meaning "without any reference to time") programs never wait on one another; they are not synchronized (serialized) with one another. Remember the "someplace box" and the missing owl (the missing answer maker) in Fig. 3.2 representing messaging passing (asynchronous communication)? Messaging between programs is message passing, which is asynchronous.

There are both synchronous and asynchronous choices possible for processing, as shown in Fig. 3.6. Applications can, of course, have a mixture of both synchronous and asynchronous processes. The top of

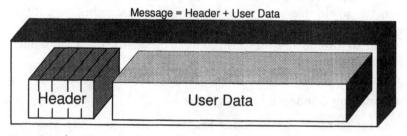

Message = Header + User Data

Header

User Data

Figure 3.5 A message.

Fig. 3.6 illustrates synchronous processing and the bottom illustrates asynchronous processing. Synchronous processing is much like a pass-the-baton operation in a relay race. Whoever has the baton is in action; the others are waiting.

The example in Fig. 3.6 shows four programs (A, B, C, and D) operating in a time-dependent (synchronous, with blocked/suspended* program operations) manner at the top of the diagram and in a time-independent (asynchronous, with unblocked/unsuspended program operations) manner at the bottom of the diagram. The synchronous choice at the top shows program A connected to program B, which is in turn connected once to program C and once to program D. This is a tree information flow pattern, as discussed in detail in Chap. 5. The line graph shows that the serial sequence of executions is:

A executes, remainder wait, and then control goes to B

B executes, remainder wait, and then control goes to C

C executes, remainder wait, and then control returns to B

B executes, remainder wait, and then control goes to D

D executes, remainder wait, and then control returns to B

B executes, remainder wait, and then control returns to A

---

* Blocked operations are discussed in Sec. 3.2.2.1.

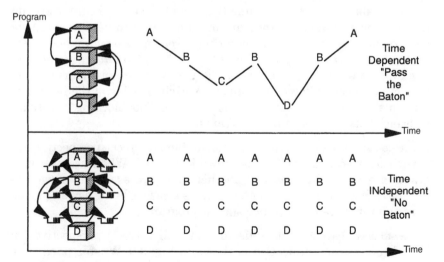

Figure 3.6   Processing choices.

The asynchronous choice at the bottom shows that no programs are connected together, except indirectly through queues.

A executes constantly, no programs wait, messages queues are filled and emptied

B executes constantly, no programs wait, messages queues are filled and emptied

C executes constantly, no programs wait, messages queues are filled and emptied

D executes constantly, no programs wait, messages queues are filled and emptied

The asynchronous choice accomplishes the same flow of information, but in a parallel, nonserialized manner, thus saving time.

3.2.2.1   Three asynchronous perspectives.   There are at least three different perspectives in common use for the term *asynchronous:*

User-to-provider (across interface)

User-to-user (singular)

User-to-users (plural)

In all three perspectives, the programs involved are time independent from one another.

User-to-provider (across interface).   A more unusual (and incorrect) perspective for asynchronous communication is shown in Fig. 3.7. When the user (above the interface) and the provider (below the interface) do not wait on one another, they are often declared to be asynchronous with respect to one another operating across an asynchronous interface. A better way to express this is to say that a *nonblocking* (nonsuspending) interface exists between the user and the provider and the user is not blocked (forced to wait) with each crossing of the interface. This is a case of being asynchronous (actually, nonblocking) across an interface between dissimilar programs.

User-to-user (singular).   The most usual (and correct) perspective for asynchronous communication is shown in Fig. 3.8. When two user programs are not waiting on one another, they are executing asynchronously with respect to one another. This is a case of being asynchronous above an interface between two similar programs.

User-to-users (plural).   A less usual (but still correct) perspective for asynchronous communication is shown in Fig. 3.9. When a particular messaging program can operate with one partner program in a manner

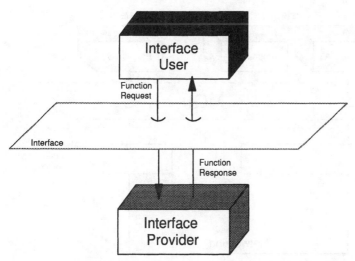

Figure 3.7 Asynchronous communication between interface user and interface provider.

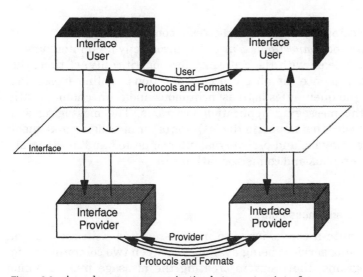

Figure 3.8 Asynchronous communication between two interface users.

unrelated to the way it operates with other partner programs, then the user program is often said to be operating asynchronously with respect to each pairing of itself and a partner program. This is a case of being asynchronous above an interface between one program and two (or more) similar programs.

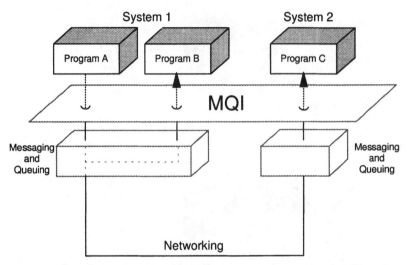

Figure 3.9 Asynchronous communication between two (or more) pairs of interface users.

3.2.2.2 Asynchronous and connectionless communication. The terms *asynchronous* and *connectionless* are often incorrectly used interchangeably. The first deals with the criticality (or noncriticality) of time; the second with presence (or absence) of connections. Online messaging (messaging and queuing) is both asynchronous and connectionless with respect to the messaging application programs. The messaging and queuing software that supports the MQI (or other messaging and queuing interface) may be, and probably is, interconnected and interoperating in a synchronous and connected fashion.

3.2.3   Queuing and messaging

Queuing is quite a convenience to have associated with messaging. Without queuing services being available between two communicating partner programs, the program creating the message must assume that the program digesting the message is available to receive it from a shared storage area of some type within a reasonable amount of time before the message might be replaced with another message.

In some cases, robotics devices are controlled by messaging without queuing, whereupon the signals are expected to be used as fast as they become available. Printer driver programs sometimes use messaging without queuing to send print streams to a printer device with the expectation that, with the proper time delay between messages, the printer device can keep up. Meanwhile messages find their way, often in a multihop manner, between printer driver and printer device.

Queuing concepts are discussed in the next chapter. Again, messaging means asynchronous communication or communication without waiting, regardless of the associated use or nonuse of queuing.

Chapter

4

Queuing Concepts

Queuing is a NO-CONNECTION communication choice. Queuing partners are not directly connected; they communicate only through queues.

Queuing is a *time adaptation* technique used for saving information until the intended message receiver is ready to receive it, be that just nanoseconds, milliseconds, or many minutes away. Entities are indirectly communicating, and each is operating at its own preferred or maximum speed unaffected by the others.

This chapter explains queuing concepts from two perspectives: first generally (as between persons) and then in more detail (as between programs).

4.1 General Queuing Concepts

For purposes of simplicity, let us assume (as in the previous chapter describing messaging concepts) that we are faced with 20 questions that need 20 answers. Let us further assume that the questions are not related to each other and that they can be asked and answered in any sequence by any qualified person. We are then free to choose a method for getting the 20 answers, using both the principles of messaging (from the previous chapter) and the principles of queuing (from this chapter).

4.1.1 Twenty questions, no queue, and one answer maker

One possible method for getting our answers is to associate ourselves with a qualified person who can quickly answer all of the questions. Then we can ask our questions as fast as we can and immediately get the answers, as shown in Fig. 4.1. We could even have another person collect our answers. This is messaging without queuing.

4.1.2 Twenty questions, one queue, and one answer maker

Using a "shoe box" to represent a queue storage space, as shown in Fig. 4.2, it is interesting to explore combinations of one or more queues taken with one or more available answer makers, as shown in the following sections, namely:

one queue, one answer maker

two queues, one answer maker

two queues, two answer makers

three queues, two answer makers

to see the resultant differences in queuing effect.

If we wish to collect all of the answers at one time after they are ready, there need not be a queue for the answers, as shown in Fig. 4.1. However, we may not wish to interrogate the answer maker to see if any or all of the answers are ready. We could insist on a combined queue for the questions and answers, as shown in Fig. 4.2, and we could search for some answers in the answer part of the question/answer queue without bothering to ask the answer maker about the status of our questions.

Now we should be able to ask our questions independently of getting our answers. The question maker and answer maker can operate independently, filling and emptying the combination question/answer queue. We would naturally expect the answers to be available in the same sequence that they were asked, even if the answer-maker person did not prefer to (or could not manage to) answer them in the same sequence.

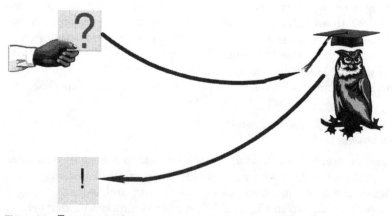

Figure 4.1 Twenty questions, no queue, and one answer maker.

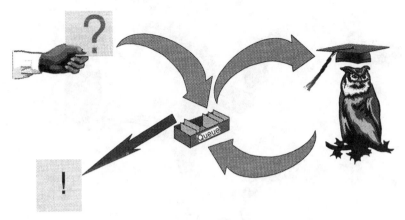

Figure 4.2 Twenty questions, one queue, and one answer maker.

The question maker could choose to ask all 20 questions before looking for any answers, or the question maker could choose to ask 10 questions and then begin looking for the first 5 answers and then proceed with the remaining 10 questions. All sorts of choices between asking questions and looking for answers are possible, but much depends upon the answer maker's ability to quickly answer and to answer in sequence.

We can now see one of the major benefits of queuing, namely, over-lapping of question-maker and answer-maker time. Of course, the answer maker must be allowed to get a little ahead of the question maker, or the answer maker must be much quicker than the question maker to allow the maximum amount of overlap of time.

Queuing can reduce the amount of time required to complete the answers to the 20 questions by unserializing the entire process and eliminating most of the time that the question maker must wait on the answer maker, and vice versa. Perhaps all of the questions can be answered (and digested by the question maker) within 8 minutes instead of 10 minutes from the earlier example. Yet, there must be more room for improvement.

4.1.3 Twenty questions, two queues, and one answer maker

Another possible method for getting our answers is the same as in the previous section but with two separate queues, one for the questions and one for the answers, as shown in Fig. 4.3. There is now no contention between question maker and answer maker for accessing the combination answer/question queue (box).

We would naturally expect the answers to be available in the same sequence that they were asked, but the answer-maker person could

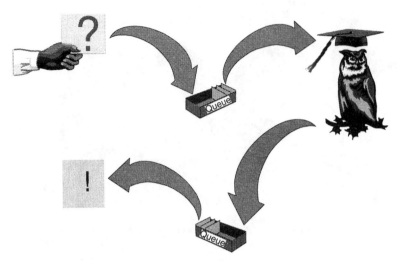

Figure 4.3 Twenty questions, two queues, and one answer maker.

choose to answer them in a different sequence. There may be a need to pause and arrange the questions in sequence by difficulty before answering them. Rarely do all questions (regardless of their type) take exactly the same amount of time to answer (some are short, others long), even if the answers are always instantly available in the answer maker's mind. In addition, some questions do require more thought than others.

4.1.4   Twenty questions, two queues, and two answer makers

Yet another possible method for getting our answers is the same as in the previous section with two separate queues, one for the questions and one for the answers, but with two answer makers instead of one, as shown in Fig. 4.4.

It is natural to expect that answers are harder to create than are the questions. This means that much of the delay might be caused by the overworked answer maker. We can use two answer makers to speed up the process.

Two answer makers and queuing can reduce the amount of time required to complete the answers to the 20 questions by not only unserializing the entire process but also making the answer-maker time (20 seconds/person) more balanced with the question-maker time (10 seconds/person). Now questions and answers both take 10 seconds each on the average, since there is twice as much "horsepower" (and half as much time) used for answering. Perhaps the total time for asking, answering, and digesting the 20 questions can be reduced to 5 minutes

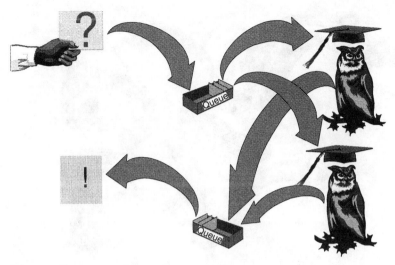

Figure 4.4 Twenty questions, two queues, and two answer makers.

(from the original 10 minutes, serially). Now we have some reasonable improvement but everyone is contending for some of the same queues.

4.1.5 Twenty questions, three queues, and two answer makers

A further possible method for getting our answers is as described in the previous section with two answer makers but using three queues, as shown in Fig. 4.5.

By maintaining separate question queues but a combined answer queue, we can speed up the process perhaps a bit more. In this way there is no contention for accessing the question "box" and the answer makers can operate in parallel, independent of one another, to speed up the answer-making process. The answer queue is accessed only whenever an answer is created by either answer maker, and these accesses might tend to be distributed.

4.2 Program Queuing Concepts

The exact same principles discussed in the preceding general queuing concepts section apply equally well to queue-based communication between programs. Making good choices for how many queues, how (or if) they should be shared, the number of messages allowed to accumulate in each (a *threshold*), the information flow patterns between queues, and many other related choices is directly involved in good messaging design (as described in Chap. 10).

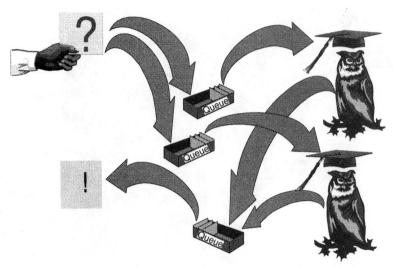

Figure 4.5 Twenty questions, three queues, and two answer makers.

4.2.1 A message queue

A queue is a storage area for saving information (messages) in an ordered manner for later retrieval. Each queue has a queue descriptor control block.

Why are queues necessary? Because timing is never perfect between computers and people and devices. Because not all things happen at exactly the right time. Some programs execute faster than others, have more work to do, or require more logic to complete processing. Information flows in spurts, comes in various lengths and at varying speeds from many directions to many directions. Many things are kept waiting in a completely serialized process where programs take turns waiting on each other to execute. Serialized processes do not need queues; they tolerate waiting on one another.

4.2.2 Some advantages of queues

There are, of course, both advantages and disadvantages for connections and queues. Some of the advantages of queues that are not quite so apparent include:

- Queues introduce unlimited varieties of information flow patterns (as described in Chap. 5) promoting

  new application designs

  integration of old and new applications on different platforms

  deserialization of existing processes introducing parallel operations

- Queues drive down the requirement for large numbers of connections (as described in Chap. 7) causing

  reduced networking setup and operation expense

  simpler application programming

  separation of communication and application environments (as described in Chap. 7)

- Queues allow easy sharing of a data stream of messages (as described in Chap. 6) for such purposes as

  concurrent processing of single messages

  load balancing (with feedback implied)

  load distribution (regardless of load and with no feedback implied)

- Queues separate queue management from program management allowing

  quick substitution of programs

  independent control of queues and programs

  trigger queues to help bring programs alive when necessary

4.2.3 Synchronous telephone conversation

By analogy, an ordinary telephone conversation is a synchronous (time-dependent) set of processes with connections and without queues, as shown in Fig. 4.6. Both parties are talking during the exact same time period, on a rotation basis (usually). Persons naturally take turns waiting on each other to speak or answer in the natural course of conversations (except for some less well mannered individuals who might not take turns waiting and speaking).

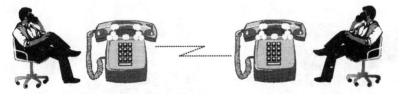

Successful call depends upon:

Constant communication line availability

Constant people availability

Common, defined language

Figure 4.6   Connection-based telephone communication.

4.2.4 Asynchronous telephone messaging

By analogy, an ordinary telephone answering machine recording and playback is an asynchronous (time-independent) set of processes with queues and without connections, as shown in Fig. 4.7. Several messages may be recorded before you hear any of them. You may be absent from the phone or busy using the phone while (phone) messages are queued.

4.2.5 Independently operating programs

As explained in Chap. 3, each messaging program operates independently from all other messaging programs. The logic within each application program is not necessarily intertwined with the logic of others; that is, each program executes independently of every other and does not, by the very nature of messaging, have to wait on any other.*

The partner program (or programs) are never specified† with messaging; only a queue name is specified for input or for output, as shown in Fig. 4.8. Unlike conventional connection-based communication programming, both predictable and unpredictable pairings of programs are easily accommodated with queue-based messaging.

Connection-based (that is, conversational and synchronous) programs are dependently operating programs; they take turns waiting on each other. Queue-based (that is, messaging and asynchronous) programs are independently operating programs; they never wait on each

---

* This does not mean, however, that application programs could not be written in such a manner as to force themselves to wait upon each other.

† The queue name could, of course, be indicative of a program name or set of program names.

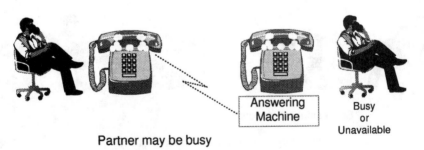

Partner may be busy

Partner may be absent

Messages can be picked up at later time

Figure 4.7 Queue-based telephone communication.

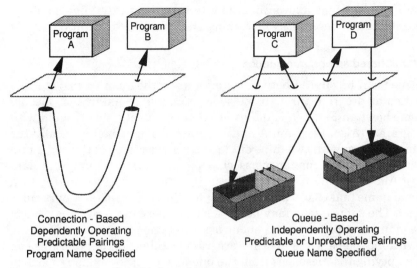

Connection - Based
Dependently Operating
Predictable Pairings
Program Name Specified

Queue - Based
Independently Operating
Predictable or Unpredictable Pairings
Queue Name Specified

Figure 4.8   Dependently and independently operating programs.

other, unless they do so with application intent and control. Again, remember the "someplace box" and the missing owl (the missing answer maker) in Fig. 3.2 for message passing (asynchronous communication)? The question maker and the answer maker can operate independently.

4.2.6   Separated queue and program management

In general with messaging, queue management and program management are divorced or separated. Queue manager programs create, control, and delete queues. Program manager programs load, start, control, stop, and deactivate programs. Trigger queues (discussed in Chap. 11) help associate queues to programs. Queue and program managers are separate. This has the distinct advantage of allowing message queuing (and moving) to operate entirely independently from program loading (and starting/stopping). Messages are delivered even if some programs are not operational, whether intentionally or unintentionally. There are many varieties of association possible between programs and queues. Messaging and queuing software provides queue management but leaves program management to normal operating systems or special-purpose application systems. It is often the case that application design or choice by computer installation procedures dictates when and if application programs are available during any particular time of the day or at a particular application system location. Separation of queue and program management allows programs to be temporarily unavailable or absent without necessarily disturbing an application's operation.

4.2.7 Shared and private queues

Queues can be shared between programs or they can be private to a particular program through the separate handles and queues mechanism shown in Fig. 4.9. Handles are temporary labels associated with programs. (Also, Program A and Program C might be identically the same program with two different labels for two copies of the same program.) The queue manager associates queue names to program handles. Allowing only a single program handle to be associated with a queue name makes it a private queue for that program (at a particular time). The same queue may later be shared between programs. One of the main reasons to share a queue is so that one program can be filling it with messages while another program or collection of programs can be emptying the messages from the queue. One of the main reasons to have exclusive use of a queue is so that a queue can be completely filled with messages or completely emptied of messages before another program gains access to the queue.

As reflected in Fig. 4.9, the programs lie "above" the queue manager and are controlled by a program management system of some sort, for example, an operating system task supervisor. The queues lie "below"

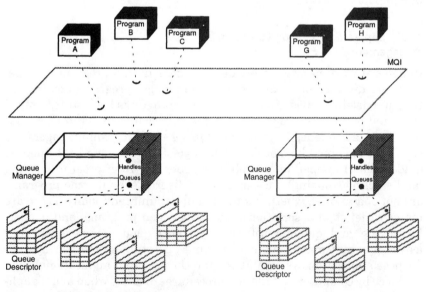

Figure 4.9 Shared and private queues.

the queue manager and are controlled by the queue manager. In order for queues to be filled and emptied regularly and for programs to locate their queue-based input and output areas, there must be an association made between queues and programs in messaging and queuing. The queue descriptor often (but not always) can be used to identify which programs are using the queue and how they are using it (input, output, shared, and so forth). Messaging and queuing products can vary widely in the technique chosen for making the program–queue associations.

4.2.8 Local and remote queues

Queues can be local (found in the local application environment) or remote (found in some remote application environment). See Fig. 4.10. The messaging application program does not distinguish between local and remote queues. The message queue manager moves messages to remote queues automatically when necessary without requiring information from the application program. For example, queue Q2 is a remote queue while queue Q1 is a local queue unbeknownst to Program A in Fig. 4.10. Program A is not necessarily aware of the existence of Programs B and C. Program B could even be replaced with Programs, D, E, and F all reading the same queue Q1, assuming the application logic permitted such.

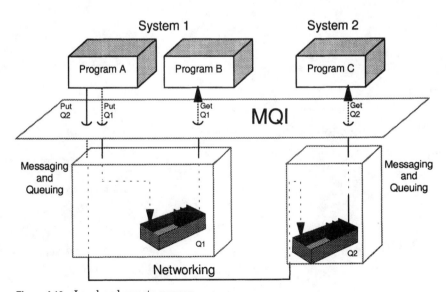

Figure 4.10   Local and remote queues.

4.2.9 Dynamic and static queues

Queues can be dynamic (not predefined) or static (predefined). Dynamic queues are for the convenience of programs that need messages stored only for a short duration of time, during unusual circumstances, or for many other reasons that do not justify the permanent existence of a queue. Chapter 11 offers a complete discussion of dynamic and static queues.

2

The Analysis

This part of the book provides hints on analyzing how and why messaging is so helpful in application development and so economical in application execution.

In order to more easily understand messaging, it is important to analyze how many ways information can flow among programs and how different processing models function. Messaging is convenient for all information flow patterns and many processing models.

As explained in Chaps. 3 and 4, messaging is an availability adaptation technique and queuing is a time adaptation technique.

The study of information flow patterns among application programs is both interesting and vital to a better understanding of the importance of messaging. Messaging accomplishes information flow patterns that are difficult or impossible with more traditional techniques.

Messaging and queuing processing models of many types are in practice already. Many of these models result in faster application deployment, simpler maintenance, easier integration of existing and new applications, less demand for highly skilled programmers, better utilization of computer resources of all kinds, less expensive business operations, and more efficient company operations.

This part of the book describes:

- *Information flow patterns*
- *Messaging models*

5

Information Flow Patterns

The patterns by which information flows among application programs are interesting to study and vital to understand in order to optimize application designs. It is not just the flow of information between exactly two programs that is important but also the flow among all of the programs. Few applications consist of just two programs by themselves; there are usually many.

The purpose of this chapter is to stimulate thinking about new patterns of information flow that become possible with online messaging. With queue-based online messaging, any and all information flow patterns are possible. With traditional connection-based communication (nonmessaging), there are a considerable number of flow patterns that are either impossible or very difficult to achieve.

The client-server flow pattern has become a particularly important flow pattern used as the basis for much information processing, particularly where it is especially economical and convenient to consolidate computing services and data at a server location so that programs at many client locations can share these server resources (for example, printers and files). Online messaging is well suited not only to the popular client-server pattern, but to all possible patterns. Furthermore, online messaging is well suited for the integration of human procedures, device operations, and computer program operations.

5.1   Information Flow

In general, information is generated (created) at some point (or points) and absorbed (processed) at another point (or points). Information can take the form of a request, a response, a one-way signal, or a one-way tidbit of data. Information can flow unidirectionally (one-way only) or, more commonly, bidirectionally (in both directions) between two programs.

5.1.1   Application

An application is a set of programs that collectively serve some useful business, administrative, scientific, or other purpose. An application is a set of application programs (and sometimes includes human, device, and other operations). In this book, we are concerned primarily with just the set of programs that constitutes an application.

The application can be subdivided into programs, as shown in Fig. 5.1. (The terms *application* and *application program* are often used interchangeably.) The programs collectively accomplish the application purpose. In general, some programs are subordinate to others. Some deal with human interfaces or device interfaces, but many deal only with other programs. However, all programs need to communicate with at least one other program (except for the larger, monolithic programs of many decades past).

5.1.2   Application program

An application program is one of the programs within an application. (See Sec. 5.1.1.) Application programs interoperate in a tightly coupled or loosely coupled manner, as shown in Fig. 5.2, being nondistributed or distributed, respectively. You will notice that not all programs are flowing information to all others; the arrows show information flow patterns that leave some programs at the edges while others are in the middle of an overall flow pattern.

If a subset of programs is nondistributed (for example, link-edited together), then the programs are tightly coupled within this subset. If a subset of programs is distributed (for example, interconnected with channels, networks, buses, and such), then the programs are loosely coupled within this subset. Programs must be first *interconnected* before

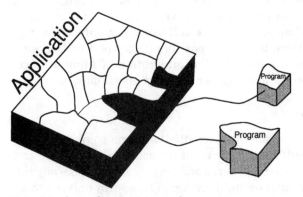

Figure 5.1   An application is a set of programs.

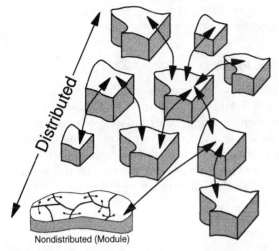

Nondistributed (Module)

Figure 5.2   An application program is part of an application.

they can *interoperate*. Interconnection is a prerequisite for interoperation, whether that connection is tight or loose.

Much academic work has been accomplished on the subject of information flow patterns between tightly and loosely coupled programs, but, unfortunately, little of this work has been accumulated and evaluated inside the computer industry or by computer users at large. Flows among operating system components constitute a particularly well-researched area of study. The flows among operating system components are not unlike flows among application programs. Online messaging opens up the old secrets and internal efficiencies of operating systems to application designers and programmers.

5.1.3   User data (among application programs)

User data flows among application programs across interconnections. Some programs are directly (adjacently) interconnected, while others are indirectly (nonadjacently) interconnected. The user data belongs to the messaging application program.

5.1.4   Control data (among systems)

Control data flows among systems programs across interconnections, often sharing connections with user data flows and being intermixed with user data flows. The control data belongs to the messaging and queuing software and to other systems. It is not accessible or interesting to the messaging application programs.

5.2 Basic Information Flow Patterns

Patterns of flow among application programs are of basic and composite types, as shown in Fig. 5.3. Basic patterns are the simplest building blocks, while composite patterns are aggregates of simple patterns.

In Fig. 5.3, the basic flows are shown in the left column and the composite flows are shown in the right column. For simplicity, only unidirectional (one-way) flows are shown; however, bidirectional flows are more common. Basic patterns in the left column include:

- Transfer
- Relay
- Split
- Join

Composite patterns in the right column include:

- (Extended) dialog
- Tree
- Chain
- Lattice

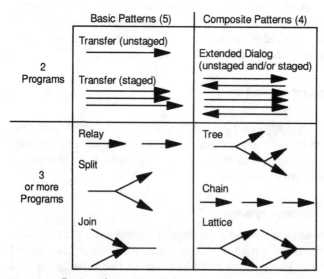

Figure 5.3 Basic and composite information flow patterns.

Basic flow patterns involve exactly two or exactly three programs. Composite flow patterns involve two or many more programs. The basic and composite flows are discussed in the following sections. Both the connections choice and the queues choice are discussed for each flow pattern.

5.2.1 Transfer

A transfer of information involves exactly two programs and is accomplished either in an unstaged manner or in a staged manner.

5.2.1.1 Unstaged transfer. The simplest flow pattern is an unstaged transfer of information whereby all of the information is transferred with a single operation (request from a program). *Unstaged* means one-staged.

A one-way message is an example of an unstaged transfer, as shown in Fig. 5.4. Sending arrival/departure information to a display device in an airport is a simple example of a one-way message flow. A status inquiry message is another simple example of an unstaged transfer, since the entire inquiry is contained within a single message. The inquiry is what is transferred. When connections are used for unstaged transfer, the connection is always transporting just the one message.

To accomplish the equivalent using queues, the queue must be restricted to containing a maximum of one message, as Fig. 5.4 shows. The queue is either empty or it has the allowed one message inside. Unstaged transfers with queues are most effective whenever the two programs are running at almost the exact same speeds and with a steady flow of messages between them of a rather constant length. In

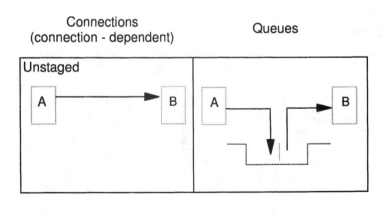

Figure 5.4   Unstaged transfer between programs.

general, a queue size restricted to a maximum of one element is not very satisfactory. It does, however, represent what happens when using a connection for unstaged transfer (that is, lower utilization of the connection over time, since connection usage alternates with program usage).

5.2.1.2 Staged transfer. The next simplest flow pattern is a staged transfer of information whereby all of the information is transferred only by multiple operations (requests from a program). *Staged* means multistaged.

An ordinary file transfer is an example of a staged transfer, as shown in Fig. 5.5. Sending a printout to a printer is a file transfer. The segmentation and reassembly of a single record or message is another example of a staged transfer where program A segments in stages and program B reassembles in stages. With connections, the segmentation usually completes before the reassembly begins, while with queues the reassembly can overlap the segmentation process.

With connections, staged transfer requires the serial operations of programs and connections proceeding generally in a manner such as program A, the connection, program B, the connection, program A, and so forth. Program A sometimes waits on program B; program B sometimes waits on program A.

With queues, staged transfer is the perfect fit. One program fills the queue as fast as it can, while the other empties it. The programs do not wait on each other.

5.2.2 Relay

A serial, unstaged transfer among three programs is a *relay* whenever one program transfers to a second, which transfers to a third. Sending

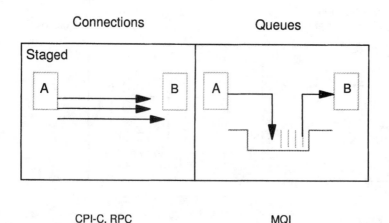

Figure 5.5   Staged transfer between programs.

a message between two PCs through a LAN gateway is a simple example of a relay flow.

A relay of information involves exactly three programs and is accomplished either in a point-to-point manner or in an end-to-end manner.

5.2.2.1 Point-to-point relay. A point-to-point relay involves two protocols: one protocol between the first and second programs in a series and another (same or different) protocol between the second and third programs.

5.2.2.2 End-to-end relay. An end-to-end relay involves one protocol between the first and third programs in a series. The second (middle) program is acting in a transparent manner (for example, like a gateway).

5.2.3 Split

A split of information involves exactly three programs and is accomplished either in an unedited manner or in an edited manner. Sending a single electronic funds transfer (EFT) message to both a debit program and a credit program is a simple example of a split flow.

5.2.3.1 Unedited split. A parallel, unstaged transfer among three programs is an unedited split if the first program simultaneously transfers the same exact information to two other programs. The message is simply copied (replicated).

5.2.3.2 Edited split. A parallel, unstaged transfer among three programs is an edited split if the first program simultaneously transfers the same exact information to one of the two other programs but transfers different information derived from its original input (or created from scratch) to the other of the two programs. The message is *not* simply copied (or replicated); it is edited (or modified to some degree).

5.2.4 Join

A join of information involves exactly three programs and is accomplished either in an unedited manner or in an edited manner. Collecting data, from several locations or entities, for the preparation of a report is a simple example of a join flow.

5.2.4.1 Unedited join. A parallel, unstaged transfer among three programs is an unedited join if one program collects identical information from two other programs.

5.2.4.2 Edited join. A parallel, unstaged transfer among three programs is an edited join if one program collects nonidentical information from two other programs.

5.3   One-way (Open) and Two-way (Closed) Flows

Flow patterns can be open or closed, one-way or two-way, respectively. Open flows are sometimes called nonreturning flows (the loop is open); closed flows are sometimes called returning flows (the loop is closed).

5.3.1   One-way (open) flow type

An open flow means that no message (or set of messages) is expected to be paired (from the opposite direction) with the outgoing message. That is, the message is one-way only. One-way flows are a specialty of online messaging. Open flow patterns are becoming commonplace in PC-based programming with window-based design. They are, however, quite uncommon in more traditional, older program designs.

A supervisor can issue a directive from a window on his or her PC screen and have the directive appear in another window on the PC screen of the subordinate (or subordinates) who are the object of the directive. A one-way flow suffices. Similarly, robotic signals are predominantly one-way messages, except when the robotic device needs to provide feedback.

Open flows are one-way flows, as shown in Fig. 5.6. The Sales Recording program is flowing information in a nonreturning manner into Queue 1; it expects no returning message. The Sales Reporting program is flowing information in a nonreturning manner from Queue 1; it expects to return no message to the Sales Recording program.

Open flows may branch or remain as a single sequence, as explained later in this chapter.

5.3.2   Two-way (closed) flow type

A closed flow means that a message (or set of messages) is expected to be paired with the outgoing message. That is, the message is part of a mes-

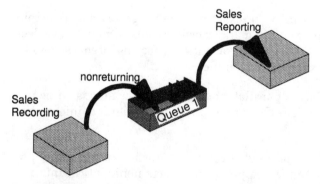

Figure 5.6   One-way-only message flow.

sage pair (or collection). Closed flow patterns are extremely common in more traditional, older program designs. For example, an inquiry message paired with its response message constitutes a closed flow.

Messaging programs emptying and filling queues between them can constitute a closed flow, as shown in Fig. 5.7. Program A is filling Queue 1 (in a direction toward program B) while Program B is filling Queue 2 (in a direction returning to program A). The flow is two-way between programs A and B.

It is usually more efficient (because of contention and queue discipline choices) to have separate queues for messages moving in different directions.

5.4   Composite Information Flow Patterns

Composite flow patterns involve two or more programs and are composed of multiple basic flow patterns in all sorts of permutations and combinations. Composite flow patterns can be open (one-way, nonreturning) or closed (two-way, returning) just as basic flow patterns can.

5.4.1   Dialog

Dialog is an ISO-defined term for combinations of staged and unstaged, closed (two-way) flows. Dialogs are of either the short or the extended type.

5.4.1.1   Short dialog.   Short dialog refers to an ISO-dialog that persists only for a short number of interchanges, for example, two basic transfer flows, staged or unstaged.

5.4.1.2   Extended dialog.   Extended dialog refers to an OSI-dialog that persists for an extended number of interchanges, for example, four or

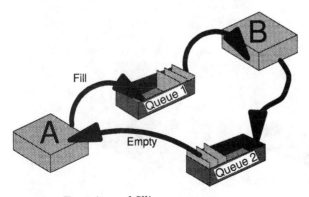

Figure 5.7   Emptying and filling queues.

five basic transfer flows, staged or unstaged. Extended dialog is the most general case of information exchange between two programs.

5.4.2 Tree

A tree structure consists of a series of split basic flows (and constitutes the classical *nested* programming structure).

5.4.2.1 Simple tree. A simple tree has exactly one split basic flow. The split can be into two or more branches. The flow can be open or closed, but it is usually closed (that is, a returning flow).

5.4.2.2 Compound tree. A compound tree has multiple split basic flows in a nested fashion. Each split can be into two or more branches. The flow can be open or closed, but is usually closed (that is, a returning flow).

5.4.3 Chain

A chain structure consists of a series of relay basic flows. A chain structure is much like a degenerate tree where every split has only one branch.

5.4.3.1 Open chain. An open chain consists of a series of nonreturning (one-way) relay basic flows. An open chain has a nonreturning message flow, as shown in Fig. 5.8. The loop remains open.

5.4.3.2 Closed chain. A closed chain consists of a series of nonreturning (one-way) relay basic flows where the last relay flow returns to the first

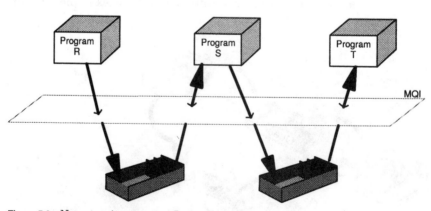

Figure 5.8 Nonreturning message flow.

program. This return is either direct (unnested) or indirect (nested), as explained in the following paragraphs describing nested and unnested closed chains.

Nonreturning flows can connect the last program to the first program to form a closed chain, as shown in Fig. 5.9. The loop becomes closed.

Type 1 closed chain (nested). One type of closed chain, called a nested closed chain, reverses the execution sequence at the end of the chain so as to return to the first element of the chain through all of the intermediate elements. This is actually a special case of an open chain in that the last and first programs are not directly "attached" in the flow pattern. It is, however, usually described as a closed chain. In some sense, it is also a special degenerate case of a closed tree with all branches having only one branch.

Type 2 closed chain (unnested). One type of closed chain, called an unnested closed chain, skips all of the intermediate elements in the chain and returns directly to the first element in the chain. This is the purest case of a closed chain, illustrated in Fig. 5.9.

5.4.4 Lattice

A lattice structure consists of a series of split basic flows followed by join basic flows.

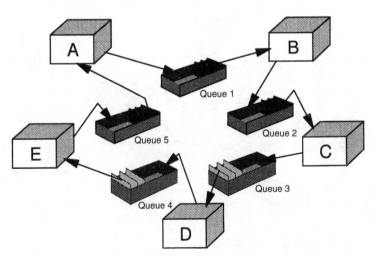

Figure 5.9   Closed chain message flow.

5.4.4.1 Simple lattice. A simple lattice has exactly one split basic flow followed by exactly one join basic flow, as shown in Fig. 5.10. In order for the join operation to be accomplished, the split operation before it in the information flow pattern must place a correlation ID into the messages so that they can be associated with one another when the join is to be performed. In addition, the number of messages to be collected at the join step must be indicated in the split step.

5.4.4.2 Compound lattice. A compound lattice has multiple intermixed split and join basic flows, as shown in Fig. 5.11. In a compound split-join flow pattern, the number of split operations and the number of join operations required as a result must be resolved over the entire flow pattern. In Fig. 5.11, program D is collecting in the join operation three messages with a particular correlation ID, two from Queue 4 and one from Queue 3.

Compound lattice flows are particularly helpful in work flow management solutions that involve not only program operations but also human and device operations.

Online messaging provides an excellent choice for handling lattice flows of all types; other nonmessaging choices are much more difficult (almost impossible) to use in lattice flows.

5.5   Client-Server Information Flow Patterns

The client-server flow pattern (see Ortali [1994] in the Bibliography) is a special case of a closed chain with exactly two programs. There are lots of implications associated with this flow pattern, for example, con-

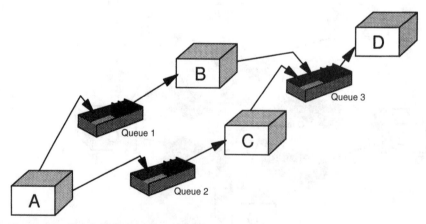

Figure 5.10 Simple split-join message flow.

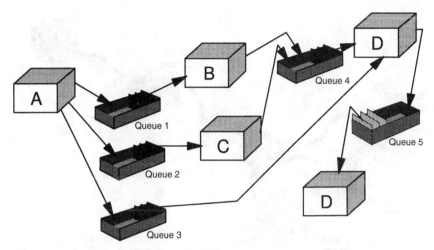

Figure 5.11   Compound split-join message flow.

figurations of a small client machine and a larger server machine and configurations of multiple clients and a single server.

The special case of a two-element closed chain is the basis for all client-server (requestor-server, requestor-responder) flow patterns, as shown in Fig. 5.12. Superimposed over a client-server configuration can be many two-element, closed chains consisting of the client and the server where the server is a single server. In fact, the most common case is a single server with many clients simultaneously requesting services. In Fig. 5.12, many insurance agents (at client locations) are requesting insurance quotations from a single server (at a server location) that has access to a single copy of insurance data. Queue 1 is being used in lieu of separate connections between each client and server. In this way, the server can be emptying the queue as fast as possible without having to service a multitude of connections on a round-robin or other basis. In addition, any servicing priorities can be based upon a queue element rather than upon a particular connection, which makes for faster priority handling.

A server can play both the role of server and the role of client to other servers, in a nested fashion. In one direction, the server is a server; in the other direction, the server is a client.

Again, online messaging is well suited to client-server flow patterns, especially when the sheer numbers of logical connections become expensive and difficult to manage. Client-server flow patterns can be embedded into the middle (or edges) of larger composite flow patterns that are mixtures of all types of flow patterns, simple and complex.

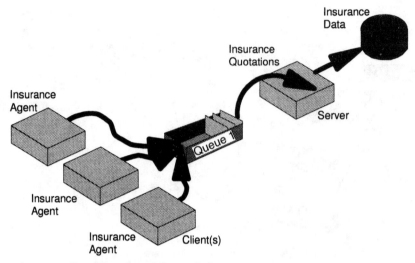

Figure 5.12   Client-server special case chain.

5.6   Transactional Information Flow Patterns

A transaction is a collection of messages with a definite beginning and a definite ending. Transactional flow patterns involve two or more programs and endless combinations of basic and composite flow patterns.

Transaction manager programs usually control the registering of the start and end of a transaction as well as all of the actions that take place between the start and end, including such things as recovery, security, and restart.

Application enablers such as IBM's CICS and IMS manage transactional information flow patterns of all sorts, usually being predefined flow patterns. (See Lamb [1993] and Wipfler [1989] in the Bibliography.)

5.7   Information Flow Pattern Summary

The information flow patterns are summarized as shown in Fig. 5.13. In this summary diagram, the transfer, relay, split, and join basic flows are shown under the A and D columns that show one-way flows. The chain, tree (diverge and converge), and lattice composite flows are shown under the B, C, and E columns. Some observations can be made from this chart:

- The return flow on a closed tree is a series of join flows.
- The relay flow is a special three-element case of a chain flow.
- The reverse tree flow is ideal for report compilation.
- Mixtures of flow patterns are necessary for ideal application solutions.

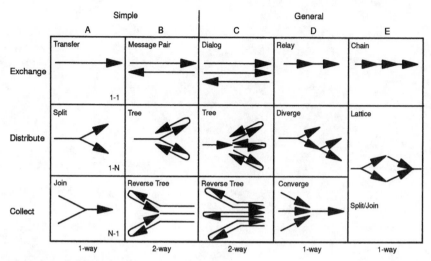

Figure 5.13   Information flow pattern summary.

There are many other observations that can be helpful in using online messaging to accomplish unconstrained information flows in application design.

This chapter has provided the opportunity for examining many flow patterns that have been sometimes ignored, discounted, or hidden from design thinking (for example, closed chains and compound lattices). Now you have the chance to break old design thinking habits and create application flows that better suit business and other procedures without being limited by what the underlying communication software can support easily and naturally. You can escape to messaging. In addition, you can better understand the principles of the more traditional and heavily used flow patterns.

6

Messaging and Queuing Models

There are many application models supported by messaging and queuing middleware, including not only those with client-server information flow patterns but also those with any and all information flow patterns. In fact, most application models (transaction processing and others) are supported quite easily by messaging and queuing, and quite a few application models (all message-driven processing and some work-flow management) are supported only by messaging and queuing.

Simple models, in varying combinations, always constitute the more sophisticated models described later in this chapter.

6.1 Range of MQI Applications

There is a wide range of MQI applications, from simple messaging through sophisticated messaging. Some applications use the MQI directly and some use the MQI indirectly through an application *platform,* which is itself an MQI program, as shown in Fig. 6.1.

The MQSeries family of messaging and queuing products provides support on many IBM and non-IBM operating systems, operating with many networking protocols, and allowing many programming source languages. See App. C for a description of the MQSeries product family as of this writing.

You will notice that application platform writers make the messaging and queuing application design and development even easier by providing yet one more cushion between the application programmer and networking. This extra cushion often encourages the use of smaller, simpler program modules and segments by removing many commonly used functions from the application program.

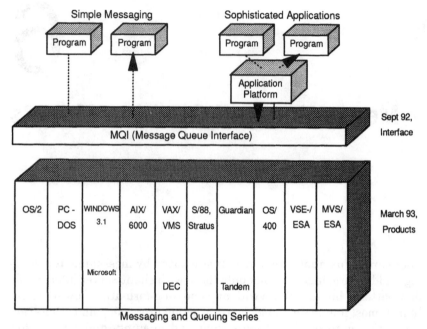

Figure 6.1   Range of MQI applications.

6.2   Simple Messaging Models

Simple messaging models involve application programs using messaging and queuing products directly. Both programs use the MQI to access messaging and queuing services.

6.2.1   A simple program pair

Messaging between exactly two programs (pairs of programs) is the simplest case, as shown in Fig. 6.2. The programs of the simple pair may both be in the same application environment or in different application environments. Messaging and queuing provides the cross-platform communication in a location-transparent manner. That is, Program A in System 1 does not know if the message it places in a queue will be read by Program B in the same system (System 1) or by Program C in a different system (System 2). Remember, the MQI always requires the specification of a queue name* and has no provision for the specification of a program name.

---

* Technically, the queue name can, however, be optionally qualified with the name of a queue manager.

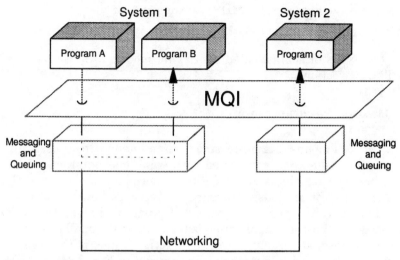

Figure 6.2   Cross-platform communication with MQI.

6.2.2   A simple program set

Messaging among more than two programs (sets of programs) is simply the aggregate of messaging between program pairs, using simple information flow patterns (as described in Chap. 5).

6.3   Sophisticated Messaging Models

Sophisticated messaging models involve application programs using message and queuing products indirectly, through platforms of many types provided by many different software packages.

6.3.1   Stacks of middleware

Application platform or middleware can serve as "helper code" to application programs, helping the application program deal with something more complex, for example, operating systems, database managers, and networks. Middleware can be stacked upon messaging and queuing middleware; that is, platforms (for example, CICS and IMS subsystems or work-flow managers) can be stacked upon messaging and queuing middleware.

6.3.2   Types of sophisticated messaging models

There are two general types of sophisticated messaging models, one newer type and one older type:

Message-driven processing (MDP), the newer type

Transaction processing (TP), the older type

These two types of sophisticated messaging models are discussed in the following sections.

6.3.3   Message-driven processing (MDP) models

Message-driven means that processing is stimulated by the presence of a message in a queue as contrasted with traditional interactive processing between terminals and programs where programs constantly read and write connections to terminals. Messages flow in an unsolicited manner to the programs instead of programs having to solicit the messages. A business application is broken up into small execution sequences identified through a workflow script, which is interpreted and executed by a workflow manager as shown in Fig. 6.3. The IBM Flowmark product is an example of software supporting the MDP model.

Message-driven processing models can be regulated or governed by three types of control:

Message monitor

Queue monitor

Probabilistic

These are described in the following sections. All three types of MDP model control operate to some degree with an absence of state control information in the programs (or program managers); that is, most MDP processes (programs) are stateless (without state tables).

- Business application is broken into discrete functional elements, the MDP processes
- Business application is satisfied by specific execution sequence of the individual MDP processes

  —Processes are started or resumed when messages arrive on a queue

  Hence, the term Message-Driven Processing
- The execution sequence, and its management and control, is defined in a *workflow script*
- The workflow script is interpreted and executed by a *workflow manager*

Figure 6.3   Message-driven processing.

6.3.3.1 Stateful message, stateless process. State information can be stored inside a message and can be totally absent within a process. This contributes to the fastest possible restart of processing whenever a process becomes unavailable, since the message can have indicators of the degree to which the message has been processed and the name of the process that deserves to see this message next.

Digital Equipment Corporation and Tandem, Incorporated have been very successful in providing software that operates with the stateful-message/stateless-process technique.

High-volume, high-performance (HVHP) traffic. Extremely high volumes of traffic and high performance are often thought to be mutually exclusive. High volume with average performance is easy to imagine; conversely, low volume with high performance is also easy to imagine. Having both high volume and high performance is more difficult to imagine and, indeed, to achieve. Messaging helps achieve HVHP traffic objectives.

Fastest process recovery. The fastest recovery possible occurs whenever no connections and no state tables need be recovered following a process (program) failure.

6.3.3.2 Message monitors. A message monitor is a control program that monitors every movement of a message through a messaging system and network. Message monitor and application program executions alternate so that messages thread their way from application to monitor to application and back to monitor. The monitor decides the application message flow pattern among programs (and queues), as shown in Fig. 6.4.

Message monitors provide for absolute control of message flow, but often at the expense of performance if traffic volumes are high and processor speeds are low.

Script-driven processing. Message monitors often use *script* data files to control message flow patterns. In this way, processing is driven by what is written in the script file. Patterns can be changed as required without changing (or recompiling) programs. Small programs are simply used (and reused) in endless combinations with all forms of the information flow patterns described in Chap. 5.

Programming without programming languages. With script-driven processing, systems and programming analysts can alter the contents of script files to change the message flow patterns among application programs

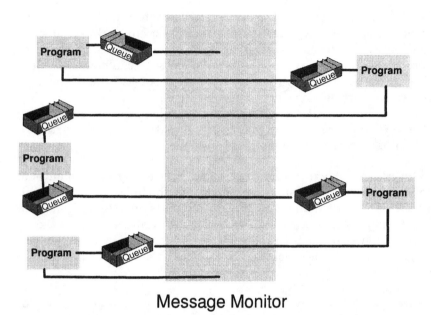

Message Monitor

Figure 6.4   Message queue monitor.

(and queues). New programs can be added, old programs can be deleted, and sequences can be rearranged, all without the use of conventional programming languages.

6.3.3.3 Queue monitors. A queue monitor is a control program that monitors only message queues so as to know how to perform program management. Queue monitors bridge the gap between independently operating queue managers and program managers. If a queue is reaching, or has already reached, a threshold, then the queue monitor can signal a program manager to load and start another copy of a program to process that queue (assuming that application logic allows multiple copies of the same program to be accessing a particular overfilling queue). If a queue remains empty for an extended time, the queue monitor can signal a program manager to deactivate programs accessing that queue. If a queue has no program associated with it, the queue monitor can signal a program manager to load and start the associated program or programs.

Program management. Program managers manage programs; queue managers manage queues. Program and queue management need not be independent, but having this independence allows a wider choice of program management to be coupled with messaging and queuing systems.

Load balancing. Load balancing can be performed on a distributed basis or on a centralized basis. Messaging can easily be used in both cases, as shown in Fig. 6.5, in which programs B, C, and D are identical copies of a program located either in a single system or in different systems. All of the programs are sharing the (message traffic in the queue) load in a balanced manner. When the message traffic load decreases, perhaps program C by itself can handle the load.

Load distribution. Load distribution differs from load balancing, in the most widely accepted way of thinking, by its lack of feedback. Load balancing implies feedback from the processing points; load distribution assumes no feedback and unconditionally distributes messages to particular selected processing points.

6.3.3.4 Probabilistic messaging. Probabilistic messaging is messaging with no queue monitors and with no message monitors. Traffic flows from queue to queue through programs in a natural manner, using chance and expecting a reasonable probability that most programs will have queues to empty and queues to fill most of the time.

Probabilistic messaging involves some degree of risk in that queues are expected to be emptied and filled with some regularity by a collection of messaging programs that may or may not continue to operate properly, continuously, and without terminating abnormally or normally. Messaging traffic and associated applications are often, however, so steady state in nature that probabilistic messaging is economical and sensible.

Predictable traffic. Traffic can be of a predictable or unpredictable sort with respect to origination points, message lengths, message directions, volumes/time period, application mix, criticality, and source-target pairings.

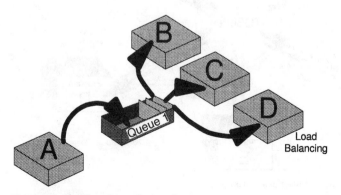

Figure 6.5   Traffic volume fluctuations.

Predetermined source and target. Some message traffic originates regularly at particular *message generation* points and terminates regularly at particular *message absorption* points for processing. That is, the origination and destinations (generation and absorption points) are paired (associated with one another) in a predetermined manner.

6.3.4   Transaction processing models

Transaction processing can be performed in an asynchronous manner as well as in a synchronous manner.

6.3.4.1   Transaction monitors. Transaction monitors are similar to message monitors and queue monitors except that they monitor entire sequences of messages through queues and through programs. They assist in guaranteeing the successful completion of each transaction.

There are many ways to access MQSeries messaging and queuing, as shown in Fig. 6.6. Programs B and C in the diagram are communicating directly through MQSeries products. Program A (a CICS-based program) can communicate through the CICS environment (using the Resource Manager Interface or RMI facility of CICS) to program C (a non-CICS program). Similarly, Program D (an IMS-based program) can communicate through the IMS environment to program C (a non-IMS program).

Now, let's extrapolate to communication between programs A and D, which is also possible on a queue basis through MQSeries.

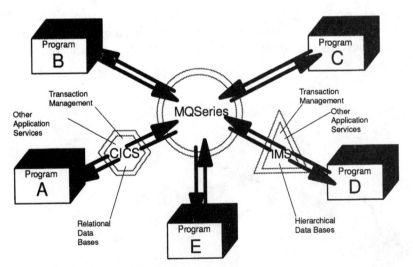

Figure 6.6   Ways to access MQSeries products.

CICS, MQI, and other non-CICS programs can all intercommunicate with the help of queues and the MQSeries product family, as shown in Fig. 6.7. You will notice from the diagram that CICS and MQ operate in entirely separate MVS/ESA address spaces. CICS is not part of MQSeries; MQSeries is not part of CICS.

The very significant function that MQSeries offers is the interoperation of existing CICS programs with newer (or older) non-CICS programs on an automatic basis through MQSeries.

Business and system transactions. There are at least two perspectives on transactions: one from the point of view of the originator of the first message (the initiator of the transaction) and the other from the point of view of the processor of one or more of the messages.

- *Business initiator.* Human-initiated transactions are often described as business transactions incorporating not only computerized operations but also clerical operations, including human think time.

- *System processor.* Subsystem-controlled transactions (for example, CICS transactions) often are involved in processing only a subset of all of the messages within a transaction. For example, a business transaction may consist of one CICS transaction, followed by three clerical operations serially, followed by an IMS transaction.

Terminal managers. Terminal managers control nonprogrammable terminal equipment, for example, IBM 3270 display stations or teletype printers.

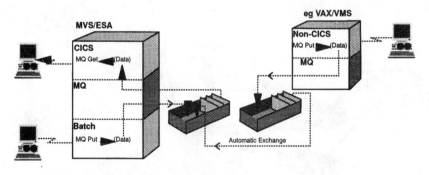

- Gives CICS Access to Messages from:
  - Other Regions within MVS
  - Other Platforms outside MVS

- Allows other Platforms access to CICS without using the CICS API

Figure 6.7 CICS, MQI, and non-CICS.

Database middleware. Database managers are middleware between an application program and a database.

Operating system middleware. CICS, IMS, and other application platforms provide middleware between an application program and a raw operating system service. These platforms also provide other types of middleware services.

Transaction integrity. Some transactions must be completed with a very high degree of integrity and control, especially if they involve the updating of multiple databases that should remain consistent.

6.3.4.2 Process monitors. Some monitors deal exclusively with processes (that is, programs or tasks) and their control. Process monitors monitor programs, not messages or queues.

6.3.4.3 Queued transaction monitors. Some transaction monitors deal exclusively with queued transaction messages, as shown in Fig. 6.8.

New designs and greater business productivity are possible with the messaging and queuing model and the MQI, as shown in Fig. 6.9. In this example, a loan officer has the entire loan processing procedure accomplished with a closed, unnested chain (as described in Chap. 5). Without MDP design and messaging, the bank loan officer might likely perform the many steps of this example in a clerical fashion involving himself and other persons. With MDP design and messaging, the bank loan officer must determine both the applicant's eligibility and the

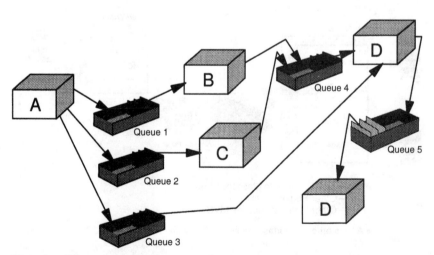

Figure 6.8 Compound split-join message flow.

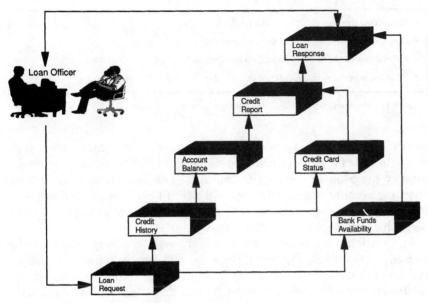

Figure 6.9 New designs, greater business productivity.

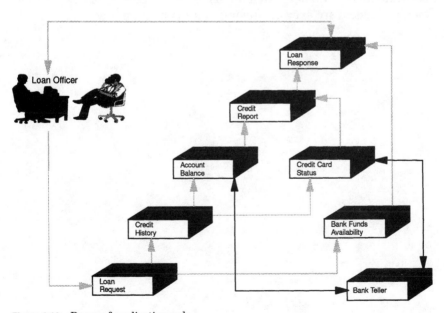

Figure 6.10 Reuse of application code.

- Simplifies the design of complex business solutions
- Encourages the design and development of reusable processes
- Allows processes to be located as appropriate in a heterogeneous network
- Enables processes to be deleted or added quickly as business changes

Figure 6.11   Business advantages of message-driven processing.

bank's ability. The loan request program *splits* the information flow into one flow for the applicant and one flow for the bank. The credit inquiry program in turn splits the applicant information flow into a checking account balance inquiry and a credit card history inquiry. The loan response program *joins* (collects) all of the information to complete the lattice (two splits plus two joins) flow.

Application code, through the use of messaging and queuing, can be reused, as shown in Fig. 6.10. Here, both the bank teller (see bottom right corner of Fig. 6.10) and the loan officer (see top left corner of Fig. 6.10) are communicating using the account balance and credit card status programs.

The business advantages of MDP (message-driven processing) are as shown in Fig. 6.11.

Messaging and queuing models are being recognized and utilized more and more to accomplish more efficient computing. Messaging and queuing models are ever increasing in numbers and in value.

3

The Design

This part of the book describes how to design messaging applications. In order to efficiently and effectively use messaging software, it is necessary to first consider and then properly integrate messaging into any and all application designs where it is appropriate. It is all too easy to overlook places where messaging could (and should) be used and also all too easy to hang on to old assumptions about what information flow patterns are possible. With messaging, all information flow patterns are possible and new, more powerful, more inclusive applications become possible. Messaging promotes intermixing new applications, intermixing existing applications, and intermixing new and existing applications.

Messaging provides asynchronous programming and offers time independence so that logical connections between application programs never exist and serialized executions of application programs are naturally avoided. Programs do not have to take turns waiting on each other and are not constrained by how they communicate with information flows.

Application environments and communication environments are of many diverse types. Some application environments function without any communication environment, but most application environments are interconnected by one or more communication environments.

There are many communication models supported by networking software from many vendors. Queue-based communication is an alternative to connection-based communication. Messaging and queuing is queue-based; conversations and remote procedure calls are usually connection-based.

Distributed applications need distributed communication services. Directories, data definition and conversion, security, resource recovery, and systems management can be either optional, light duty, or extremely vital.

New application development tools are required for messaging. For example, messages are seen as flowing between queues and not between programs. Programs are associated with queues and not with connections.

This part of the book describes:

- *Application and communication environments*
- *Communication models*
- *Distributed computing services*
- *Approaching messaging application development*

Application and Communication Environments

Application environments and communication environments are of many diverse types. This chapter describes both types of environments and their relationships as these relationships relate to messaging and queuing.

Messaging and queuing is one of the forms* of *communication middleware* that sits between an application environment and a communication environment to *isolate applications from communication networking*. This chapter first discusses application environments and then communication environments.

7.1 Applications and Application Enablers

Applications do the useful work for which computers were, and still are, intended. That is, applications apply the computer (or a collection of computers) to some useful purpose or purposes. Spreadsheet programs and game programs are both examples of applications. Balance inquiry and cash dispensing are simple examples of applications performed by a computer within an automated teller machine (ATM). There are tens of thousands of application programs, many operating on the principles of messaging and queuing.

Application enablers help applications do their work more easily and with more efficiency, more control, and more dependability; that is, they enable (or help) the execution of the application programs. The

---

* See Chap. 8 for a complete discussion of several forms of communication middleware access, including CPI-C, RPC, and MQI.

IBM's Customer Information Control System (CICS)™ is one example of an application enabler that is used extensively in many environments. Middleware* of various types can be contained within or outside of application enabler packages. Communication middleware is but one form of middleware; database and many other forms of middleware exist. Messaging and queuing (communication) middleware is found outside of the CICS enabler package but can be accessed by programs from within the CICS application environment.

Applications and application enablers are collections of computer programs that make the computer useful to humans, either directly or indirectly. Humans and devices interact with computers to accomplish work or play through applications and application enablers. Humans use computers to varying degrees, based upon their computer skills, interest in computers, and job responsibilities.

The primary types of devices used with computers are:

Input (keyboard, mouse, image scanner, and more)

Storage (hard file, diskette, CD-ROM, tape cassette, and more)

Communication (high-speed channels, local area networks, wide area networks, and more)

Output (display, print, sound, motion video, and more)

Input, output, and storage devices are attached primarily to a single computer, although they often can be shared among computers; communication devices are between computers.

Applications tie together information from some or all of these types of devices as well as from persons and programs. Applications are generally recognized by humans, while application enablers are generally hidden behind the scene, helping out the applications. The following sections discuss first the applications and then the application enablers. Applications can, of course, operate without application enablers.

7.2 Applications

Applications usually involve programs, information, and signals, along with devices and persons. All of these entities can be distributed across many locations or found at a single location. Programs, devices, and persons are often said to be end users of a computer system, even though "end user" most often implies a person.

---

* The term *middleware* is discussed more fully in Sec. 7.3.1.1.

7.2.1 Humans, computers, and information

Notwithstanding the fun derived from computer games, people generally treasure their computers because of the value of the information that can be derived from them or stored within them and the extreme utility of having them around. A relationship between humans, computers, and information is shown in Fig. 7.1. Information goes into and comes out of computers through the actions of humans. Of course, computers can also operate without the actions of humans (for example, in controlling chemical processes or the flight of an airplane).

7.2.1.1 Humans. Humans need information for personal, business, and other reasons, and computers help greatly when people search for or create information. In the same way that programs need to queue information, humans often need to queue (save, sequentially or otherwise) information.

7.2.1.2 Computers. Computers, through programs, create, analyze, summarize, store, and retrieve information very quickly. Humans use input devices to access a computer and use output devices to get information, as shown in Fig. 7.1. Storage devices save information.

7.2.1.3 Information. Information is an asset that is recognized more and more as a very important and valuable commodity to be protected and nurtured. Information can be derived from data through computer

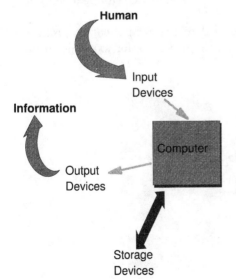

Figure 7.1 Humans, computers, and information.

programs. Information can be derived by programs through the simple retrieval of a single piece of stored data, by associating many pieces of freshly created data with or without retrieving previously stored data, by analyzing signals and events, and by many other means. The terms *information* and *data* are often used interchangeably.

A message can contain a single piece of data, many pieces of data, or only a portion of one piece of data. Application programs must be able to interpret what is stored within each message within a message queue. Messages and message queues are explained in Secs. 3.2.1 and 4.2.1, respectively.

7.2.1.4   Information flow.   At the upper left corner of Fig. 7.2, a human interacts with the computer through one or more input devices (for example, a keyboard, a mouse, a magnetic strip reader, or a scanner). At the middle left of the diagram, information can be returned (for example, through a display device, a printer, or an audio device). At the bottom of the diagram, information is both stored and retrieved. All of these operations happen through computer software programs (executing, in most cases, inside the computer processor shown in the center of Fig. 7.1) and as the dark box in Fig. 7.2. The "S/W" in the oval of Fig. 7.2 means "software program": either input, output, storage, or application software program. There are, of course, many situations where special-purpose software is operating entirely separate from the main computer within processors that are part of the input, storage, and output devices. That is, device access software can be housed both in the device and in the main computer. The device can be a computer (or set of computers) as well.

Figure 7.2 demonstrates the central role that application programs play in our use of computers. The human input (for example, from a

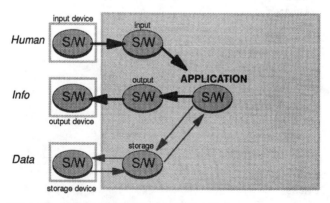

Figure 7.2   Human input and computer output.

keyboard or mouse input device) is delivered to the application program or perhaps, equivalently, the application program solicits the human input from the input device. The next action depends entirely upon the application program. Consequently, we have placed the application program (or collection of programs) in the middle of Fig. 7.2. It may or many not be necessary to store or retrieve data from the storage device before delivering information to the human through the output device (for example, a display screen or a sound speaker).

Information can flow among programs, devices, and persons in endless combinations of patterns, as explained in Chap. 5. Messaging and queuing makes all information flow patterns, including client-server and peer-to-peer, much easier to accomplish.

7.2.2 Programmers and application software development

The application types of computer software (for example, those programs required for business purposes) are developed by programmers, as shown in Fig. 7.3. The programmer, using the software development tool programs loaded into the computer, creates a collection of programs and stores them within a program library on a storage device (usually a disk) for later use. This program library is separate from normal data stored on a computer. Programs are executable by the computer; data is usually not executable, but some data (control blocks and control files) can control the execution of programs. During the creation of these application programs, the programmer uses the information provided through output devices to "see and touch" his or her programs. Application programs can be retrieved from the storage devices, executed, and tested to see that they work properly.

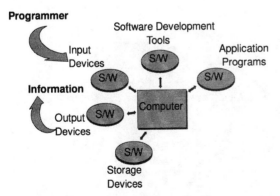

Figure 7.3 Programmers and application program development.

Figure 7.4 demonstrates the central role that software development tools (usually collections of many different software packages) play in the creation of application programs. First, the programmer, using an input device (for example, a keyboard) and a selected text editor from the tools collection (or from operating system software), writes the source programming language statements and stores them on a storage device. Next, he or she uses a language translator tool (a compiler or assembler utility program) to transform one or more source language files into one or more program load modules, which are then stored in a ready-to-run program library on a storage device. During the program development cycle (code, compile, test, and such), the output devices provide information about the application programs being created for later production use. It is also possible to write a program that is not translated from one language to another but is, instead, interpreted and executed in its original form by a program interpreter, but this form of program execution is very inefficient except for small programs or programs that do not need to execute quickly.

7.2.3 Application and storage programs

Application programs can use *storage helper* programs to assist them in using storage devices, as shown in Fig. 7.5. Entire database systems exist for the sole purpose of assisting applications to access data on storage devices. Hierarchical, flat, relational, indexed, and other data organizations are all supported by various database systems.

In general, storage devices can contain both nonexecutable data and executable programs. Programs and information are both stored. Programs can be distributed, information can be distributed, or they can both be distributed.

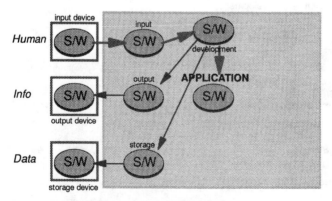

Figure 7.4 Application program development.

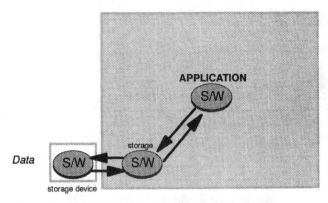

Figure 7.5   Applications and storage programs.

Often the questions become:

Do we distribute only programs or only data? or both?

Should data be located very near the programs?

Should data master files be disjointly partitioned across locations?

Should identical programs exist at multiple locations?

How can data master files be kept consistent (identical within some time period)?

Many storage programs allow for accessing information stored on remote systems, in which case the storage programs themselves use communication programs (discussed in the next section) much as any ordinary application program would. This is known as remote data access or distributed data systems.

7.2.4   Application and communication programs

At first, application programs dealt directly with communication devices, but this resulted in very difficult programming, special skills, and much training of many programmers. Later, communication helper programs became available to the application and enabled easier communication between computers, as shown in Fig. 7.6. For example, the IBM OS/2 Communications Manager contains networking programs for allowing PC-based application programs to communicate across LAN communication devices with other PC-based application programs.

7.2.5   Application development and production environments

The application development and production environments, when combined, result in a collection of software as shown in Fig. 7.7. This dia-

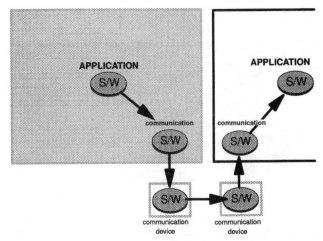

Figure 7.6   Applications and communication programs.

gram shows the application development environment superimposed over the application production environment. Application development tools are of value only within the application development environment. In a similar way, communication devices are usually of value only with the application production environment, since most application development (except for testing) is accomplished within a single local operating environment. The source environment where an application is developed need not be the same as the target environment where an application is in production.

The application programs are central to application processing, as shown in Fig. 7.8. The left of the diagram shows humans receiving information and storing data through the use of the application programs. The left of the diagram shows helper programs of various sorts

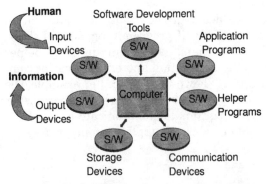

Figure 7.7 Application development and production environments.

assisting the application program in the center of the diagram. General helper programs at the top right, storage helper program at the bottom left, and communication helper programs at the bottom right all assist the application program to operate in the local application environment (and to interoperate with other application environments).

The application production environment is the environment that supports the continuous or continuing execution of application programs. It does not include the development environment that makes the execution of applications possible.

Applications do useful work with the help of application enablers, which are discussed in the next section.

7.3   Application Enablers

Application enablers are special collections of software that assist, help, or enable applications to accomplish their work. Most programs written recently or now being written depend upon one or more application enablers and cannot operate without the assistance provided by the enablers. Obviously some programs, new or older, can operate without enablers, whereby they either provide enabler-type services for themselves or do without.

In the past, application enablers have also been called platforms, subsystems, application environments, and application systems. Anything that enables applications to execute can be classified as an application enabler. In the earliest days of computing, the application programmer had to write all of the programs necessary to enable his or her program to execute; now there are many helper programs called enablers to relieve the application programmer of some of the pro-

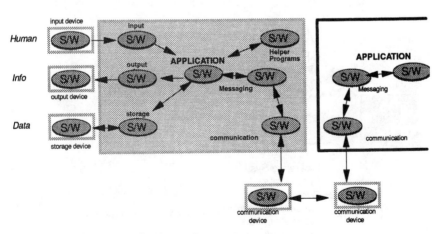

Figure 7.8   Application production environment.

gramming burden to ensure that the application can not only execute but execute with efficiency and dependability.

Specifically, many application enablers help applications with access to output, storage, and communication devices.

7.3.1   Application and helper programs

Application programs benefit greatly from helper programs that perform routine but often complex functions such as accessing data storage devices and communicating over networks. The lower right corner of Fig. 7.9 lists some common helper programs that have been developed over the last several decades.

Operating system helper programs make it much easier to use basic operating system functions. Database helper programs make storing and retrieving data quick, easy, and reliable. Communication networking helper programs (such as IBM's MQSeries™ family of products) enable communication within and between applications without regard for the type of networks being utilized. Transaction control helper programs allow applications to define a starting point and an end point for a series of messages that constitutes a transaction. IBM's CICS and IMS™ are two examples of transaction control helper programs. Information flow helper programs assist applications in associating information flows. Information flow patterns are discussed in Chap. 5.

Other types of helper programs (for multimedia operations, desktop publishing, library research, and such) are being created constantly. The application programs are central to application processing, as shown in Fig. 7.10. With an abundance of helper programs now available, it is much easier to write application programs.

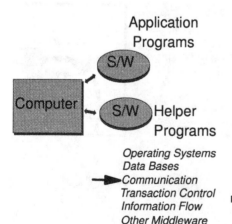

Figure 7.9   Helper programs.

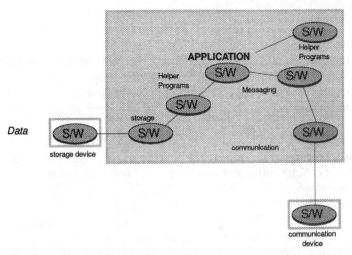

Figure 7.10   Helper programs and devices.

7.3.1.1 Early middleware. The very earliest example of middleware (that is, software "in the middle") was a computer operating system (a system to help operate the computer even before computers were interconnected by networks) that cushioned the programs from hardware devices and eliminated the need for constant human button pushing and punched card hopper filling and emptying operations. This middleware was never labeled as middleware, but it was, indeed, in the middle between humans and devices and between humans and programs.

7.3.1.2 Application platforms. Application platforms provide helper code (a cushion) between the application program and something else more difficult. Application platforms have been called application enablers, application services, application subsystems, application support, and application environments. "Platform" is a rather imprecise term in that some platforms are associated primarily with a particular operating system, others with a particular networking protocol, and still others with a particular programming language. "Platform" is, however, a widely used and familiar term.

Having application programs deal directly with communication devices is very costly and inefficient. First, large masses of computer programmers must be taught the complexities of the communication devices. Next, large numbers of application programs must contain network-specific communication logic and large-sized error recovery procedures. The very sensible solution has been to reduce the demand for most programmers to understand networking and to trim away the communication code from application programs that should be dedicated solely to application purposes (not communication purposes). Hence, today fewer programmers need to understand networking and fewer programs need to execute networking of their own. Skill demands are reduced for masses of programmers and execution efficiencies are improved for masses of programs. Application programmers can now be dedicated to their purpose (that is, application development) instead of to communication programming (that is, to a nonproductive and costly purpose).

Application programs can now use helper programs to assist them in using communication programs (and communication devices), as shown in Fig. 7.11. Messaging and queuing is helper program software for queue-based networking. SNA logical unit of type 6.2 (LU 6.2) is helper program software for connection-based networking. Some application enablers, such as IBM's CICS and IMS subsystems, provide access to both queue-based and connection-based helper programs.

Each set of application programs can access other sets of application programs, within the same application or within another application, through networking programs and communication devices, as shown earlier in Fig. 7.8. Application programs may choose to use the messaging middleware shown in the middle of Fig. 7.10 or bypass it and use communication software directly through connection-based APIs. When using networking directly through connection-based APIs, program names (and networking parameters) are specified, but when using networking indirectly through queue-based APIs, queue names (and no program names or networking parameters) are specified, as shown in Fig. 7.12. In the figure, Program A specifies Program B (plus networking parameters) for connection-based communication or Queue J7 for queue-based communication. Connection-based and queue-based APIs are thoroughly discussed in Sec. 8.3.

7.3.2 Distributed application environments

Applications are collections or sets of application programs, as shown in Fig. 7.13. Some of the application programs can be tightly coupled, as shown in Fig. 7.14, in a nondistributed application or portion of an application. There is no networking involved in the traditional sense of

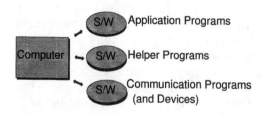

Figure 7.11 Helper programs and communication.

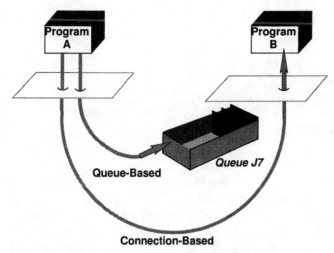

Figure 7.12   Connection and queue specification.

communication networking with communication devices. Programs
are interacting within a single load module or across computer address
spaces within a single processor. The arrows show interactions where
programs are paired with other programs. Some programs interact
with only one other program; others interact, in a paired fashion, with
multiple other programs.

Some examples of applications that are often nondistributed include
simple desktop publishing, inventory control within a single shop, and
a retail point-of-sale terminal with cassette tapes to record sales transactions for later transmission to a central processing site.

A single application can also be distributed across many environments. Some of the application programs can be loosely coupled, as
shown in Fig. 7.15, in a distributed application or portion of an appli-

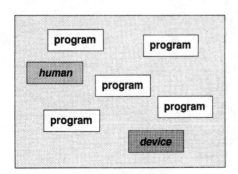

APPLICATION

A collection of programs
(and human and device
operations) that serves
some useful purpose.

Figure 7.13   Application definition.

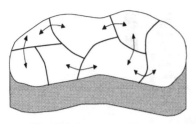

Figure 7.14   Nondistributed application.

Nondistributed (Module)

cation. The programs within the application can be interconnected in multiple manners. Messaging and queuing can be used to accomplish communication between some application programs, while other methods are used between other application programs. All too often in the past, a single method of communication was chosen exclusively for all communication programming within an application or between applications. Sometimes this was an easy choice because there was only a single network available for use for application intercommunication. Today, with lots of network choices available in most computer installations, the choices are not so easy to understand and make.

Simplifying Fig. 7.18 into Fig. 7.16, we can see that application programs must depend upon communication devices for communication between programs. Application information flows first into a communication device, then into another communication device before flowing

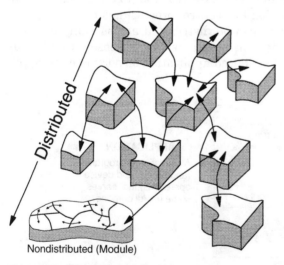

Nondistributed (Module)

Figure 7.15   Distributed application.

eventually into the target application, as shown in Fig. 7.16. The application programs are logically connected, through communication devices (one or more), as shown in Fig. 7.17.

7.3.2.1 Communication middleware. Middleware to shield developers from networking has been rather slow in its evolution. The large variety and extreme complexity of most communication networks has had a stalling effect on efforts to create communication middleware.

As shown in Fig. 7.18, application environments can be "cushioned" with middleware or they can exist in a very primitive form where programs have lots to do other than those things that deal directly with the purpose of the application. Communication middleware takes over from application programs the role of dealing directly with networking programs. Applications can now deal with networks indirectly through communication middleware, such as that provided by messaging and queuing, as shown in Fig. 7.19.

Applications do not need to do networking; middleware can do it instead. Logical connections can exist between communication middleware as well as between application programs, as shown in Fig. 7.20. The physical connections are between the communication devices. Ignoring the physical communication devices and physical

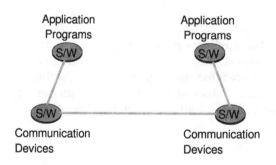

Figure 7.16   Application-network information flow.

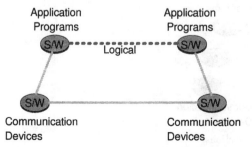

Figure 7.17   Logical connection between programs.

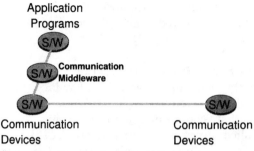

Application
Programs

Communication
Middleware

Communication   Communication
Devices   Devices

Figure 7.18   Communication middleware.

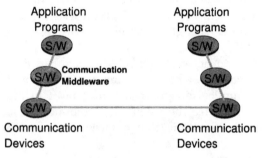

Application   Application
Programs   Programs

Communication
Middleware

Communication   Communication
Devices   Devices

Figure 7.19   Distributed communication middleware.

connections between them, the applications and middleware both can have their separate logical connections, as shown in Fig. 7.21. The logical connections between applications are separate from the logical connections between application helper programs, such as messaging and queuing helper programs. MQI programs have no logical connections, only queues.

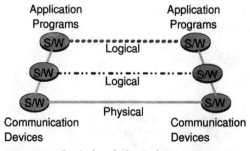

Application   Application
Programs   Programs

Logical

Logical

Physical

Communication   Communication
Devices   Devices

Figure 7.20   Logical and physical connections.

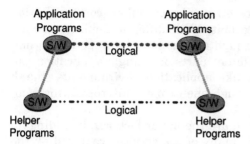

Application
Programs

Application
Programs

Logical

Logical

Helper
Programs

Helper
Programs

Figure 7.21   Application and helper logical connections.

7.3.3   Applications and networks

Applications need networks; networks support applications. Application environments and communication environments are both of many diverse types. Communication middleware, which may or may not be part of an application enabler, sits between an application environment and a communication environment to isolate applications from networking. Entire applications and parts of applications within different application environments can interoperate through a collection of communication environments, as shown in Fig. 7.22. In the diagram, "A" within the oval shape means "application environment" and "C" within the oval shape means "communication environment." At the top of the diagram, the A environments communicate through a single C environment. At the right side of the diagram, the A environments communicate through two C environments.

Each application environment can be associated with one or more communication environments. Each communication environment can

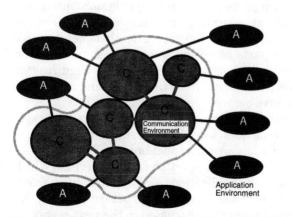

Figure 7.22   Interoperating application and communication environments.

interoperate with one or more other like or unlike communication environments. Separate applications within different application environments can interoperate through one or more interoperating communication environments. Different parts of a single application can also be in different like or unlike application environments (thank goodness) and interoperate through one or more interoperating communication environments.

Operating systems quite often are the major factor in bounding an application environment, since many of the services required by the application are provided only by the operating system. For example, an application developed with the OS/2 operating system is naturally separated from an application developed with the Unix operating system because of the basic differences between the operating systems. In addition, operating systems usually have great affinities with particular communication network types. For example, the various UNIX-type operating systems have an affinity with the TCP/IP network type, and the OS/400 operating system has an affinity with the SNA network type.

Choices for programming source languages, macro languages, CALL libraries, and compile-time options, as well as choices for processor hardware and run-time facilities, also bound an application environment. Messaging and queuing is a natural and easy choice for interoperations among application programs from different application environments, regardless of whether the programs are within a single application or in separate applications. Messaging is a queue-to-queue mechanism that substitutes for a program-to-program mechanism. It is easier to adapt queues to different environments than it is to adapt programs to different environments. Queues are always nonexecutable; programs are always executable.

If two applications (or two parts of the same application) are developed not only with different operating systems but also oriented toward different network types, then interoperation between the two applications or application parts becomes, of course, less natural and more difficult (but certainly not impossible). Such a case requires two nonadjacent application environments to be interoperating across two adjacent communication environments, as shown in Fig. 7.23.

Applications enablers help applications in their use of input, storage, communication, and output devices, as well as helping with the control of the application itself. In some cases, application enabling products are packaged to include application support and other services, which are explained later in this book. Applications and application enablers interoperate with the help of lower-layer infrastructure (that is, the communication environment), which is discussed in the next section of this book.

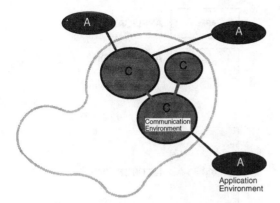

Figure 7.23   Interoperating applications or application parts.

7.4   Communication Environments

Communication environments support application environments, as explained in the previous section.

Many different groups in the computing industry have been involved during the last decade in developing computing reference models—some models for operating systems, some for databases, some for application systems, and some for communications networking—but only recently have efforts begun in earnest to combine these various models into a single, more complete, yet simpler reference model. Such a generalized model can be based easily upon established networking models.

There are, indeed, many different technologies available in today's marketplace that provide solutions for a variety of networking problems, but to comprehend how these technologies function, one must start with a reference model. One of the best reference models that exists today for networking is the OSI Reference Model. We will utilize this reference model for all discussion throughout this book in a slightly simplified format.

Figure 7.24 relates the terminology used in this chapter (and in the remainder of the book) to an approximation of the terminology used in the ISO's OSI networking model. Referring to the right side of Fig. 7.24, the application programs (the APPL box in the diagram) conditionally depend upon the application support programs (the SUPP box in the diagram), which in turn conditionally depend upon the networking programs (the NETG box in the diagram). That is, sometime during their execution, application programs may require application support programs (or in some cases they may not). Likewise, sometime during their execution, application support programs may require networking programs (or in some cases they may not).

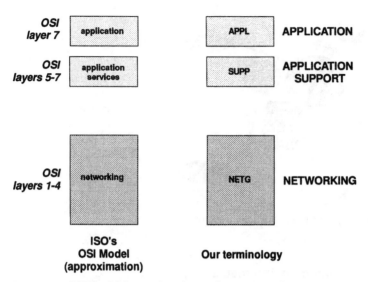

| OSI layer 7 | application |
| OSI layers 5-7 | application services |

| APPL | APPLICATION |
| SUPP | APPLICATION SUPPORT |

| OSI layers 1-4 | networking |

| NETG | NETWORKING |

ISO's
OSI Model
(approximation)

Our terminology

Figure 7.24   OSI model (approximation) and our terminology.

Referring to the left side of Fig. 7.24 and speaking loosely or as an approximation, the "bottom" (four layers) of the OSI model accomplishes networking; the "top" (three layers) of the OSI model accomplishes application and application services (which can be considered part of application enabling). In the pure OSI networking model, application services are defined to be entirely within layer 7 (at the bottom of the application layer). We have stretched the OSI application services definition to include those communication middleware and other services provided by OSI layers 5 and 6.

We are striving in this book to relate our single open networking model (that is, our generalized reference model) to other models—for example, to established application system models and networking models at large. Relative to networking, the simple model shown in Fig. 7.24 is extremely similar to the more detailed IEEE-POSIX Open System Environment (OSE) model,* as shown on the extreme left side of Fig. 7.25, using just the "communication entities" (our networking), "application platform entities" (our application support), and "application entities" (our application) parts of the OSE model.

---

* You will find the POSIX-OSE model described in the P1003.0/D16.1 draft dated October 1993, sponsored by the Portable Applications Standards Committee of the IEEE Computer Society. Many other computing models are also similar to the POSIX OSE model, but, when simplified, most models reduce to our simple model with respect to networking and ignoring the "people" and "information exchange" entities represented in the complete OSE model.

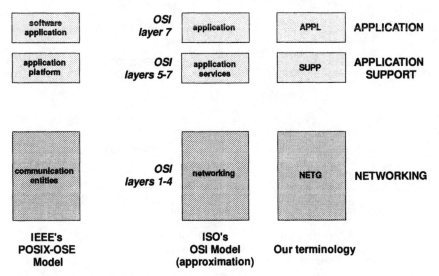

software application	OSI layer 7	application	APPL	APPLICATION
application platform	OSI layers 5-7	application services	SUPP	APPLICATION SUPPORT
communication entities	OSI layers 1-4	networking	NETG	NETWORKING

| IEEE's POSIX-OSE Model | ISO's OSI Model (approximation) | Our terminology |

Figure 7.25   OSE, OSI, and our terminologies.

The OSE model (see the extreme left of Fig. 7.25), although not, strictly speaking, a layered model, shows communication entities (or networking) programs being separated from software application programs by application platform programs. We have labeled this separator software simply "application support" (SUPP) in our simple reference model. Other models, similar to OSI (an older model) and OSE (a newer model), have similar labeling. (See Halsall [1988] in the Bibliography.)

The OSI networking layers (layers 1 through 4) can be subdivided, approximately in half, into transport networking (layers 3 and 4) and subnetworking (layers 1 and 2), as shown in Fig. 7.26, thus making four layers instead of three in our simple model. These four layers (or software program groups) provide the basis for the discussions in the remainder of this book. The "bottom" two layers can always be regarded as just networking.[†] The "top" two layers can always be regarded as just the application environment. Application and application support (boxes) constitute the "top" of the model; transport network and subnetworking (boxes) constitute the "bottom" of the model.

Figure 7.27 identifies this useful equation used throughout the remainder of this book:

$$COMPUTING = APPL + SUPP + TPORT + SNETG$$

[†] The subequation NETG = TPORT + SNETG causes the original three layers (boxes) to be expanded into four layers (boxes) in our simple model.

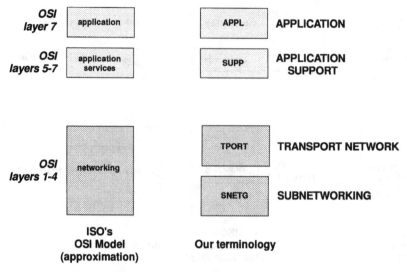

ISO's
OSI Model   Our terminology
(approximation)

Figure 7.26   OSI model (approximation) and our terminology, expanded.

where each of the four layers (software program groups) interacts only with its adjacent layers (above and below). The diagrams in the remainder of this book use the short labels "APPL," "SUPP," "TPORT," and "SNETG" for the four layers of our computing model.

Collectively, the four layers constitute computing,* as the equation reflects. In our diagrams, the top two layers are closely associated with one another, as are the bottom two layers. Messaging and queuing is in the bottom layer of the top two layers.

---

* Computing can be accomplished at various points in time by using either all of the elements on the right side of the computing equation or just some of them. That is, computing involves, at various times, just applications (APPL), or applications and application support (APPL and SUPP), or applications and application support and networking (APPL, SUPP, TPORT, and SNETG).

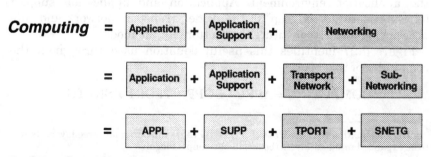

Figure 7.27   Some computing equations.

Referring to our previous discussion of communications helper programs in Sec. 7.3, the layers of our simple model can be depicted as
shown in Fig. 7.28.

All discussion is from the perspective of a particular application's
required application support and communications infrastructure on a
system. Briefly, the four major groupings previously identified are as
follows:

- *Application.* Abbreviated APPL-A for a particular application A
  under consideration. Note that this application can be user-developed
  to satisfy a particular business requirement, or it can be a more
  generic, prepackaged, off-the-shelf purchased application, such as an
  X.400 mail application or an X.500 directory server.

- *Application support.* This has two divisions, called the *application
  programming interface* (abbreviated API-A for a particular application
  A) and *application support* (abbreviated SUPP-A). The API and the
  application support are closely tied together and are chosen by the
  application programmer based upon the requirements of the application. Examples of application programming interfaces include CPI-C
  (Common Programming Interface for Communications), RPC (Remote
  Procedure Call), and MQI (Message Queue Interface). Depending
  upon the API selected, the application services may be quite different.
  For instance, CPI-C usually utilizes SNA logical unit 6.2 (LU 6.2) services, which includes the protocol flows between two applications for
  establishing a conversation, exchanging data, ensuring commitment
  of resources, and terminating a conversation. RPC does networking
  through program stubs that are customized for each application program and then attached (linked), and RPC usually operates over
  TCP/IP protocols. MQI provides queue-to-queue communication, in

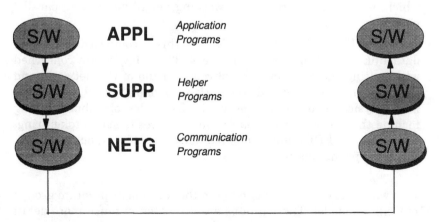

Figure 7.28   Application programs and networking.

lieu of direct program-to-program communication over a dedicated logical connection, and is a form of communication middleware with resource commitment and assured message delivery. MQI operates over LU 6.2, TCP/IP, and other networking protocols.

- *Transport network.* This includes the critical *transport* and *network* OSI layers, and is abbreviated TPORT-A for a particular application A. These two layers work closely together to ensure that user data is transmitted with a predictable level of service from the source node to an end node, perhaps through a set of intermediate nodes. Depending upon the specific protocol chosen, these layers provide such functions as optimal route determination, error detection and recovery, and congestion control. Examples of transport protocols include TCP/IP and SNA APPN. Each of these protocols utilizes unique addressing structures, protocol flows between peer transport layers in end nodes, and routing protocols between intermediary nodes. Please note that, throughout this book, the term *transport protocol* will refer to the combination of these two OSI layers (unless explicitly identified as OSI layer 3 or 4) to match commonly used nomenclature in the computer industry. (See Cypser [1991] in the Bibliography.)

   Also note that, historically, the application support and transport network have been very closely tied together, and the selection of a particular API forced the selection of a particular network protocol—or, conversely, a programmer was forced to select an API based on the currently supported transport protocol in the network. For instance, Remote Procedure Call (RPC) and the TCP sockets interface are both closely tied with the TCP/IP transport protocol, and would be an application programming interface of choice for a TCP/IP-based transport network; however, if a CPI-C-based application might solve a particular business requirement, then SNA transport would have to be added to this TCP/IP network to support the CPI-C-based application—which, without multiprotocol networking, might necessitate parallel networks.

- *Subnetworking.* Abbreviated SNETG, this is composed of the OSI data link control and physical layers. These layers are concerned with getting the data onto the physical media of the network, and then getting it reliably and efficiently from one physical node to the next physical node in the network. Examples of subnetworking include the various local area network choices (such as token ring, Ethernet, and FDDI) and wide area network choices (such as SDLC, X.25, and frame relay).

This was a very brief introduction to the four major program groups of our simple model. These groups are discussed in much more detail throughout the remainder of the book.

7.5   Some Fundamental Networking Concepts

The fundamental concepts necessary for an easier understanding of the relationship between applications and networking are:

Stacks of software

Switching points

Model layers (or software program groups)

Application (APPL)

Application support (SUPP)

Networking (NETG)

Transport Network (TPORT)

Subnetworking (SNETG)

You will recognize these concepts as being related to the computing equation introduced earlier (that is, Computing = APPL + SUPP + TPORT + SNETG). The four layers are stacked, switching points exist between each layer, and each layer serves some purpose. These concepts are discussed in the following sections.

7.5.1   Stacks of software

As evidenced by both the POSIX-OSE computing model and the OSI networking model,* it is often convenient to think of groups of computer programs as being stacked,† one group on top of another, on top of another. Figure 7.29 illustrates this stacks of software concept for:

- Application
- Application support
- Transport network
- Subnetworking

software program groups or layers.

---

* The OSI networking model is mostly concerned with layering of software and interoperations between like layers and is not so much concerned with interoperations between adjacent unlike layers; that is, OSI is concerned more with protocols and less with interfaces. Furthermore, OSI is mostly concerned with consolidating techniques (protocols and formats) into a single technique and disallowing dissimilar techniques.

† Each group of programs has its own set (or sets) of formats (information bit patterns) and protocols (information exchange rules); hence, part of or all of this stack is often referred to as a protocol stack, whereby a format stack is also implied. Of course, each protocol stack is composed of a particular collection of protocols stacked upon one another; not all possible protocols are included.

Layers, Levels, Stacks, Groups

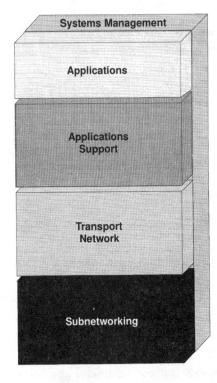

Figure 7.29   Stacks of software.

An *application program interface* (API) is a mechanism that defines interoperation between the application (APPL) and application support (SUPP) program groups. The API is a contract of sorts whereby the application software (above the API) requests particular functions and the application support software (beneath the API) supports the interface by interpreting the requests (program calls), executing them, and returning the results across the API to the application software.

The application (APPL) group of programs is often divided into particular subgroups or application sets or *application suites,* according to the purpose of each application. The application support (SUPP) group of programs is often divided into several general types of application support, according to what the support does for the application.

A *network access* is a mechanism that defines interoperation between the application support (SUPP) and the networking (TPORT and SNETG) program groups. The network access operates in much the same way as the API but instead operates between a networking user (for example, an application support program) and a networking provider (for example, a transport network program).

The networking (TPORT and SNETG) group of programs is often divided into particular types of networking, according to the functions provided and the communications medium being utilized.

The various subgroups of the aforementioned main groups (layers) of programs are discussed in more detail later in this chapter.

Lots of different technologies exist for interoperations between these groups (and between subgroups within these groups). This chapter identifies many of these technologies, particularly those in the networking groups (that is, in the transport network and in the subnetworking program groups). (See IBM [GG24-4338] in the Bibliography.)

Together, the transport network (TPORT) and subnetworking (SNETG) constitute the networking (NETG) program group with its network access mechanism. Access to networking is almost always through the transport network so that the transport network access and the network access, as shown in the middle of Fig. 7.30, are usually synonymous. Subnetworking is usually accessed only indirectly through a transport network.

In most of the figures within this chapter, you will see the networking group of programs replaced by two groups of programs, the transport network group and the subnetworking group, so that the stack of software will appear as in Fig. 7.30 (that is, with the NETG box expanded into the TPORT and SNETG boxes).

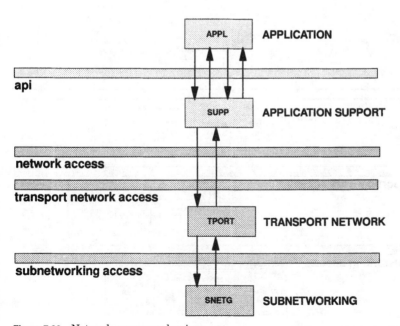

Figure 7.30   Network access mechanism.

7.5.2 Switching points

The four program groups (or model layers) discussed previously are separated by three dividers between the four groups, as shown in Fig. 7.31. In the best case, these group-separating dividers (or APIs or access points) allow for a selection of software beneath each in such a way that each divider serves as a switching point. Each software program group (layer) is a collection of programs. The programs within each group can interact, directly and indirectly, with programs in other groups above and below it. For example, in Fig. 7.31, application program A7 is being supported by application support program S14, which in turn is supported by transport program T5, which in turn is supported by subnetworking program W14. This might, for example, represent an invoice printing application program using the OSI-FTP application support program to transmit an invoice batch over a TCP/IP transport network using a LAN between the invoice printing application program and the invoice printing device.

Because of the manner in which software has been developed (and packaged) for each of these software program groups, there are some affinities between particular types of programs within each layer. For example, a particular type of application support might be available only with a particular type of transport network, which is available with only two types of subnetworking. It is the very rare case that all

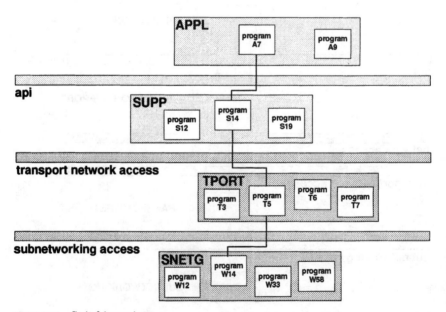

Figure 7.31   Switching points.

layers are found within a single software package; usually the layers are distributed across many software packages where each package may span layers or exist entirely within a layer.

The next sections discuss the four layers (application, application support, transport network, and subnetworking) of our model in more detail. Transport network (TPORT) and subnetworking (SNETG) are sometimes discussed collectively under networking (NETG).

7.5.3 Application definition

A computer *application* is best described as a collection of programs (often intermixed with human and device operations) that serves some useful purpose. For example, a credit inquiry application and a high-way traffic control application are both serving a particular useful purpose (that is, getting information about credit worthiness, and controlling automobiles and trucks on a highway system, respectively).

Figure 7.32 illustrates the general concept of an application. A collection of computer programs does the information processing while interacting with one or more humans and devices (input devices, data storage devices, printing devices, display devices, communications devices, and such). Humans very often instigate interactions among the parts of an application, for example, whenever a human interacts with a workstation to print a report.

For our purposes within this book, an application is simply a collection of application programs.

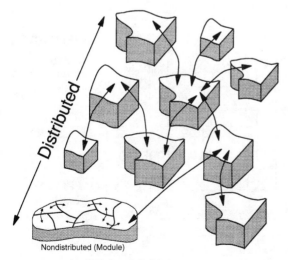

Nondistributed (Module)

Figure 7.32   Application definition.

7.5.3.1 Nondistributed application. A *nondistributed application* is best described as an application existing entirely within one location. As shown in Fig. 7.33, all of the programs of a nondistributed application are at the same location (for example, within one computer processor at location 1). Similarly, all of the application support is at one location (for example, within one computer processor's software collection). No networking is required. (Figure 7.33 is an alternate form of Fig. 7.14, which shows a nondistributed application with no networking.)

Interoperations can be application-to-application through application support (see flows 1, 2, and 3 at the middle left part of Fig. 7.33), or, by contrast, simply application-to-application support and back (see flow 4 at the bottom of Fig. 7.33). Of course, much of the application processing can be executed within each program.

7.5.3.2 Distributed application. A *distributed application* is best described as an application existing within two or more locations. As shown in Fig. 7.34, some of the programs of a distributed application are at one location (for example, within one computer processor at location 1), while others are at another location (for example, within one computer processor at location 2). Networking is required between the locations, and the network access mechanism must be used by the application support programs at each location. Figure 7.34 is an alternate form of Fig. 7.15. In this example, program A1 (within application A) is communicating with program A6 (also within application A but at a different location). The application is distributed across locations, or is a distributed application.

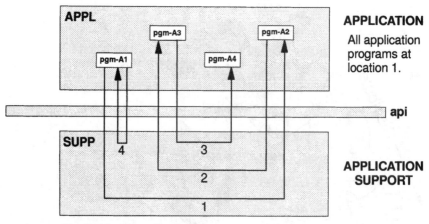

Figure 7.33   Nondistributed application.

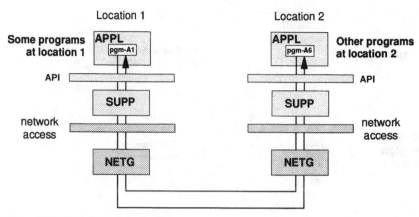

Figure 7.34   Distributed application.

In the remainder of this book, the label "APPL-A*x*" is often used in figures and in text as a short label for "application *program* A*x*." For example, APPL-A4 should be read as "application program name A4." Similarly, APPL-A (without the number) is usually meant to denote the entire application collection of programs. For example, APPL-A should be read as "Application A, consisting of many programs."

7.5.4   Application support definition

Application support is best described as a set of software services useful to an application. As shown in Fig. 7.35, application support is generally composed of multiple software packages, each offering a particular set of helpful functions. Application support comprises helper programs containing functions (grouped together into services) that can be used by ordinary application programs. Application support is simultaneously available to multiple application programs and saves each program from having to support itself. Individual application programs access functions within services within application support through one or more application program interfaces (that is, interface to and from application support), commonly called just APIs.*

Application support is usually packaged so that multiple APIs, one or more for each package, are required to access all of the functions available within the total aggregate of application support.

---

* In a reciprocal manner, application support programs are accessing individual application programs through the same API.

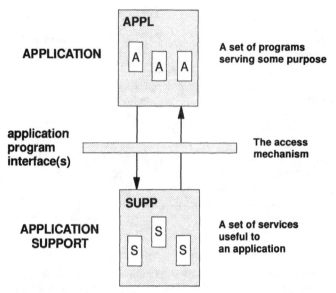

Figure 7.35   Application support definition.

7.5.4.1   Types of application support.   Among the many types of application support (SUPP) software are those types of a more global, less local nature, such as:

Communications (that is, networking)

Standard applications (that is, industry utilities)

Distributed services (that is, sets of common application functions)

which are shown in Fig. 7.36 and discussed in this chapter, and those of a less global, more local nature, such as:

Presentation services (that is, human/device interfaces)

Data access services (that is, information storage/retrieval methods)

Application services (that is, application control functions)

Object-oriented services (that is, evolving object management functions)

which are discussed later in this chapter.

Individual application programs almost never need all of the application support available; most application programs use application support selectively and repetitively; some application programs never need any of it. The following sections describe the more global (more network-oriented) application support previously identified.

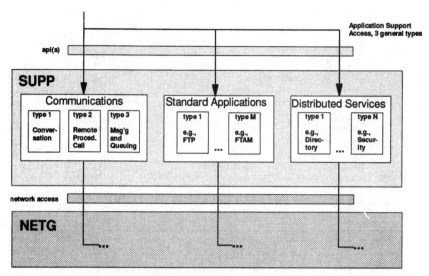

Figure 7.36   Some types of application support.

7.5.4.2 Communications support. Communications support programs support communication within and between applications. The communications support programs provide application-to-application (commonly called either program-to-program or queue-to-queue) communication, as shown in Fig. 7.37. There are many styles of communications support, providing differing functions and differing degrees of functions. Figure 7.37 shows queue-based application support (specifically, messaging and queuing), which differs from connection-based application support. Both queue-based and connection-based application support, however, use networking for intercommunication. These are discussed and contrasted in Chap. 8.

7.5.4.3 Standard applications support. Standard application programs support communication within and between applications or end users, for example, e-mail between humans (X.400 or SMTP) or file transfer (FTAM or FTP) between systems, as shown in Fig. 7.38. These are applications that have become "standardized" through constant and heavy usage over a long period of time by large numbers of organizations. This program group essentially constitutes a tried-and-tested set of industry utility programs for moving information (for example, large files) across networks.

7.5.4.4 Distributed services support. Distributed services programs provide ancillary support for communication within and between applications. These services are typically used selectively (for example,

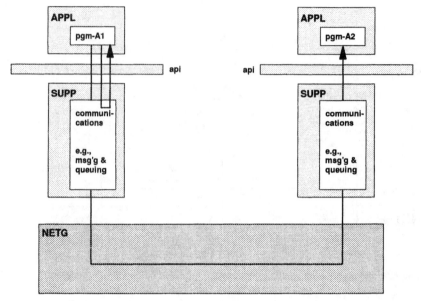

Figure 7.37   Communications support.

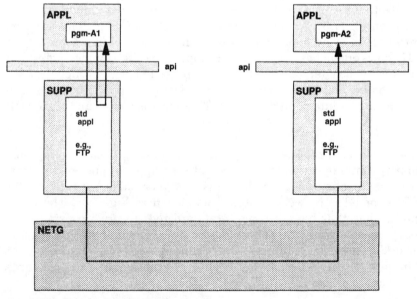

Figure 7.38   Standard applications support.

directory or security) or as an enhancement to normal application processing (for example, transaction management or resource recovery). See Fig. 7.39.

Distributed services are, for the most part, in a constant state of evolution in the computer industry. There are no uniform sets of distributed services available yet, except for such consortium offerings as OSF-DCE. Distributed computing services are discussed in detail in Chap. 9.

7.5.5 Networking definition

Networking is best described as a set of software services accomplishing communication between computer systems. Individual application support programs access functions within services within networking through the network access mechanism or set of mechanisms (see Fig. 7.40). A particular program S from within the application support group selects a particular program T from within the networking group. Program T provides the networking; program S uses the networking.

7.5.5.1 Networking componentry.   Networking (NETG) is composed of two parts:

- Transport network (TPORT)

- Subnetworking (SNETG)

as shown in Fig. 7.41. That is, NETG = TPORT + SNETG.

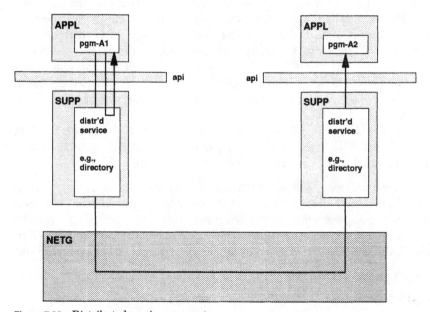

Figure 7.39   Distributed services support.

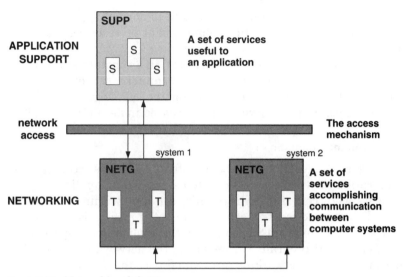

Figure 7.40   Networking definition.

The TPORT box is composed of several selectable transport network types with the implication that only one is used at a time. Likewise, the SNETG box is composed of several selectable subnetworking types with the implication that only one is used at a time.

In general (but based on product availabilities), a network-using (SUPP) program can mix and match types of transport networks and types of subnetworking, as shown in Fig. 7.42. In this example, TCP/IP was chosen for the transport network and LAN was chosen for the subnetworking. It is as if a giant rotary switch exists between the SUPP and TPORT layers such that, by rotating the dial, the SUPP can choose exactly one of the TPORT choices. Similarly, the TPORT can choose exactly one of the SNETG choices. Of course, this rotary switch choosing is based solely upon available networking products that offer these devices.

The number of mix-and-match combinations between SUPP, TPORT, and SNETG available today is limited, but this number is destined to grow with more choices becoming available.

Transport network definition.   A transport network is best described as a collection of networking programs that exchange information between and among adjacent and nonadjacent computer systems using a variety of available communications media. The transport network (TPORT in Fig. 7.43) is composed of selectable transport network types, including:

- SNA transport
- APPN transport

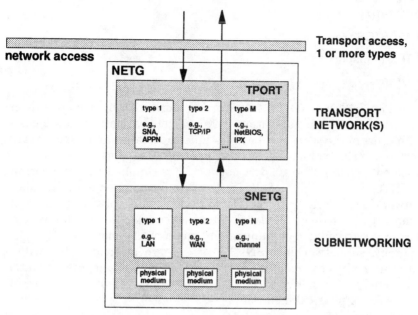

Figure 7.41   Networking componentry.

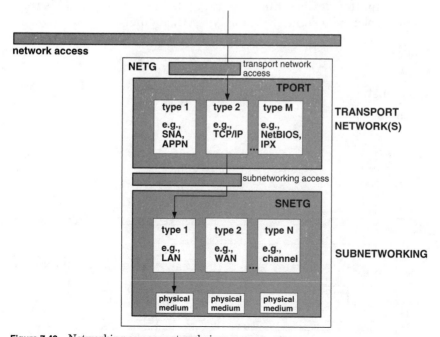

Figure 7.42   Networking componentry choices.

- TCP/IP transport
- NetBIOS transport
- IPX transport
- Others

In Fig. 7.43, a NetBIOS transport network is being used to support conversational communication between applications. In most cases, both users of the transport network must select the same type of transport network. In Fig. 7.43, both application support programs have selected a NetBIOS transport network. For simplicity, subnetworking (SNETG) is purposely not shown. In a similar manner, many of today's networking diagrams purposely exclude subnetworking to promote simplicity. A *network cloud* is often substituted for subnetworking across a particular communications medium, as shown in Fig. 7.44. Thereby, subnetworking (SNETG) layers are interconnected with a particular physical medium and transport network layers are interconnected with a network cloud, which is composed of subnetworking layers and a physical medium.

Subnetworking definition. Subnetworking is best described as a collection of networking programs that exchange information between immediately adjacent (or logically adjacent) physical communications/computing devices. Subnetworking (SNETG in Fig. 7.45) is composed of selectable subnetworking types, including:

- LAN
- WAN
- Channel
- Others

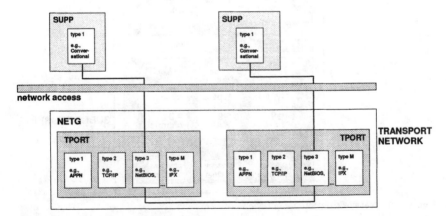

Figure 7.43 Transport network definition.

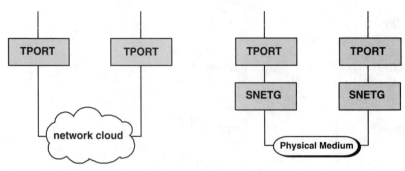

Figure 7.44   Transport network cloud and subnetworking medium.

In this example diagram, LAN (local area network) subnetworking is being used to support a TCP/IP transport network. A very good, simple introduction to the physical components of networking can be found in Derfler (see Bibliography), parts 3, 4, and 5.

7.5.5.2 Repeaters and multiplexors. Repeaters and multiplexors span subnetworking (SNETG) stacks "on the bottom side," as shown in Fig. 7.46. In this way, subnetworking stacks are connected to one another through a physical medium, which is most often not discussed in subnetworking descriptions.

Also, for the sake of consistency, we will use the following definitions.

Repeaters. A repeater operates at the physical layer (OSI layer 1), simply extending the physical characteristics of the network by regenerating signals so that the optimum performance in terms of signal quality and distance can be achieved. Repeaters can sometimes also

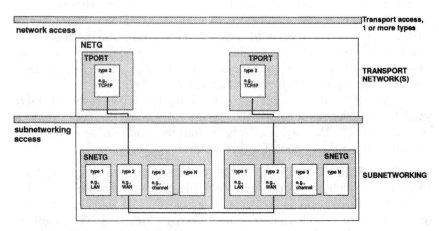

Figure 7.45   Subnetworking definition.

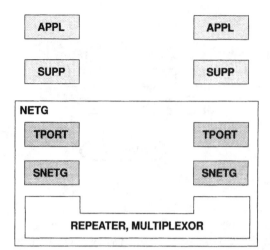

Figure 7.46 Repeaters and multiplexors.

provide media conversion from one type to another, for example, from fiber optics to copper.

Multiplexors. A multiplexor operates at the physical layer (OSI layer 1), taking data bits from several devices and interleaving this data onto a single physical link. Such methods as time-division multiplexing (TDM) or frequency-division multiplexing (FDM), may be used along with statistical division multiplexing (SDM) approaches to maximize the number of devices that can share a single link. Multiplexors manage the bandwidth available on the serial link and, hence, are often called *bandwidth managers*.

7.5.5.3 Bridges, routers, and gateways. Aside from the reference model, some basic concepts and terminology must be defined so that particular technologies can be better understood. For instance, terms such as "bridge," "router," and "gateway" keep reappearing in various types of literature relating to networking, but the definitions of these terms may not be consistent. Figure 7.47 illustrates the "formal" definition of these terms with regard to the OSI model.

In general, *bridges* span subnetworking (SNETG) layers, *routers* span transport network (TPORT) layers, and *gateways* span application support (SUPP) layers. Bridges, routers, and gateways all serve the same general purpose of interconnection of software.

Bridges. Bridges effectively "melt" two LANs of like types together. In general, bridges handle protocols like-to-like, not like-to-unlike. Bridges operate at the media access control (MAC) sublayer of the data link control (DLC), OSI layer 2, as shown in Fig. 7.48. They connect

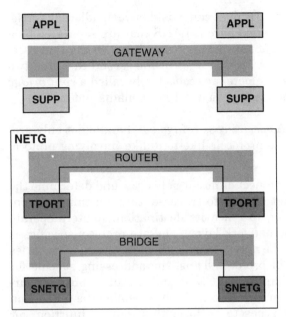

Figure 7.47   Bridges, routers, and gateways.

two LAN segments, and forward frames from one LAN segment to the other. Minimal processing is needed for this processing, and bridges can be quite efficient. Since all the processing takes place at the OSI layer 2 and does not involve the higher layers, bridges are often referred to as being *protocol-independent*. A bridge does not care if it transports TCP/IP, DECNet, or OSI protocols, which all operate at OSI layers 3 and above.

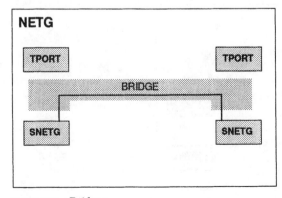

Figure 7.48   Bridges.

Local and split bridges.   Bridges can exist between adjacent LANs, or they can exist between nonadjacent LANS that are separated by a WAN (or other non-LAN subnetworking technology). In the first case, the single bridge is called a *local bridge*. In the second case, the two bridges (one local and one remote) are collectively called a *split bridge* (one part in one LAN and another in different, nonadjacent LAN).

Protocol converters.   A *special-case bridge* can handle not only protocols like-to-like, but also protocols like-to-unlike (involving protocol conversion).

Routers.   Routers interconnect different networks, and determine the optimal path for a packet of data to traverse, through this interconnected series of networks, to reach its destination. Routers operate through OSI layer 3 (the network layer), and are protocol-specific, as reflected in Fig. 7.49. At the network layer, the network address varies according to the choice of protocol. Thus, the addressing scheme for SNA will be different than for DECNet, and a router must be purchased for each of the desired protocols. Routers enable the connection of different subnetwork types to perform this routing function, for instance, connecting a LAN to a frame relay wide area network for a specific OSI layer 3 protocol.

A *brouter* is a combination of a remote bridge and a router, performing the functions of both. Often it will route the protocols that it knows how to route, and bridge all other protocols.

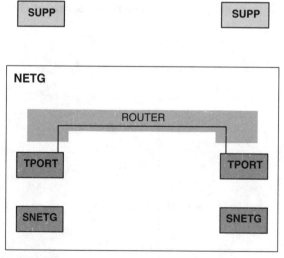

Figure 7.49   Routers.

Gateways. The most basic (and earliest) concept of a gateway is that entity (software and hardware, combined) which funnels traffic from one network to another. It is that point where networks touch. The networks are usually of different types, for example, a collection of devices (that is, a device network) attached to a communications network. In the middle 1970s (and even today), devices in a subscriber network could be hidden (addresswise) behind the gateway where they were collectively attached to a service network. Today with LANs, the LAN gateway is the funnel point for all LAN devices to access non-LAN devices (for example, between LANs and WANs) or to access LAN devices on a LAN of a different type.

Gateways generally operate above the network layer (layer 3) and may involve all layers. The key to a gateway is conversion. A gateway will convert all layers up through the layer at which it operates. A common type of gateway is the application gateway, which is very application-specific and which converts all seven layers as needed, as reflected by Fig. 7.50. An electronic mail gateway, which takes office mail from one type of electronic mail system (such as IBM PROFS) and converts it to another mail format (such as DEC All-In-One), is an example of this type of gateway.

There are gateways that provide *protocol conversion* from one session and data stream type to another, such as VT100 to 3270. These gateways usually operate at the higher OSI layers.

Another special type of a gateway is a transport gateway spanning transport network (TPORT) layers, where different transport facilities

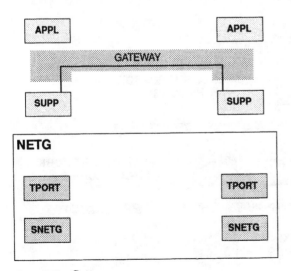

Figure 7.50 Gateways.

are matched to each other. Examples of transport gateways include a NetBIOS application transported over a TCP/IP network, or a TCP/IP application run over an SNA network. These gateways perform a relay function, where the original application information is relayed over one or more transport protocols to its final destination. There is usually extra code provided at the transport level to compensate for the change in transport protocol and to permit the relaying of the application information through the network, without impacting the application program itself.

Technologies can be grouped by layers

Subnetworking

Transport network

Application support

as shown in Fig. 7.51.

In Fig. 7.51, programs within the application program group (layer) are labeled program A, indicating an application program. Similarly, programs within the application program group are labeled program S, indicating a support program. In the transport network program group, programs are labeled program T, indicating a transport network program. In the subnetworking program group, programs are labeled program W, indicating the "wire" (or equivalent physical communication medium).

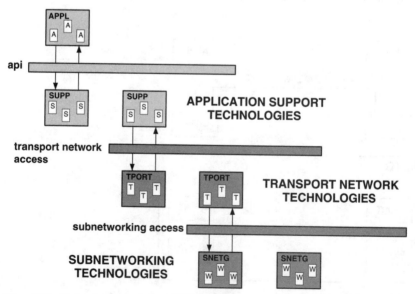

Figure 7.51 Technologies by layers (SNETG, TPORT, SUPP).

7.5.6 Direct and indirect networking

Ideally, only application support programs, and not ordinary application programs, are involved in networking, as shown in Fig. 7.40. There are cases, as shown in Fig. 7.52, where ordinary application programs deal directly with networking instead of indirectly with networking through application support. These cases are often to effect extremely high efficiency and performance at the expense of application program simplicity, portability, and maintainability.

Indirect networking saves enormous amounts of application development time and complexity. It shields the developer from the complexities of networking.

7.5.7 Simple networking

Simple networking involves just a single network interconnecting two systems, within each of which are application programs, as shown in Fig. 7.53. With simple networking, program A1 and program A2 within the same application intercommunicate (indirectly through the SUPP layer) with great simplicity and efficiency across a single network of some particular type. The networking is used by the SUPP software and provided by the TPORT software. The application support is used by the APPL software and provided by the SUPP software.

7.5.8 Separated (private) networking

Separated networking involves two (or more) disjoint networks each interconnecting two (or more) systems. Each separate network has its

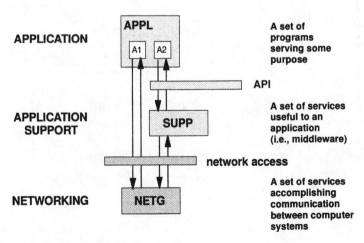

APPLICATION — APPL — A1 — A2 — A set of programs serving some purpose

API

APPLICATION SUPPORT — SUPP — A set of services useful to an application (i.e., middleware)

network access

NETWORKING — NETG — A set of services accomplishing communication between computer systems

Figure 7.52   Direct and indirect networking.

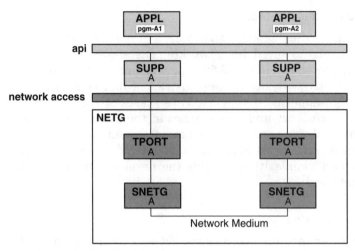

Figure 7.53 Simple networking.

own *private* or separate connections, even if several of the networks are of the same type (for example, two separate NetBIOS networks). Nothing is shared, logically or physically.

In Fig. 7.54, program A1 and program A2 are intercommunicating across a network A, while program B1 and program B2 are intercommunicating across a different network B.

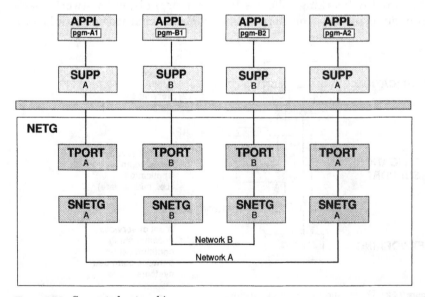

Figure 7.54 Separated networking.

7.5.9 Combined (shared) networking

Combined networking involves two (or more) separate networks operating as one combined network. In Fig. 7.55, program A1 and program A2 are intercommunicating across a network A/B, while program B1 and program B2 are intercommunicating across the same shared network A/B.

7.6 Communications Models Galore

Many communications* networking models have been created in the past several decades. Standards bodies have them; computer vendors have them; universities have them; everyone has them. See Fig. 7.56. Each of these models serves as a thinking tool to organize the myriad components of a typical communications system or set of systems. Since messaging and queuing software sits on the edges of both application software and communication software environments (as discussed in Chap. 7) and between the environments as middleware, it is useful to consider how messaging and queuing relates to other communication middleware varieties within various communications models.

---

* The plural term "communications" is often used interchangeably with the singular term "communication" in the computer industry, especially when it is used as an adjective referring to systems (plural).

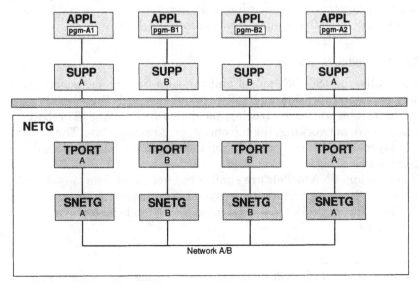

Figure 7.55 Combined networking.

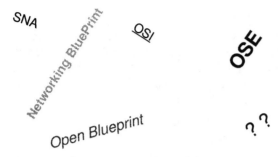

Figure 7.56   Communications models galore.

Some models are depicted as layered into a stack of distinct layers somewhat like the floors of an office building; other models are depicted as layered into a set of concentric shells somewhat like the skins of an onion. Still other models have no layers at all and describe only the association between groups of software in a multidimensional sense.

7.6.1 OSE model

One of the easiest-to-understand models is the IEEE-POSIX* Open System Environment (OSE) model, as shown in Fig. 7.57, that includes not only communication and application software entities but also human and device entities. The OSE model acknowledges that interfaces are necessary between programs and other programs, between programs and devices, and between programs and people. Messaging and queuing is a set of programs between ordinary application programs and the networking programs that control communications devices.

7.6.2 OSI and SNA models

Historically, the ISO-OSI model and the IBM-SNA model concentrate solely on communication and software entities with the emphasis on communication software layering, as shown in Fig. 7.58. The bottom six layers are networking; the top one layer is applications. The bottom three layers are physical networking; the next three layers upward are logical networking.

The OSI and SNA models are similar but not equal. The lowest three layers (1/2/3) of each model correspond to physical networking, the next three layers (4/5/6) to logical networking, and the top layer (7) to applications.

---

* Institute of Electrical and Electronics Engineers' Portable Operating System Interprocess Execution.

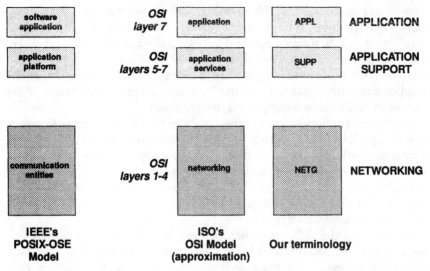

Figure 7.57   IEEE-POSIX Open System Environment (OSE) model, simplified.

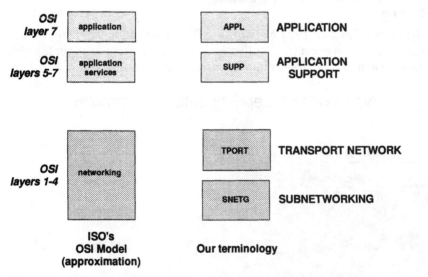

Figure 7.58   ISO's OSI and IBM's SNA model approximations.

7.7   Networking Blueprint

In March 1992, IBM introduced a computing model that incorporated the SNA networking model and many other models, in a superimposed manner, to reflect the abundance of networking protocols and the desirability for open systems networking. This "Networking Blueprint," as shown in Fig. 7.59, subdivides the networking environment into transport network and subnetworking. It also subdivides the application environment into application and application support. Thus the seven-layered OSI and SNA (and other) models have been absorbed into a four-layered blueprint. Application program interfaces separate the application and application suport layers, giving applications their access to application support and networking.

Basically, subnetworking corresponds to the bottom three layers of the OSI or SNA model, which provide for communications between two physically adjacent communications devices separated by a communications link or other physical connection medium, such as a LAN medium. The transport network corresponds to layer 4 of the OSI or SNA model and provides logical paths between adjacent or nonadjacent communications devices.

The Networking Blueprint has two major versions, the original version as it was introduced in March 1992, as shown in Fig. 7.59, and a generalized version introduced in September 1992, as shown in Fig. 7.60.

7.7.1 Original March 1992 Networking Blueprint

The original March 1992 Networking Blueprint identified three examples of APIs (between the applications and application support layers), one for each of the three communications programming styles.

Networking Blueprint (featuring examples)

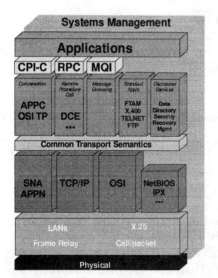

Applications

Multivendor
Application Support

Multi-Protocol
Networking

Subnetworking

Systems Management

Figure 7.59 Original Networking Blueprint.

Networking Blueprint

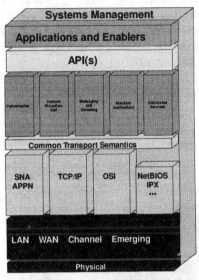

Applications

Multivendor Application Support

Multi-Protocol Networking

Subnetworking

Systems Management

Figure 7.60   Networking Blueprint.

The standard applications and distributed services portions of the application support layer were represented without example APIs. Some representative examples of the types of application support were included.

The March 1992 Networking Blueprint also identified some specific representative subnetworking technologies, but some technology groups were not represented.

7.7.2   Generalized September 1992 Networking Blueprint

The generalized September 1992 Networking Blueprint identified a nonspecific *set of APIs* (between the applications and application support layers) for accessing all parts of application support. This was more generic to reflect the many varieties or APIs that were already being used to access application support of the types identified.

The types of application support were generalized by removing the representative examples of support. Application enablers (collections of application helper programs) were added to the applications layer to reflect the availability of this type of software.

The September 1992 Networking Blueprint also identified broader categories of subnetworking technologies to reflect the many types of communication subnetworks.

7.8   Open Blueprint

In April 1994, IBM introduced an applications and communication combined model that not only incorporated the Networking Blueprint from two years earlier, but also expanded the applications environment that is discussed earlier in this chapter. The "Open Blueprint," as shown in Fig. 7.61, shows the application environment at the top and the communications environment at the bottom.

The Open Blueprint and many other communications models include both connection-based and queue-based communication methods.

Open Blueprint

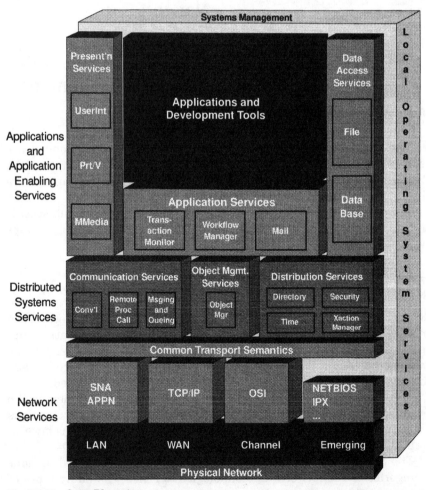

Figure 7.61   Open Blueprint.

The Open Blueprint is a superset of the Networking Blueprint. The Networking Blueprint focused on the communication environment; the Open Blueprint details the application environment but also includes the Networking Blueprint in total. Stated differently, the Networking Blueprint simplified the representation of the application environment; the Open Blueprint expands the representation of the application environment.

The Networking Blueprint recognized, but did not fully elaborate on the application environment. The Open Blueprint expands (explodes and subdivides) the application environment (the applications and application support layers) to reveal the software machinery available to an application program.

The three major Open Blueprint componentry groups are:

- Applications and Application Enabling Services

- Distributed Systems Services

- Network Services (same as the NETG layer in our simple model)

7.8.1   Open Blueprint application environment

The top two componentry groups constitute the Open Blueprint application environment.

7.8.1.1   Open Blueprint applications and application enabling services.   The applications layer is labeled *applications and application enabling services,* which is subdivided into subgroups:

- Application development tools, shown as part of the applications

- Presentation services

- Data access services

- Application services

The application enabling services surround the applications in Fig. 7.61 in a U-shaped manner.

Application development tools.   Application development tools were not detailed in the Networking Blueprint but have been added to the Open Blueprint. Application development tools deal with creating and changing an application.

Presentation services.   Presentation services were not detailed in the Networking Blueprint but have been added to the Open Blueprint. Presentation services deal with input/output between programs and persons, or between programs and input/output devices. Presentation services include services for user interface, printing/viewing, and multimedia operations.

User interface. The user interface presentation services deal with presenting data and information to human beings through various devices. These interfaces can be of the newer graphical user interface (GUI) type or the older keyboard and other types. The same user interface presentation services *receive* data from, as well as *present* data to, human beings through devices and programs.

Print / view. The print/view presentation services deal with printing and displaying information.

Multimedia. The multimedia presentation services deal with preparing information for multiple forms of media, including voice, video, image, and data.

Data access services. Data access services were not detailed in the Networking Blueprint but have been added to the Open Blueprint. Data access services deal with storage and retrieval of information to and from a storage medium. Data access services include both file and database access.

File. File data access services deal with direct access to individual data files.

Database. Database data access services deal with access to collections of data files having hierarchical, relational, or object-oriented organizations.

Application services. Application services were not detailed in the Networking Blueprint but have been added to the Open Blueprint. Application services deal with helper programs that provide functions useful in controlling applications, their processing, and their data.

The standard applications of the Networking Blueprint have been absorbed into the application services of the Open Blueprint. These are applications that have become "standardized" through frequent and heavy usage over many years by many computing installations.

Application services include transaction monitors, workflow managers, and mail systems.

Transaction monitors. Transaction monitors* control transaction processing such as that accomplished by such transaction processing systems as CICS and IMS.

---

* See also *transaction manager* under Distribution Service in the Open Blueprint.

Workflow managers. Workflow managers deal with the integration and control of work activities accomplished by persons, devices, and programs.

Mail systems. Mail systems deal with the distribution of notes, documents, and files between one person and one or more other persons.

7.8.1.2 Open Blueprint distributed systems services. The application support layer is labeled *distributed system services,* which is subdivided into subgroups (of services):

- Communication services, for communication within and between applications

  Conversational

  Remote procedure call (caller-callee)

  Messaging and queuing

- Object management services, for object-oriented application processing

- Distribution* services, for services useful to a distributed application

These application support or distributed systems services are described in the following sections.

Communication services. The communication services are the leftmost three stacks of the application support layer of the Networking Blueprint. The choices include programming with CPI-C, RPC, and MQI. Communication services deal with communication within and between application programs.

The communication services are subdivided into conversational, remote procedure call, and messaging and queuing choices.

Conversational communication. Conversational communication, introduced with APPC and SNA LU 6.2 in 1981, provides connection-based, program-to-program communication using CPI-C, APPC, EXEC-CICS, or other similar interfaces and SEND, RECEIVE, CONFIRM, and other verbs. TCP/IP's sockets interface provides a lower-level conversational interface.

Remote procedure call communication. Remote procedure call communication, introduced with various RPCs (remote procedure calls) from SUN, OSF-DCE, and other sources in the middle 1980s, provides

---

* Note the distinction between the *distributed system* services componentry group and the *distribution* services within that group.

connection-based, program-to-program communication using both CALL and RETURN statements (CALL in the caller program, RETURN in the callee program).

Messaging and queuing communication. Messaging and queuing communication, introduced with MQI (Message Queue Interface) and the MQSeries product in 1992, provides queue-based, queue-to-queue communication using MQPUT and MQGET verbs.

Object management services. Object management services were not in the Networking Blueprint but have been added to the Open Blueprint. Object management services deal with object-oriented processing, an emerging technology.

Distribution services. Distribution services were labeled "distributed services" in the Networking Blueprint. Distribution services deal with services required by distributed applications. Among these distribution services are directory, security, time, and transaction manager.

Directory. Directory services provide a means to locate programs, persons, and other entities. Both centralized and distributed directories provide location-independent mapping information between names and names, between names and addresses, and other mappings.

Security. Security services provide access control, authentication, and other functions.

Time. Time services provide the time of day with a consistent value across geographical regions in a variety of formats.

Transaction manager. A transaction manager* provides coordination control in the starting, execution, and ending of a transaction where one or more programs and systems are involved (for example, the synchronization of operations between transaction monitors, database managers, and distributed transaction programs).

Standard applications. The standard applications part of the application support layer of the Networking Blueprint has been moved up one layer in the Open Blueprint. Standard applications (such as mail and file transfer) are part of the distributed systems services of the Open Blueprint; they are part of the application services of the Open Blueprint.

---

* See also *transaction monitors* under Application Services in the Open Blueprint.

Networking Blueprint

Open Blueprint

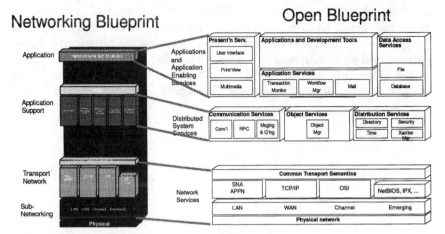

Figure 7.62   Networking and Open Blueprints.

7.8.2   Open Blueprint communication environment

The bottom componentry group (network services) constitutes the Open Blueprint communication environment. The communication environment is identical to the lower portion (transport network and subnetworking layers) of the Networking Blueprint.

The Open Blueprint network services componentry group corresponds to the aggregate of the transport network and subnetworking layers of the Networking Blueprint. These services include:

- Transport network

  SNA, APPN, TCP/IP, OSI, NetBIOS, IPX, . . .
- Subnetworking

  LAN, WAN, channel, emerging
- Physical media

7.8.3   Evolution from Networking Blueprint to Open Blueprint

In summary, the Open Blueprint (introduced in April 1994) evolved from the Networking Blueprint (introduced in March 1992) as an expansion (and refinement) of the application aspects (environment) of the Networking Blueprint, as shown in Fig. 7.62. The left side of the diagram shows the Networking Blueprint; the right side shows the Open Blueprint. Network services are identical between the blueprints. Distributed system services are similar between the blueprints. Applications and application enablers are expanded with much more detail in the Open Blueprint. (See IBM [GC23-3808] in the Bibliography for general and technical information, respectively.)

8

Communication Models

This chapter explores selected communication models and their communication middleware, including messaging and queuing. It is quite important to realize that any particular application should be allowed to communicate with itself or with other like or unlike applications using more than just one communication model. The communication model that is best suited to each application situation should be chosen so as to save money, time, and effort.

8.1 Communication

Communication involves the exchange and/or distribution of information, as shown in Fig. 8.1. Exchange implies pairs; distribution implies groups. Exchange implies two-way interchange; distribution implies one-way shipment and delivery.

Program   **Program**

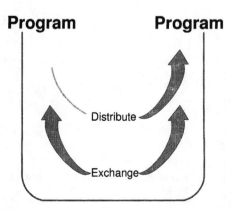

Figure 8.1 Exchange and distribution concepts.

In general, programs, devices, and persons can be involved in communication, as explained in Chap. 7 and as shown in Fig. 8.2. Persons access programs through devices. Persons and programs directly access devices but in different manners.

Software program communication generally involves the exchange of information within, between, and among software programs of all types, as well as between and among devices, people, systems, and networks. Communication between programs and devices is sometimes accomplished through *device drivers* or *access methods,* which control devices (input, output, storage, communication, and others). Communication between persons and devices (input and output) is usually accomplished through written device operating procedures. Messaging and queuing deals with software program communication (see bottom of Fig. 8.2); person-to-person communication (see top of Fig. 8.2) is discussed in App. D.

8.2 Connection-based and Queue-based Communication

Connection-based and queue-based communication differ primarily with regard to *availability and time.* Connection-based communication *demands* constant availabilities and a coincident time period; queue-based communication *tolerates* varying availabilities and unmatched time periods. Furthermore, connection-based communication occurs by pairing programs two at a time, whereas queue-based communication occurs by both grouping and pairing programs. This section compares and contrasts connections and queues as communication choices.

8.2.1 Some perspectives on communication methods

A few simple examples can serve to illustrate the basic distinctions between connection-based and queue-based communication methods.

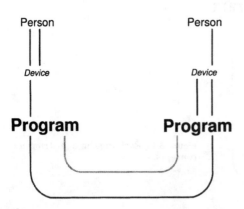

Figure 8.2 Communication through programs, devices, and persons.

8.2.1.1 Connection-based, Morse code signaling. In some sense, the now-obsolete military communication based upon Morse code signals by light or sound is an example of connection-based communication. Both parties were necessarily constantly available (at least the signaling clerks were) and the communication occurred during the same time period. The connection was the light or sound path.

8.2.1.2 Connection-based, person-to-person conversing. Likewise, the ordinary face-to-face (or telephone-to-telephone) human social and business communication of today is usually a connection-based communication (at least in more polite circles). Both persons are available (physically, if not mentally) and they are conversing during the exact same time period. The connection is the visible and audible air (or audible telephone) path between the humans.

8.2.1.3 Queue-based, paper tape correspondence. In some sense, the now-obsolete "torn-tape" military and commercial communication based upon round holes punched in a continuous paper tape wrapped around a turning reel is an example of queue-based communication.* The party destined to receive the encoded information from the paper tape and the party who originated the encoded information[†] need not be constantly (or at all) available during a particular same time period. With paper tape-based teletype keyboards, transmitters, receivers, and printers, the communicating humans did not have to be attending the printers and keyboards at the same moments in time (but the recipient could, by choice, watch the printer and detect the originator's keystrokes in many cases). There was no necessary connection between originator and recipient.

8.2.1.4 Queue-based, person-to-person correspondence. Likewise, the ordinary human social and business communication of today need not be a face-to-face (or telephone-to-telephone or computer-to-computer), connection-based communication. Information exchange and distribution can be by paper or electronic mail,[‡] handwritten instructions or notes, picture greeting cards, formal correspondence through secretaries, third-party negotiations, and all forms of indirect interaction. There is no necessary connection between originator and recipient, and

---

* See Chap. 1 for a short history of queue-based communication.

[†] The communicating paper tape reader (hole-sensing) and paper tape writer (hole-punching) devices were, of course, operating necessarily in a connection-based manner across a telephone line, leased or switched.

[‡] See App. D for a very thorough explanation of the distinction between mail messaging as provided by X.400-type electronic mail (e-mail) systems and the online/real-time messaging as provided by messaging and queuing.

both parties can operate (originate or receive information) by choice during different or overlapped or coincident time periods.

8.2.2 Connections, physical and logical

Connections are either physical or logical in form. Connections always have both a local end and a remote end. Physical connections are tangible (for example, copper wire or optical fiber cable). Logical connections are intangible (for example, storage areas in two different processors that are associated with one another for receiving and transmitting information between a pair of programs). Logical connections (often called sessions) are a logical subdivision of the physical connections (often called data links or just links), as shown in Fig. 8.3.

Logical connections between programs are based upon one or more physical connections between and among two or more communications devices. One method used for subdividing a physical connection is to allocate equal time periods on a round-robin basis for each logical connection's data. For example, with six logical connections, every sixth second (or smaller time unit) would belong to any given connection and the data segments would be interspersed in rotational order. This method has the obvious disadvantage that, unless all of the logical connections have a steady stream of data, some time slots will be wasted with no data from a logical connection. Many time- and resource-sharing techniques exist for subdividing physical connections into logical ones.

8.2.2.1 Links are physical connections.

Links are physical connections between adjacent communications devices, as shown in Fig. 8.4. For a local area network (LAN), physical links can consist of twisted/shielded copper wire pairs or some other short-distance medium. For a wide area network (WAN), physical links can consist of telephone facilities or some other long-distance medium. Collectively, physical links interconnect communications devices to form a physical network, as shown in Fig. 8.5.

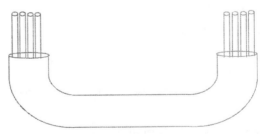

Figure 8.3 Logical sessions and physical links.

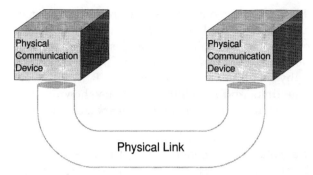

Figure 8.4 Physical links interconnect physical communication devices.

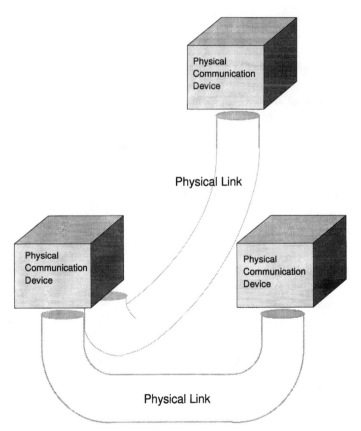

Figure 8.5 Collectively, physical links interconnect physical networks.

8.2.2.2 Sessions are logical connections. Sessions are logical connections that result from subdividing a physical link connection (or series of physical link connections), as shown in Fig. 8.6. Session connections interconnect logical network software, as shown in Fig. 8.7, and they can span one or more physical links, as shown in Fig. 8.8. Session connections can share some links and not share others, as shown in Fig. 8.9. Collectively, sessions form *logical transport networks*, as shown in Fig. 8.10.

8.2.2.3 Conversations are also logical connections. Conversations are logical connections that result from a reserved private usage of a ses-

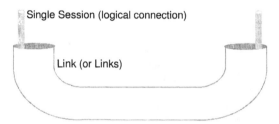

Figure 8.6 Session connections are logical connections.

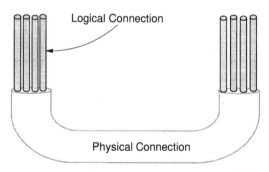

Figure 8.7 Session connections interconnect logical network software.

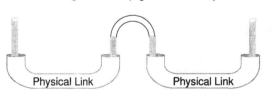

Figure 8.8 Session connections can span links.

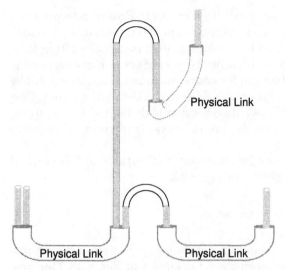

Figure 8.9 Session connections share some links.

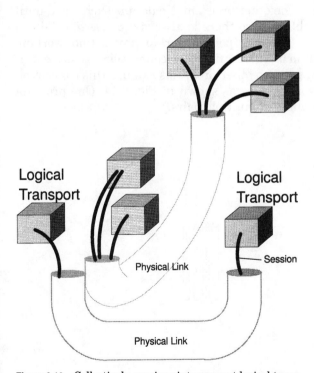

Figure 8.10 Collectively, sessions interconnect logical transport networks.

sion logical connection, as shown in Fig. 8.11. Program-to-program conversations occur over logical unit-to-logical unit sessions. Sessions are prebuilt and maintained by the logical unit in a pool waiting to be dedicated to a program pair. Ideally, the pool of session connections is always larger than the demand for conversations so that there will be no delay or pause for building an additional LU-to-LU session. Sessions are not subdivided; the entire session is dedicated for the duration of a conversation (seconds, hours, days) to a connection-based program pair.

Collectively, conversations interconnect applications to form *logical application networks,* as shown in Fig. 8.12.

8.2.3 Connection-based communication assumptions

When using connections, programs communicate in pairs across a private logical connection between them, as shown in Fig. 8.13. One program sends data into one end of the connection; the other program receives data out of the other end of the connection. The private logical connection is established by one of the programs of the program pair with the cooperation of the other program. No data is exchanged until the connection is established, so there is always a pause to establish the logical connection at some point prior to any actual working exchange or distribution of information. The connection belongs exclusively to the program pair; no other programs can use this connection while both programs are active, as shown in Fig. 8.14. One program can, however, be a member of multiple pairs.

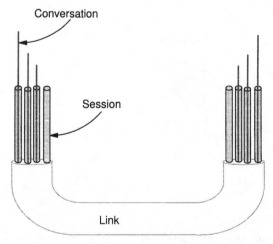

Figure 8.11 Logical conversations and logical sessions.

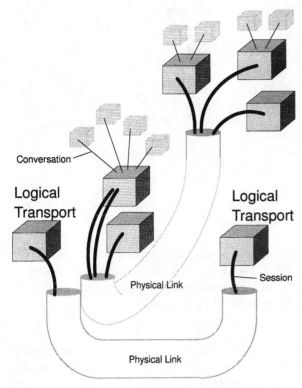

Figure 8.12   Collectively, conversations interconnect logical application networks.

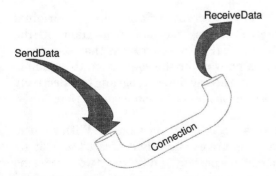

Figure 8.13   Connection-based communication.

8.2.3.1   Availability. Both source and target programs must be constantly available, as well as the logical connection between them. If either of the programs or the connection between them fails, then communication ceases.

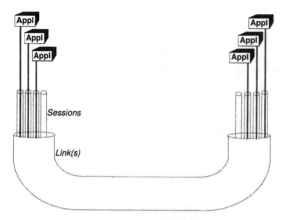

Figure 8.14   Conversational communication.

8.2.3.2 Predictability. Reading a connection carries with it a certain guarantee of predictability about what kind of data will be read. Hence, logic within an application program to screen out unsolicited or invalid traffic can be minimized. Program pairings are prearranged and predictable.

8.2.3.3 Delivery integrity. Across a connection, the information sent is always either delivered at the other end of the connection or the sender is notified of nondelivery. This function eliminates lots of programming logic that could otherwise be necessary within large numbers of application programs.

8.2.3.4 Delivery sequence. Delivery of information in the original sequence is assured. With connections, information is taken off the connection in a first-in first-out (FIFO) manner such that sequences are always preserved. There is no way for the connection to scramble the information sequence or for the receiving program to selectively skip information packets (records or messages) on a connection.

8.2.3.5 Privacy. Privacy across a connection is assured. With connections, both ends of the connection are exclusively dedicated to a single pair of programs, thus assuring privacy. Only the two programs assigned to the two ends of the connections can access the connection.

8.2.3.6 Utilization. In actuality, utilization of a logical connection can seldom be anything but extremely low (below 10 percent), because, if either of the two programs is executing, then the connection cannot be in use. The convenience of an available connection usually far outweighs the poor utilization of the connection.

8.2.3.7 Efficiency. Connection-based communication promotes efficiency in that helper programs that manage the creation of connections and that monitor the connections to assure that they are operational are themselves very efficient.

8.2.3.8 Control. Programs (of any type) may choose either logical or physical connections (or both), as shown in Fig. 8.15. Program A is using a logical connection to communicate with Program R but a physical connection to communicate with Program S. Program R is a member of both the A-R and A-S program pairs.

8.2.4 Queues, local and remote

Queues are either in a local system or in a remote system. As discussed in this book, queues are always logical.*

8.2.5 Queue-based communication assumptions

When using queues, programs communicate with queues in lieu of connections, as shown in Fig. 8.16. Queue-based programs do not specify the target *program* name; they specify the target *queue* name instead. The target queue can, however, be indicative of the target program through the use of trigger queues as explained in Chap. 11. If the target queue is actually remote, then the message goes first into a local transfer queue whereupon it is moved automatically to the remote queue, as shown in Figs. 8.17 and 8.18. One program puts a message into a queue; another program gets a message from that queue, as shown in Fig. 8.19. Logically there is a single queue even if a transfer queue is actually involved.

---

* Physical queues usually refer to memory management or hardware-based storage areas.

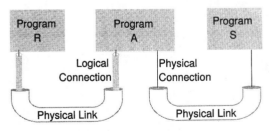

Figure 8.15 Programs may choose logical or physical connections.

Program   **Program**

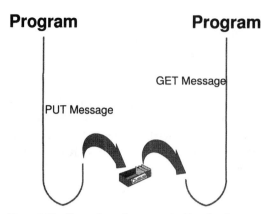

Figure 8.16   Queue-based communication, local queue.

Program   **Program**

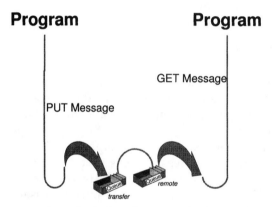

Figure 8.17   Queue-based   communication,   remote queue.

Program   **Program**

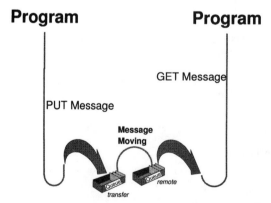

Figure 8.18   Queue-based   communication,   message moving.

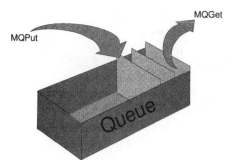

Figure 8.19   Queue-based communication.

There are never any logical connections between queue-based programs; the connections (if any) are always between the message queue managers, which do the networking for the queue-based programs, as shown in Fig. 8.20.

In Fig. 8.20, Application A put a message into Queue q8a while Application R gets the same message from Queue q8r, which is logically the same queue as q8a only with a different name. Similarly, Programs A, B, and C communicate through logical queues with Programs R, S, and T, as shown in Fig. 8.21.

Message moving machinery is necessary, as shown in Fig. 8.22. This machinery automatically shuffles queue elements (messages) to the right queue.

8.2.5.1 Availability. The target program can be busy or unavailable. Furthermore, there is no way to identify the target program in the MQI; a target queue is identified instead.

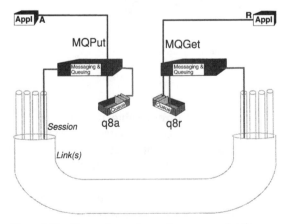

Figure 8.20   Messaging and queuing communication.

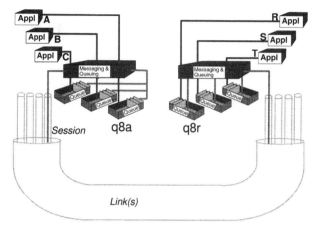

Figure 8.21   Message moving.

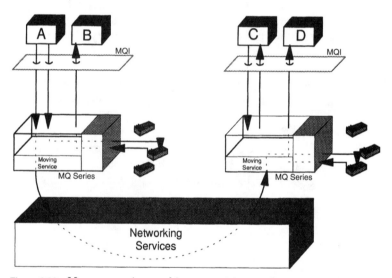

Figure 8.22   Message moving machinery.

8.2.5.2 Predictability. Unplanned and unpredictable pairings of programs can easily be accommodated. Message flow patterns are unconstrained among programs. Only by specific application design intent are program pairings and information flow patterns controlled.

8.2.5.3 Queue labels. Queues are locally, remotely, and globally named. It is usually best to separate queue names from program names.

8.2.5.4 Queue definitions. Queues are statically or dynamically named and described.

8.2.5.5 Queue access. Queues can be accessed in an exclusive (private) manner or in a nonexclusive (shared) manner. Queues can be defined as exclusive-use to force private access.

8.2.5.6 Queue usage. Queues are used as input-only, output-only, or input-output.

8.2.5.7 Delivery integrity. Messaging and queuing optionally offers assured delivery of messages to target queues. In many applications, assured delivery can be sacrificed to gain extra performance, especially with low-value messages in high volumes.

8.2.5.8 Delivery sequence. In general, delivery sequence is not assured with messaging since a single connection between program pairs is always absent.

8.2.5.9 Privacy. Queues are private only to the degree that access to the queues is controlled.

8.2.5.10 Utilization. Queues are time adapters and availability adapters (as explained in Chap. 4) and, as such, are extremely well utilized to take care of the adjustments that must be made to allow programs to run at their maximum speeds. The ideal case is to have all queues empty and all programs busy; the next most ideal case is to have some queues of messages and all programs busy; a less desirable case is to have no queues, many idle or waiting programs, and only some busy programs.

8.2.5.11 Efficiency. Queues are extremely efficient in that they help keep programs running as fast as they possibly can within the constraints of program logic. Without queues, programs often slow down, wait long periods of time on other programs, or go inactive.

8.2.5.12 Control. Queues provide for control of unlimited flow patterns, as explained in Chap. 5. Queue managers handle network control.

8.3 Program-to-Program Communication

Programs communicate with particular other programs, in a simple-tree manner as shown in Fig. 8.23, by using either connection-based communication or queue-based communication. Communication occurs between programs in pairs, either directly (for example, across program boundaries within a single load module) or indirectly (for example, across a communications network, with or without communications helper programs). A given program can, in general, communicate with some programs in a connection-based manner and, at the same time, communicate with other programs in a compound-tree manner as shown in Fig. 8.24. A guide for selecting which of the two methods (connection-based or queue-based) to use when and why is given in Sec. 8.8.

8.3.1   Interdependence

Figure 8.25 demonstrates that connection-based communication makes programs dependent upon one another, while queue-based communica-

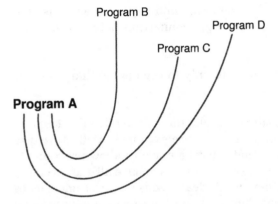

Figure 8.23   Simple tree delivery sequences.

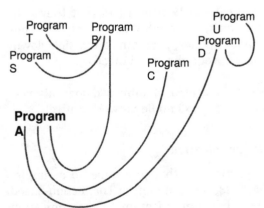

Figure 8.24   Compound tree delivery sequences.

tion allows programs to be independent of one another. Queue-based programs execute without being influenced by the execution of other queue-based programs. They execute as fast as they can and do not wait on other programs. Connection-based programs take turns waiting on each other.

8.3.2 Target specification

Connection-based programs specify program name, connection label (session ID, connection address, or such), and some networking parameters (session mode, station address, or such). Queue-based programs specify only queue name, no program name, and no networking parameters. Figure 8.26 contrasts the target specification (Program B or Queue J7) for connection-based and queue-based communication.

8.4 Three Communication Programming Styles or Paradigms

The Networking Blueprint (see IBM [GC31-7057] in the Bibliography) introduced three distinct communication programming styles (or paradigms): two connection-based and one queue-based. CPI-C and RPC (see the left side and middle of Fig. 8.27) are primarily connection-based; MQI (see the right side of Fig. 8.27) is queue-based.

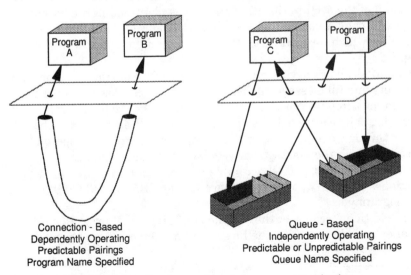

Connection - Based
Dependently Operating
Predictable Pairings
Program Name Specified

Queue - Based
Independently Operating
Predictable or Unpredictable Pairings
Queue Name Specified

Figure 8.25 Connection-based and queue-based program communication.

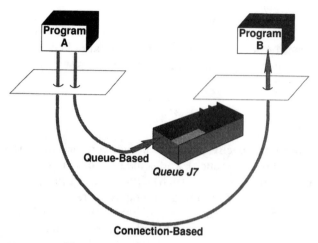

Figure 8.26 Target specification.

8.4.1 Messaging and queuing style (using MQI)

The Message Queue Interface (MQI)* was first introduced through the Networking Blueprint, which was announced by IBM in March 1992. The message queue interface is supported by a Message Queue Manager.† The MQI is approximately in the middle of the blueprint diagram, as shown in Fig. 8.28.

Figure 8.29, derived from the Networking Blueprint, shows the relationship of the MQI to the other interfaces identified in the blueprint. MQI is a companion to CPI-C and RPC. The MQI, and Message and Queuing which supports the MQI, are highlighted in the middle of Fig. 8.29.

The MQI allows asynchronous (independently operating) programs to empty and fill message queues and to exchange messages with other programs through those message queues. The MQI programs are never logically connected; they are only indirectly associated with each other through one or more of these application-type message queues through which the messages are exchanged. Queues, not connections, are shared. Thus, the MQI accomplishes "communication without connections" between application programs. See IBM (GC33-0809) in the Bibliography.

In addition, because of the simplicity of MQI, MQI programs can be smaller and simpler than the larger and more monolithic programs of

---

* See Chap. 12 for a detailed discussion of the Message Queue Interface.

† See Chap. 11 for a detailed discussion of the Message Queue Manager.

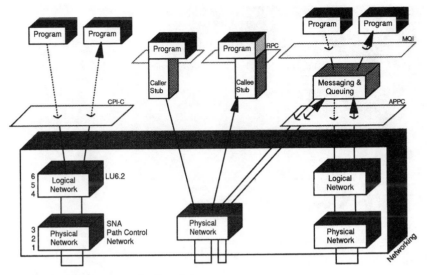

Figure 8.27 Three communication styles.

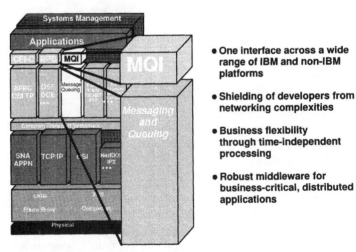

- One interface across a wide range of IBM and non-IBM platforms

- Shielding of developers from networking complexities

- Business flexibility through time-independent processing

- Robust middleware for business-critical, distributed applications

Figure 8.28 Networking Blueprint components.

the past, resulting in a lower skills demand and greater productivity for the large numbers of applications programmers who are employed to write business applications. Hence, MQI programs are

- Usually asynchronous (independently operating)
- Always connection-*in*dependent
- Mostly simple and small

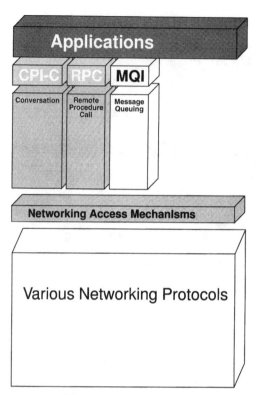

Figure 8.29   The Message Queue Interface (MQI) and Networking Blueprint.

8.4.2   Conversational style (using CPI-C or APPC)

The Common Programming Interface for Communications (CPI-C) and the Advanced Program-to-Program Communications (APPC) interfaces are both supported by an SNA logical unit of Type 6.2 (LU 6.2). The Common Programming Interface for Communications (CPI-C), along with the earlier Advanced Program-to-Program Communications (APPC) interface,* provides communication between application programs over a private logical connection called a TP-to-TP[†] conversation. The TP-to-TP conversation is supported by an LU-to-LU session, which is reserved exclusively for the two TPs for the duration of the conversation. Every pair of communicating programs has its own

---

* The CPI-C has both consistent semantics (meanings) and syntax (formats); the earlier APPC interface has consistent semantics but varying syntaxes (depending on the implementation).

[†] In SNA, TP means *transaction program* (an application program).

private conversation and any program can have multiple active conversations with other programs.

CPI-C programs tend to be synchronous (dependently operating), connection-dependent, and monolithic in structure. They are quite often involved in extended dialogs or long sequences of information exchanges where a private logical connection is vital to the logic of the application programs (for example, whenever unsolicited input could not be handled).

8.4.3 Remote procedure call style (using RPC)

The remote procedure call (RPC) is supported by stub code in the caller program and in the callee program. The remote procedure call provides communication between application programs using a call-return mechanism, usually across a private connection between caller and callee in a synchronous manner.* The caller program is suspended until the called program completes and returns a response. See Rosenberry (1993) in the Bibliography.

As previously explained, RPC programs are of two types: the caller (the calling program) and the callee (the called program). RPC traffic follows the client-server message flow model. See Orfali (1994) in the Bibliography.

The RPC caller programs tend to be synchronous (dependently operating), connection-dependent, and simple in structure. Similarly, the RPC callee programs tend to be synchronous (dependently operating), connection-dependent, but quite complex in structure since they quite often play the role of a server with many services and functions to execute and large numbers of callers to support.

8.5 Support for the Three Companion Communication Interfaces

As shown in Fig. 8.30, the CPI-C, RPC, and MQI are companion interfaces. They complement one another; they coexist and do not compete with one another or substitute for one another. A single application program is, therefore, (assuming the available facilities) eligible to use all three interfaces within a single module, choosing the interface that best suits a particular message traffic pattern or logic requirement. Or, more likely, within a single application (that is, a collection of programs) some programs use only MQI, others use only CPI-C, others use only RPC, and the remaining programs use mixtures of interfaces.

---

* Asynchronous RPC mechanisms have been proposed but not widely implemented at the time of this writing.

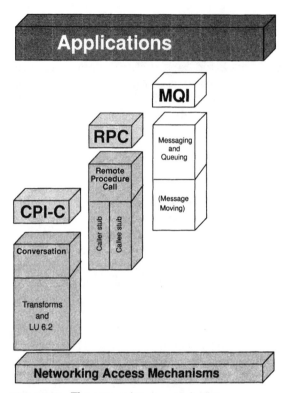

Figure 8.30 The companion communications programming interfaces.

8.5.1 MQI, an interface within layer 7

The MQI is an interface between two sublayers of layer 7 (that is, between an application sublayer and an application services sublayer of the application layer 7 of most network models).

The message queuing sublayer, the layer next to the top of the network (layer 6), is the *direct user* of the network and isolates the MQI applications from the complexities and difficulties of the communications network. Message traffic flow patterns are not necessarily fixed or preplanned since there are no connections involved among the application programs.*

The MQI is supported *"outside" a six-layered network* (by message moving above the network). MQI was born in the mixed LAN and WAN world.

_____

* Of course, unexpected message traffic must be handled by messaging programs since traffic is not flowing over private connections between programs.

8.5.2 CPI-C, an interface between layer 7 and layer 6

CPI-C is an interface between layer 7 (the application layer of most network models) and layer 6 (the top of the SNA logical unit of type 6.2, LU 6.2, within the SNA network model). Layer 7 is not subdivided; the application is using the network directly. Layers 4 and 5 (the "bottom" of LU 6.2) support SNA sessions; layer 6 (the "top" of LU 6.2) supports SNA conversations and the interpretation/execution of CPI-C.*

The CPI-C is supported *"inside"* a *six-layered network* (by LU 6.2, which is an aggregate of layers 6/5/4 and by a TCP/IP network). CPI-C was born in the WAN world.

8.5.3 RPC, an interface between layer 7 and layers 4/3

RPC is an interface between layer 7 (the application layer of most network models) and layer 4/3 (the middle of most network models). Layer 7 is not subdivided; the application is using the network indirectly through *stub code,* which is, in effect, substituted for layers 6 and 5. Stub code is compiled, once for the caller and once for the callee, and then added to both application programs. Hence, message traffic patterns must be prearranged between partner programs.

RPC is supported *"outside"* a *three-layered network* (by stub code not considered part of the network). RPC was born in the LAN world.

8.6 Unique Value of Messaging and Queuing

Messaging offers four distinct values, which are easy to understand and remember.

1. *One interface across many platforms.* A single, consistent interface (the MQI) is used in all operating systems, in all programming languages, and with all networking protocols that are supported.

2. *Shielding of developers from networking.* The application developer does not deal with networking parameters or with network programming, as shown in Fig. 8.31. The communication logic burden has been subtracted away from the application and absorbed by messaging and queuing.

   Messaging means less networking. Messaging causes a heavy reduction in the number of networking sessions required, since only

---

* A layer-7 transform can exist to map CPI-C to APPC with the use of side-tables.

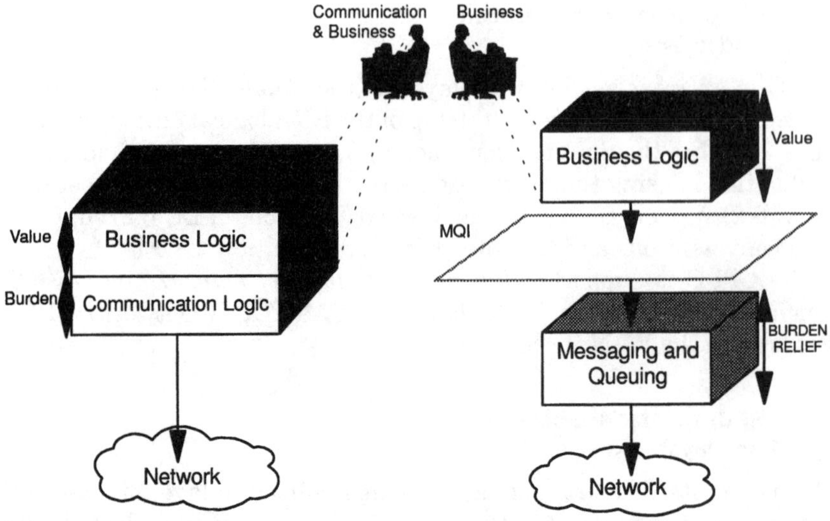

Figure 8.31   Shielding developers from networking complexities.

the queue managers are using connections, never the application programs. In Fig. 8.32, networking without messaging means every pair of communicating programs gets a connection. This can easily result in a "session explosion" when large numbers of programs must communicate. Figure 8.33 shows that messaging means less networking since messaging and queuing programs do not have connections of their own.

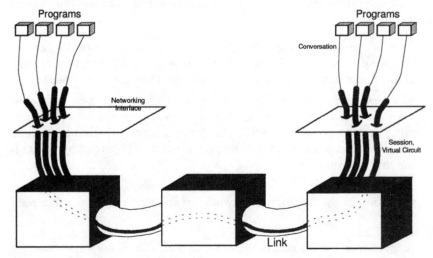

Figure 8.32   Networking without messaging.

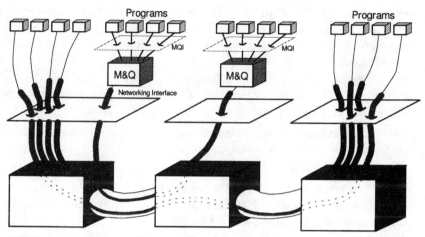

Figure 8.33 Messaging means less networking.

Messaging means less communications skills. Messaging and queuing is simpler to accomplish since the application is not involved with connections and less communication networking skill is required with queues.

Messaging means concentration upon applications. Messaging and queuing programs do not have communication programming burdens, so they can concentrate on application (business) logic.

3. *Asynchronous application programming service.* Time independence is introduced by the asynchronous programming services of the MQI. Programs do not have to take turns waiting on each other and executing serially.

Program speed differences. Programs do not all execute at the same speed; they are different in size and in function, as shown in Fig. 8.34.

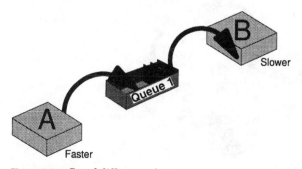

Figure 8.34 Speed differences between programs.

Program availability differences. Programs cannot be constantly available; there will be moments when they are busy and rarely will a program be unbusy at exactly the right moment to receive data arriving, as shown in Fig. 8.35.

Emptying and filling queues. Messaging programs independently empty and fill queues without knowing which other programs are filling and emptying those same queues. See Fig. 8.36.

4. *Extensive communications middleware.* Messaging provides extensive communications middleware.

Traditional communications. Traditional communications requires fully or heavily meshed networks, as shown in Fig. 8.37.

MQI communications. MQI communications requires less networking by applications, as shown in Fig. 8.38.

Application communications burden. Communications programming can be a heavy burden to application programs, as shown in Fig. 8.39. The purpose of an application program is pure application

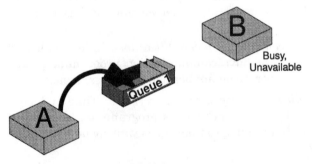

Figure 8.35   Busy or unavailable target program.

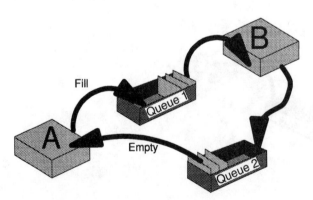

Figure 8.36   Emptying and filling queues.

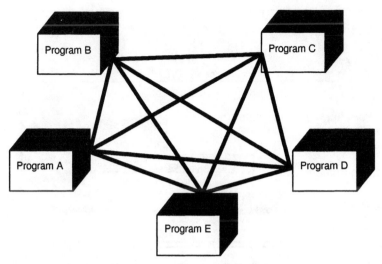

Figure 8.37   Traditional connection-based communications.

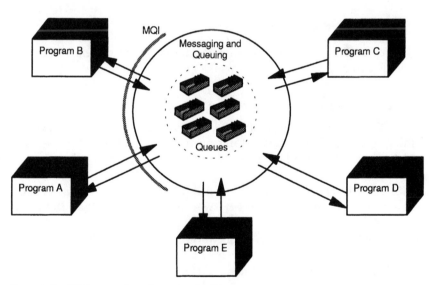

Figure 8.38   MQI queue-based communications.

programming, not communications and other programming burdens. Programs can interface directly with physical (four-layered
and three-layered) networks, directly with logical (six-layered) networks, indirectly with networks through messaging and queuing, or
very indirectly with networks through application platforms that
use messaging and queuing services themselves.

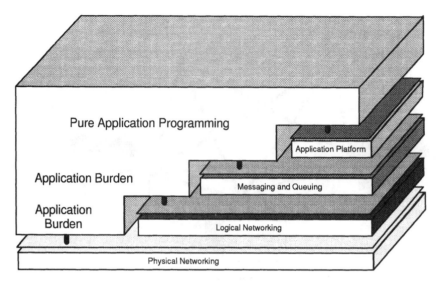

Figure 8.39   Application communications burden.

8.7   Style Selection Guide

Application developers need a selection guide to determine which one or two or three communication styles to choose, when, and why.

8.7.1   Some preliminary considerations

8.7.1.1   Software packaging affinities.   Many software packages have an affinity to a particular networking protocol, and, in some cases, will operate only with a single networking protocol.

8.7.1.2   Historical preferences.   Many computer installations have developed a preference for one or more networking protocols, especially when only a single type of network has been operational in the installation. Familiar networks are preferred networks.

8.7.1.3   Existing programming skills.   For those programmers who must use networking software directly, the level of existing communications programming skills is a very important consideration.

8.7.1.4   Duration of interoperation.   The duration of information exchanges between a given pair of programs is an important consideration for selecting communication paradigms.

8.7.1.5   Direct networking.   Not all application programs are written (or have been written) to one of the three interfaces described in the preceding sections. Many are written to do direct networking, that is, communicating directly with networks without the benefit of any communications middleware.

8.7.1.6 Designer choices. The application designer is responsible for defining the set of functions to be provided within a particular application (or set of applications frequently referred to as a *suite* of applications) and describing the structure of the logic required to interpret and execute those functions. The logic structure may be logically or physically distributed or not as the designer specifies.

The application designer's design specification may or may not take into consideration the various communication interfaces that are actually available for implementing (coding) the design. Ideally, the implementation should be made using the most suitable communication interface available, using multiple interfaces if and when necessary or desirable within a program segment or module. This implies that specification of communication interfaces might be best omitted from the design specification unless the designer realizes that some communication interface support is restricted or absent or that the communication interface support must be explicitly integrated into the design description.

8.7.1.7 Coder choices. The application coder is responsible for receiving the application design document, understanding the purpose of the application, and implementing the specified logic structure through available programming languages, program library services (macro instructions, call statements, include statements, and such), and system resources (operating systems, communication facilities, device handlers, and such). The application coder must often absorb the burden of integrating a mixture of languages, services, and resources, since these are seldom specified in the application design document but must be reckoned with at coding time.

The application coder must deal with the availability of communication interface support, regardless of whether or not the design specification refers to such.

8.7.1.8 System administrator choices. The system administrator is responsible for maintaining the programming languages, library services, and system resources upon which the application coder depends. The system administrator must install and maintain software, advertise the installation-dependent values for all types of installation parameters to application coders where necessary, and assure the integration of all software and hardware computing facilities so that these will be available to execute the application program.

The system administrator must deal with the tools and other support necessary for the application coder's use of communication interfaces at program compile time and run time. A particularly large responsibility is the assigning of local and global names and addresses.

8.7.1.9 IS/DP manager choices. The *information system* or *data processing manager* is responsible for installing, operating, and maintaining one or more computing environments from a financial point of view.

8.7.2 Connection and queue choices

Considering only the choice between connections and queues (and not choices between particular connection-based interfaces or queue-based interfaces), there are some choices that are rather obvious, as shown in Fig. 8.40.

8.7.3 Some technical considerations

Three disjoint communication styles are in practice today supported by interface *providers* and depended upon by interface *users*. The inter-

Style Selection Guide		
	Use Queues	Use Connections
Duration of Message Exchange	Short (Fewer than 10 messages)	Long (10 messages or more)
Criticality of Network Restart Time	High, need fastest restart possible	Low, delayed restart tolerated
Importance of Parallel Operations	High	Low
Variety of Information Flow Patterns	All Information Flow Patterns	Mostly Closed Trees and Chains
Dependence Upon Serialization	Very Low	High or Medium
Processing Window of Time	Heavily Utilized or Over-utilized	Under-utilized or absent
Message Traffic Volumes	Extremely High	Moderate
Relationship of Programs	Highly Independent	Highly Dependent
Programming Skill Levels	Low	High or Medium
Knowledge of Networking	Low	High or Medium
Degree of Privacy Required Between Programs	Low or None	High
Mixtures of Application Types	High or Medium	Low or Single Application Type
Mixed Old/New Applications	Many or Some	Few or None
Mixtures of Network Types	High or Medium	Low or Single Network Type

Figure 8.40 Connection and queue choices.

face users request *functions,* which are usually grouped together into sets advertised as *services;* the interface providers interpret and execute the functions and provide feedback, if necessary, to the requestor of the function. In addition, the interface provider executes lots of functions not directly related to a particular function request.

In theory, the application programmer has a choice of one, two, or three of the MQI, CPI-C, and MQI interfaces, as illustrated in Fig. 8.41, which shows the application program (the interface user) at both the bottom and top of the diagram.

8.7.3.1 MQI. Messaging and queuing is the newest communications model, which is characterized by *no logical connection between interface users.* The basic relationship between MQI user and MQI provider is that of a message PUTter/GETter and a message QUEUer/SWITCHer/MOVer service whereby messages are exchanged between MQI users.

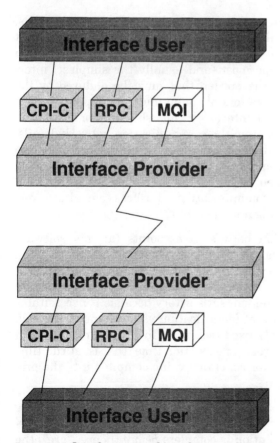

Figure 8.41   Interface user and interface provider.

8.7.3.2 CPI-C. LU 6.2 Conversations (CPI-C or APPC) is the well-known and heavily-advertised-by-IBM, level 7-to-level 6 interface supported by SNA LU 6.2 and by TCP/IP. CPI-C is the one and only direct interface to and from the SNA network. The basic relationship between CPIC user and CPIC provider is that of one program and a *helper program* (the logical unit and a CPIC-APPC transform if necessary) that locates a peer program, resulting in communication between peer programs or *co-routines,* where each program is supported by a private logical connection between the two programs dedicated to the pair of programs for the duration of a *conversation.*

8.7.3.3 RPC. Remote procedure call (RPC) is just a regular-looking program call that happens to be executed not locally but remotely instead. Hence, transparency is accomplished relative to the programmer's coding. The basic relationship between RPC user and RPC provider is that of a mainline *caller* and support to locate a subroutine *callee.*

8.7.3.4 Target identification. As with many layered network models, multiple interfaces "stacked" one on top of the other constitute the set of communication interfaces potentially available to the application programmer. Using the uppermost (and usually the simplest) interfaces significantly reduces the burden that must be endured by the application programmer, since lots of communication support can be provided beneath the highest interfaces, thus shielding or insulating the application programmer from the communication network details. In many cases, the application programmer is not disqualified from using the lower layers of communication interfaces but is discouraged from doing so. This document assumes that the application programmer is using the uppermost communication interfaces (namely, conversational, RPC, and messaging and queuing).

Resolution. Communication between two points (at any software level) involves the identification of one point by the other as shown in Fig. 8.42. The identification value used by the communication initiator to identify the partner may be compiled into the source language as a constant value, but this is seldom satisfactory because it necessitates recompiling the program if that identification value changes. Instead, a better and more commonly used technique involves a resolution of the identification value at program execution time (that is, at run time or later) from the symbolic (or constant) value compiled into the program to another usable value.

Above the interface (preresolution). When the usable ID value identifying the communication partner is compiled into the program, then preresolution occurs because resolution is complete *before* (hence, "pre") the program uses the communication interface.

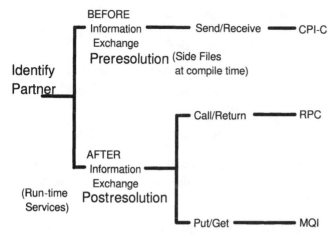

Figure 8.42   Target identification and resolution.

Below the interface (postresolution).   When only a symbolic variable or constant ID value identifying the communication partner is compiled into the program, then postresolution must occur before resolution is complete and this postresolution must be provided by the interface support as it interprets and executes (hence, "post") functions requested across the interface.

Routing (reresolution).   When the usable ID value identifying the communication partner is either compiled into the program (preresolved) or provided by the interface support (postresolved), but further resolution becomes necessary while an information unit is in transit to the communication partner, then *re*resolution must occur before resolution is actually complete. This reresolution may occur multiple times, especially in such cases as cross-network communication where each communicating program does not necessarily have access to the ID value for its partner. This technique of using a presumed target value, as contrasted with the "actual" target value, is in common practice today, particularly in interenterprise communication.

Higher-layers routing.   Routing is a two-step process whereby the route (or collection of routes) is first set up by the lower network layers and then used by the higher network layers. A route is simply a list of addresses. Each route is either complete or partial, logical or physical. Logical (as contrasted with physical) routing as it exists today occurs above layer 4, using both platform (application and communication subsystem) facilities or application facilities. See Hauzeur (1989) in the Bibliography.

Lower-layers routing.   Physical (as contrasted with logical) routing as it exists today occurs inside the network in layer 4 or below, using either software or hardware but usually in a fixed, non-application-programmable way.

End-to-end and point-to-point. Some protocols (and formats) are end-to-end while others are point-to-point. At each end are multiple stacks of protocols; at each point are multiple stacks of protocols.

Application points. Layer 7 points are typically only end-points.

Platform points. Platform points are typically only end-points but can also be midpoints.

Network points. Network points are both end-points and midpoints.

8.7.3.5 Binding choices. Binding choices (that is, association of source and target programs or queues) can be done at compile time or at run time, as shown in Fig. 8.43.

8.7.3.6 Connection expense ratios. Using a connection involves the expenses of setting up the connection, using the connection, and tearing the connection down.

Set up/tear down the connection. Before a logical connection (session) can be used, it must be set up or established. This requires a pause in the execution of the application program unless there is a spare logical connection awaiting allocation from a previously established pool of sessions. The logical unit (for example, LU 6.2) is usually responsible for tearing down a connection when it is no longer required.

Use the connection. While a connection is being used, its resources consist of control blocks and buffer space at the local and remote ends

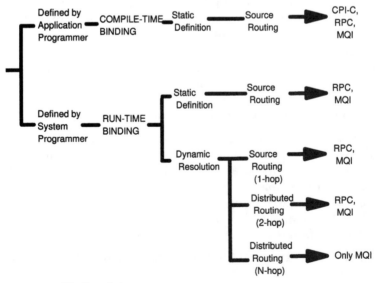

Figure 8.43   Binding choices.

of the connection as well as some time allocation for use of a physical link or sequence of links.

Pay for the connection. The ratio of setup/teardown expense to use expense for a connection is extremely important to the application designer (and coder) whenever deciding upon having a dependence upon a logical connection or not. (See Fig. 8.44.) The formula:

$$\left[\text{Expense} = \frac{(S + T)}{U} \right]$$

where S = setup
        T = teardown
        U = use

is often used by programmers and designers in making sensible choices between connections and queues. If the number of time units necessary to set up and tear down a connection is small in proportion to the number of time units necessary to use the connection, then the connection is more justified. For example, in Fig. 8.44, if 60 time units are required to set up and tear down a connection that is used for only 3 time units, then the connection is quite expensive. On the other hand, if the same 60 time units are still required to set up and tear down a connection that is used for 300 time units, the connection becomes more reasonable.

8.7.3.7 Characteristics and structures. The characteristics and structures of CPI-C, RPC, and MQI are quite different.

MQI characteristics and structures. The characteristics of the queue-based, messaging MQI are:

$$\frac{\text{setup} + \text{teardown}}{\text{use}} = \text{ratio} = \frac{S + T}{U}$$

$$\frac{50 \text{ units} + 10 \text{ units}}{3 \text{ units}} = \frac{20}{1} \quad ==> \text{expensive}$$
SHARED Connection

$$\frac{50 \text{ units} + 10 \text{ units}}{300 \text{ units}} = \frac{1}{5} \quad ==> \text{cheap}$$
PRIVATE Connection

$$\frac{0 \text{ units} + 0 \text{ units}}{300 \text{ units}} = \frac{0}{5} \quad ==> \text{very cheap}$$
NO Connection

Figure 8.44   Connection expense ratios.

- Communication is groupwise or pairwise or one-way.
- Communication is intermixed or flip-flop serial.
- Processes are frequently operating independently against queues.
- Queues replace instantiations of code.
- Messaging can be accomplished without queuing or message moving.
- Messaging and queuing can be accomplished without message moving.
- Message moving may be necessary to accomplish messaging and queuing.

MQI structure is shown in Fig. 8.45.

CPI-C characteristics and structures. The characteristics of the connection-based, conversational CPI-C are:

- Communication is pairwise.
- Communication is flip-flop serial.
- Application is dependent upon LU 6.2 (or TCP/IP) services (for example, session-outage notification).

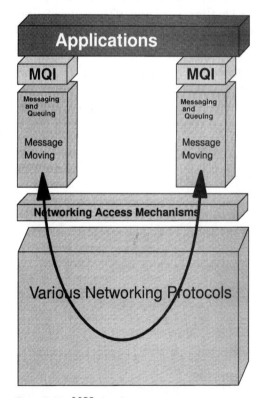

Figure 8.45 MQI structure.

- Both sets of communicating code are interdependent; requesting/ responding roles are reversible.

CPI-C structure is shown in Fig. 8.46.

RPC characteristics. The characteristics of the connection-based, call/ return RPC are:

- Communication is pairwise.

- Communication is flip-flop serial.

- Application is dependent upon RPC stubs and run-time services.

- Caller is dependent upon callee procedure (functions within services) for results.

- One chunk of code requests results; one chunk of code generates results; roles are not reversible.

RPC structure is shown in Fig. 8.47.

8.7.3.8 Flow differences and similarities. The differences and similarities of MQI, CPI-C, and RPC, with regard to their ease of handling different

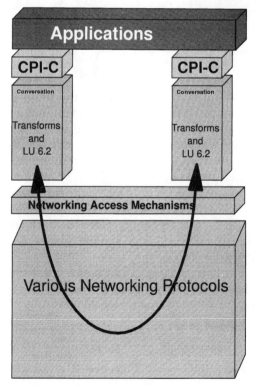

Figure 8.46 CPI-C structure.

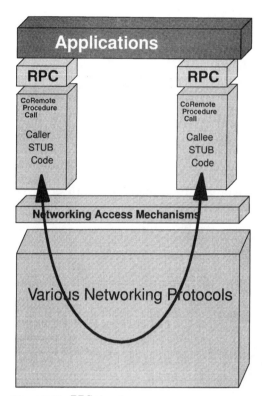

Figure 8.47   RPC structure.

flow patterns, are shown in Fig. 8.48. Some program communication is dependent (application logic-wise) upon having logical connections that are used privately between and among the communicating programs. Other program communication is, by contrast, independent of any logical connections and can be executed without any logical connections between and among the communicating programs.

The application programmer (and designer) can relax the intensity of the application program logic whenever input is predictable, comes from predictable sources, and can be solicited by connection. Without one or more private logical connections to and from a given application program, the application programmer can never be certain what will be received, where it came from, and when it will come; hence, monitor-type logic must be introduced to analyze the inbound traffic.

Today, RPC and Conversations are predominantly based upon logical connections; messaging is not.

Procedure call.   RPC uses a procedure call mechanism.

Message passing.   Conversation and messaging use a message-passing mechanism.

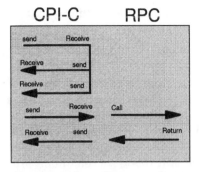 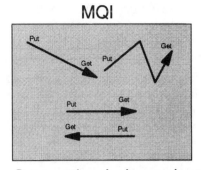

Connection-Dependent   Connection-Independent

Figure 8.48 Flow differences and similarities.

8.7.3.9 Spectrum of flows. The left side of Fig. 8.49 shows the "best, but not only" fit for the information flow shown on the right side of the figure.

One-way-only flows. One-way messages (top flow in left column) are best suited to MQI. CPI-C and RPC are founded on the expectation of constant two-way flows: CPI-C for both peer-to-peer and client-server models and RPC for the client-server model.

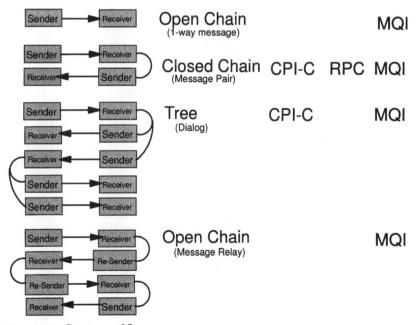

Figure 8.49 Spectrum of flows.

Message pairs. Two-way flows (second flow in left column) that do not continue are suited to any of the three styles, depending on the application requirements. This is a highly prevalent information flow pattern in most applications.

Continuous exchange flows. Extended dialogs or a continuum of flows (third flow in left column) are best suited to either conversation or messaging. RPC is not designed for the caller (client) and callee (server) to keep changing roles; the client is always the client and the server is always the server (except that the server can play the client role with nested RPCs and thus have two roles: server to one program and client to another).

Complex flow patterns. Complexly exchanged flows (bottom flow in left column) are best suited only to messaging. With messaging, both one-way and two-way flows occur in all directions without regard for any pairing of programs. With CPI-C and RPC, flows are regulated in pairs; with MQI, flows are not regulated by pairs or otherwise.

8.7.3.10 Example message flows. Figure 8.50 illustrates some basic message flows or sequences, some of which have nicknames that have become widespread in usage. The top of the figure shows five simple sequences; the bottom shows two complex sequences.

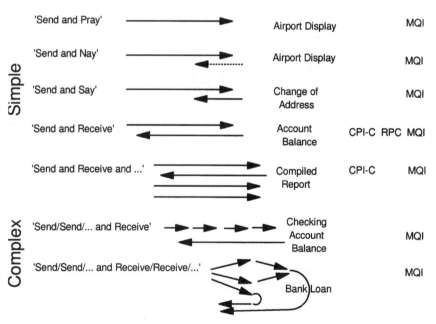

Figure 8.50 Example message flows.

The *send-and-pray* sequence is a one-way flow in which nothing can necessarily be known about what happens to the message. An example of such a sequence (of one-way messages) is a program that is feeding arrival and departure information to a display device in an airport terminal area. MQI is well suited to this sequence.

The *send-and-nay* sequence is a two-way flow that happens only when the program that digests and processes the message being sent finds something wrong with the message or has need to notify of some otherwise negative situation unrelated to the message. If nothing is wrong, the nay response is not sent. This is sometimes called exception-only responding. An example of such a sequence is based on the previous send-and-pray example: a program that periodically requests status from the airport terminal display device and finds that the display device declares itself to be nonoperational (that is, a nay response). The actual practice programs that operate arrival/departure information screens in airports intersperse in their send-and-pray sequence a *send-and-say* sequence periodically (for example, every 20 seconds) to force the display to acknowledge that all is well, or to send a nay response if something is wrong. Again, MQI is well suited to the send-and-nay and send-and-say sequences. Send-and-nay means that a response is optional (only if error exists); send-and-say means that a response is mandatory.

Another example of a send-and-say sequence is a program that sends change-of-address messages to a database expecting some type of short "say" message back.

The *send-and-receive* sequence is similar to the send-and-say sequence, except that it does not necessarily involve a question-and-answer or request-and-response sequence. An example of a send-and-receive sequence is an ATM banking program requesting an account balance (and all associated information) from a banking database.

The *send-and-receive-and* . . . sequence is an extension of the send-and-receive sequence in which message exchanges continue. Compiling a report is a good example in which the program gathering data for a report sends to many other programs and then collects all of the data into an aggregation from these many sources.

The *send/send/* . . . *and-receive* sequence is an example of a closed chain information flow (see Chap. 5). If you are requesting your checking account balance amount from an ATM banking terminal in a remote city, the ATM program may cause a request to be sent through a series of programs (across many computer networks) before the answer is returned directly to the ATM.

The *send/send/* . . . *and-receive/receive/* . . . sequence is an example of a closed lattice information flow (see Chap. 5). An example of this lattice flow is shown by a bank loan processing. The flow begins (at the left of Fig. 8.50) when a bank officer collects information from the loan applicant and then begins processing the loan application in an automated manner. The first program distributes messages to several programs (for example, credit card balance, checking account balance, and bank funds programs) and then later collects the answers from the end of the lattice that had no directly returning flows.

8.7.3.11 Work unit structures. Figure 8.51 shows work unit structures, which are actually just varieties of the information flow patterns described in Chap. 5, except that the depictions are vertically oriented instead of horizontally oriented as they are in Chap. 5.

A *tree* work unit structure allows the root Program A to request information from Program B, which farms it out (subcontracts it) to Programs C and D. Similarly, Program D farms the work out to Programs E and F. After collecting flows from Programs E and F, Program D returns information to Program B. This entire traversal of the tree is serial by nature and programs must wait for information flows to return, unless MQI is being used.

A *chain* work unit structure is serial from Program J, to Programs K, L, M, N, O, and finally back to Program J, making a closed, nonreturning chain, which is often called a *ring* in work unit parlance.

A *lattice* work unit structure is serial from Program R, through Programs S, T, U, V, W, and then back to Program R, making a closed lattice. Of course, the last Program R could have been a Program X, to make an open lattice.

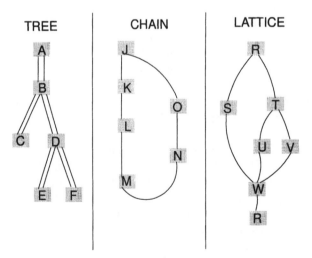

TREE   CHAIN   LATTICE

Figure 8.51   Work unit structures.

Work unit structures are simply information flow patterns used to accomplish some defined collection of activities (human, device, and program) that constitute some unit of work (for example, the processing of a single insurance claim form). Work flow is discussed in Sec. 10.7.

8.7.3.12 Some client-server relationships. Figure 8.52 shows a variety of client-server flows: the top for CPI-C, the middle for RPC, and the bottom for MQI.

CPI-C client-server flows are usually an N-element closed tree.

RPC client-server flows are usually a two-element, closed chain (nicknamed send-and-receive) flow.

MQI client-server flows are accomplished by any and all flow patterns.

8.8   Three-Way Comparison

This section provides a three-way comparison of CPI-C, RPC, and MQI.

8.8.1   Three styles comparison table

The following is a table comparing various aspects of the three communications programming styles. This table serves as a selection guide for application designers and programmers.

Comparing:	CPI-C	RPC	MQI
Primary communication method	Connection	Connection	Queue
Time sensitivity	Synchronous	Synchronous	Asynchronous
Primary networking protocols	LU 6.2,TCP/IP	TCP/IP	LU 6.2, TCP/IP
Partner availability	Constantly required	Constantly required	Never required
Connection availability	Constantly required	Constantly required	Never required
Coupling	Tight	Tight	Loose
Stateful machinery			
Stateful process	Always	Always	Yes or No
Stateful message	Never	Never	Yes or No
Blocking interface	Yes and No	Yes	No

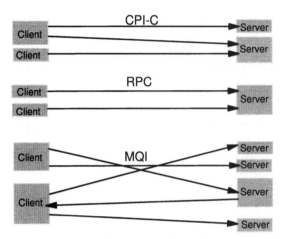

Figure 8.52 Some client-server relationships.

8.8.2   Special cases

- CPI-C special cases

  One-way bracket looks like "send-and-pray" messaging.
  Allocate / Send / Receive / Deallocate looks like RPC.

- RPC special cases

  Nested RPC can be made to look like messaging.
  RPC transport need not be session-based.
  RPC "Name Server" operation looks like relay in messaging.
  RPC caller can have multiple callees.

- MQI special cases

  Message pairs look like RPC, without a network connection (session) beneath.

  "Contextual" messaging looks like conversational communication without a session.

9

Distributed Communication Services

MQSeries products provide programming interfaces and execution time support for time-independent, distributed applications. However, any comprehensive, distributed application environment needs additional services over and above those provided by the application enabler. Distributed applications take a wide variety of forms and can be implemented in a wide variety of ways. The underlying services can, however, be grouped together under a relatively small number of headings. In this chapter, we define five such underlying services and examine them in general and in relation to the facilities offered by MQSeries.

The five classes of service are:

- Directory services
- Security services
- Data definition and conversion services
- Resource recovery services
- Systems management

Figure 9.1 shows a hypothetical distributed system containing most of the elements which will be discussed in this chapter. Several machines are connected via a network. Some of these machines hold valuable resources, which would need to be recovered in the event of failure. All systems management tasks for the entire system are carried out from a single, identified machine. Distributed communications services run on several of the machines. For example, directory and security services run on separate machines. In the sections which follow, we will examine each of these kinds of service and cover some of the issues involved in realizing practical implementations of distributed systems.

One particular implementation of some of these distributed services, namely the Open Software Foundation's Distributed Computing Environment (DCE), has received much attention in recent years. At the end of this chapter we will look briefly at its provision of distributed services. We will also identify those services for which additional products are required.

9.1   Locating the Right Services

Distributed applications consist of multiple processes which communicate with one another in order to work together on a particular problem. We will term any part of a distributed application, which executes separately in this way, an *element* of that application.

Individual application elements provide services to one another. Whether they operate on a client-server basis or as peer-to-peer partners is immaterial to this part of the discussion. In either case, at various points in the execution of the application, one element is performing work for another.

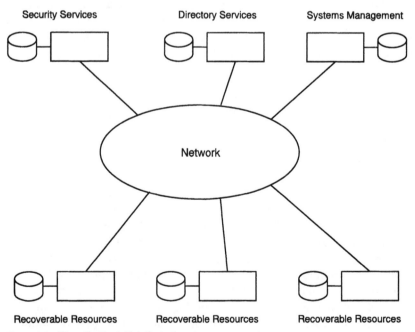

Security Services   Directory Services   Systems Management

Recoverable Resources   Recoverable Resources   Recoverable Resources

Figure 9.1   Hypothetical distributed system.

Elements may be on the same machine or different machines. In either case, some means must exist for elements to be able to locate and communicate with one another. These are two separate but related problems. First, an element needs to discover where in the network the required element is located. Subsequently, it must discover how to send information across the intervening network to that element.

The solution to the first of these problems is normally considered to be a function of the middleware. It is usually called a *name service*. The solution to the second problem is normally considered a function of the network. However, as we will see, there are occasions when the middleware layer needs to be involved in this function too. The function is known as *routing*. We will look at both of these services in the following sections.

Practical implementations of both functions involve *directories* in some form. Directories map between classes of values. Some directories map names to addresses. Others provide address-to-address mappings. Name services and routing provide two of the major uses for directories.

9.1.1 Name services

As we have already seen, elements of distributed applications need to find one another. These elements are each associated with one or more operating system processes, sometimes known as tasks. Any process in a network is associated with the machine on which it runs. A machine is identified by some network address. The precise form of this address depends on the particular network. However, in every case, a machine is uniquely identified by its network address.

Within any machine, many processes may be involved communicating over the network. In order for requests and data to be routed to the correct process within a given machine, each one is associated with some kind of local address. For example, in a TCP/IP network, each machine is associated with an *internet address*. The local address on a machine is known as a *port*. Network traffic associated with a particular machine address and port number is routed to a specific process. Some port numbers in TCP/IP are fixed by convention. The *telnet* service, which provides the ability for users to log on to a machine from a remote location, is available on port number 23, for example. These fixed assignments are called *well-known ports*. While they are very valuable for common applications such as remote login and file transfer, they are not flexible enough for general application use. Consequently, middleware products, such as DCE, need to provide facilities for dynamically associating local addresses, like port numbers, with particular application elements.

MQSeries products, on TCP/IP networks, use well-known port numbers. This is acceptable as all traffic passes between MQSeries components rather than directly between application elements. All application traffic is transferred via a single port, removing the need for complex dynamic port allocation. This reduction in the number of simultaneous connections required between machines on the network is one of the features of MQSeries products. Although only a single port is used, this does not entail any additional arbitrary limitation on throughput. This is a feature of TCP/IP.

The preceding telnet example is particularly simple. The port number is fixed by convention. The name of the machine on the network is supplied by the user. This is a remote login facility, so it is natural for users to have to specify the name of the host machine they wish to log on to. The only mapping required is between the name of the host machine and its internet address. Internet name-to-address mapping can be performed in a number of ways, the simplest being a table of host machine information stored on every machine on the network. In this case, every machine knows explicitly about all the others. Most network implementations provide a variety of ways of mapping machine names to addresses. These are, of course, examples of directories. Figure 9.2 shows a very simple name-to-address mapping. In the figure, the internet address of a machine called *hebrides.myth.com* is being looked up. The resulting address, *5.73.19.89,* is the value which TCP/IP will use to send packets to the machine in question. The usual TCP/IP implementation of this kind of directory is as a table on each machine in the network. The challenge, in managing such tables, is to keep them up to date when changes occur. Additional tools exist which allow a single copy of these tables to be distributed to other machines, on a TCP/IP network, when changes occur.

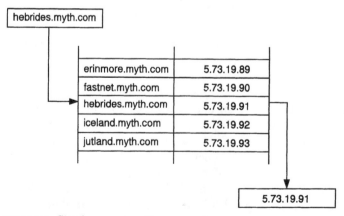

Figure 9.2   Simple name-to-address mapping.

It is, of course, possible to have some addressing information compiled into each element of an application, to allow it to access other elements without the need for a directory. This, though, is extremely inflexible. For example, it prevents parts of the application from being run on different machines at different times without the application being modified. The ability to move pieces of an application at will can greatly improve application availability. It can allow specific machines to be taken out of service for regular maintenance, without affecting the provision of the application to clients and end users. It can also be used to provide load balancing, allowing application elements to be moved from heavily loaded hosts to those which are less busy. The ability to move application elements can also form the basis for highly available systems. In this kind of system, a standby processor can take over the work load of a failed processor. For this to work, application elements from the failed machine must be restarted on the standby processor. Clients wishing to use these elements must be able to locate them on the standby processor.

The most flexible solutions to the problem of identifying application elements delay the resolution of the address of a service until the service is actually needed. This requires that some kind of mapping, relating application name and address, be available as the application executes. The mapping relies on a database which is dynamically updated as application elements start and end. This is a key factor, as it allows the distributed system to react almost immediately to changes in the state of the machines and applications that it contains. The mapping database is another kind of directory. It maps a name by which the application element, or service, is known to the addressing information required to locate it. Typically this involves a machine address and process address, as we have already seen.

In this form, a directory form is analogous to a telephone book. Application elements request information from the directory using a name. The directory is always updated when elements move, so that any element looking up another element always receives a valid address. In practice, of course, it is possible, on rare occasions, for an application to retrieve a network address, only to find that the desired element is not actually available. This might happen, for example, if the application element terminated just after its address was retrieved from the directory. Such occurrences should be rare. However, directory services of this kind do need to provide facilities to handle this situation or to allow applications to deal with it. Usually, the situation can be resolved by repeating the lookup operation.

The ability to move application elements around without affecting other elements of the same application is known as *location transparency* or *location independence*. All directories provide this ability;

however, some require more work of systems administrators than do others. For example, a directory implemented as a table on each client machine of a given service would require all client machines to be updated whenever a service moved. In contrast, a central table held in one place and made available to all clients via the network requires only a single update to be made when a service moves. Better still, a fully dynamic directory, updated by the services themselves as they start and stop, requires no intervention from systems administration staff.

It is becoming increasingly common for an application to be totally unaware of the actual network addresses of the application elements it is using. This is obviously important in insulating applications from network details. Different network types use different addressing schemes. Allowing addresses to percolate up to the application code forces the application code to be aware of the particular kind of network involved and makes it more difficult to write applications which are truly network independent. Hiding the details of the network from the application code is one important function of middleware, such as MQSeries.

Directories can hold name-to-address mappings. Some directories map machine names to machine addresses while others map service names to machine and process addresses. However, this is not the only information which a directory may hold. For example, a given service might be available from a number of machines in the network. Rather than ask for a specific instance of a service by name, an application may be able to ask for a type of service, not caring about which particular application element supplies it. Alternatively, an application might need to be very specific about a service. For example, there might be a number of application elements on the network able to offer a particular type of print service, each one corresponding to a specific physical printer. However, only one of these printers might be physically local to the current user of the application. Consequently, only one of these services is appropriate for this particular user. One way to distinguish such services is to have additional information in the directory identifying the specific resources with which they are associated. Applications requesting the service ask for a server of a particular type associated with a particular resource. In our example, the resource would be the specific instance of a printer.

Although our examples have concentrated on location of application elements, directories can be used to specify any kind of computing resource, from a machine to a database to a user or printer. Name services are always associated with a particular scope. Typically, this is a specific set of machines or some well-defined domain within a network. Within the scope covered by a particular directory, names must be unique. The set of possible unique names provided by a directory is known as a *namespace*. Name services exist which are capable of providing a single namespace large enough to encompass literally every application and machine on the planet. We will look at an example of an implementation in Sec. 9.6. These capabilities will become increasingly important as the need grows for interoperation between application elements on a global scale. In practice today, however, it is more usual for a namespace to cover a single network or a set of interconnected networks belonging to a particular organization. For example, IBM has a TCP/IP network which is used internally to provide communications among many thousands of workstations, servers, and mainframe machines across the world. The scope of each of the many namespaces involved in managing this network is typically a single physical location. However, hierarchical arrangement and segregation ensures that addresses are unique across the entire network. This is also a feature of an even bigger network known as the Internet. This network links a wide variety of organizations across the world.

9.1.2 Name services and MQSeries

Distributed applications based on message queuing use queue names to identify application elements. The location of a given queue is represented as a queue manager name. Queue manager names may be specified explicitly by an application or may be inferred from the queue name, as we shall see. Each queue manager is associated with a particular host machine. Consequently, a directory can be used to establish the network address of the host machine where a particular queue manager is running. Each queue manager maintains a set of queues, which can be viewed as its resources. A second directory can be used to relate queue names and application element names. These directories mean that even the most specific form of addressing in MQSeries, in which both the queue manager and queue names are supplied explicitly, still involves sufficient indirection to allow application elements to be moved freely.

The facilities provided by different networks and middleware layers vary considerably. MQSeries products exploit whatever the networks provide to implement their own directories. For example, the facility to have a queue name imply a particular queue manager and queue is implemented in at least two ways in MQSeries. On platforms with only basic directory support, the mechanism is known as a *remote queue definition*. A queue is defined on the local queue manager, but points to another queue on a different queue manager. Figure 9.3 shows an example of such a definition. In the table, the first column shows the name of queues known to the local queue manager. The second column shows whether these queues are actually local to the queue manager or whether they are on a different queue manager. The final column shows the actual queue name if different from that in the first column.

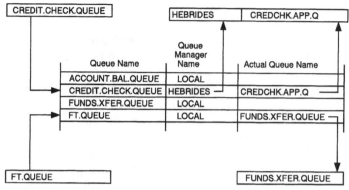

Figure 9.3   Local, remote, and alias queues.

The first entry, for ACCOUNT.BAL.QUEUE, shows a queue which is actually implemented on the local queue manager. If an application writes a message to this queue it will remain on the local queue manager. ACCOUNT.BAL.QUEUE is called a *local queue*.

The second entry is more complex. The queue, known to the local queue manager as CREDIT.CHECK.QUEUE, is actually implemented as a queue called CREDCHK.APP.Q on queue manager HEBRIDES. Messages written to CREDIT.CHECK.QUEUE will be forwarded by the local queue manager to the HEBRIDES queue manager and will be stored on its CREDCHK.APP.Q. CREDIT.CHECK.QUEUE is called a *remote queue*.

The fourth entry, for FT.QUEUE, is local to the queue manager. However, it references the actual queue FUNDS.XFER.QUEUE. This is a name-to-name mapping. FT.QUEUE is an *alias queue*. It is basically a different name for an existing queue. Aliases can be used to refer to local or remote queues. In this example, it refers to a local queue.

Remote queue definitions allow applications to refer to a queue without needing to know the name of the queue manager involved. Remote queues are effectively implicitly name-to-address mappings, since the queue manager name will map to a particular host machine address. They are implicit because the host machine address does not appear explicitly in the remote queue definition. The definitions do, of course, involve some administration. The amount required depends to a degree on the sophistication of the underlying network layer. In general, when addresses change, the remote queue definition itself can remain unchanged. It is more likely that the mapping between queue manager name and host will need to change.

Figure 9.3 shows the relationships among the various types of queue definition. However, it does not represent the actual implementation of the directory facilities. MQSeries products exploit the capabilities of the platforms on which they run in order to provide this kind of mapping.

Definitions held locally on individual queue managers all need to be updated if changes occur. The effort involved could be reduced considerably if distributed directories could be used. Such directories are accessible from any machine on the network. Common information held in such directories could be updated just once. Each machine would retrieve it as required. Emerging technologies, such as DCE, include the required facilities. These might eventually be exploited to reduce the maintenance required in managing all manner of distributed systems.

9.1.3   Routing services

The result of a lookup in a namespace is typically a network address. This is enough to identify a particular machine uniquely, but takes no account of the topology of the intervening network. The namespace defines which machine has to be contacted, but says nothing about how data should be sent to it. Getting the packets of information to the appropriate address is usually considered to be the responsibility of the underlying network. In a situation where both the sending machine and the target machine are in the same network, this is undeniably true. However, where a middleware layer allows an application to span multiple kinds of network, say SNA and TCP/IP, the middleware may have to be responsible for moving data between the networks.

The task of getting packets of data to the correct machine via intermediary hosts is known as *routing*. It is required in all but the smallest local area networks. In most practical networks of any size, it is physically impossible for all machines to be directly connected together. Machines are grouped into *domains*. The exact terminology varies from network to network, but we will define a domain as a group of machines that can directly access one another. Part of the reason for this partitioning is sheer cost; part of the reason is traffic congestion. If all machines are directly connected together, all traffic in the network will share the same physical medium. There will be an upper limit to the bandwidth of the medium and, even if it is very high, there will come a point at which so many machines are connected that the medium will become saturated. As might be expected, the more exotic media with highest bandwidth, such as fiber optic cables, are also the most expensive.

Fortunately, it is not necessary to have all machines directly connected. As long as sensible partitioning is used, grouping together machines which need to communicate a lot, it is possible to arrange domains in which satisfactory performance is maintained. Often, each domain is a single local area network (LAN). Many modern networks are arranged this way and consist of interconnected local area networks. Interconnections between these networks are of three basic kinds, namely *bridges, routers,* and *gateways.*

9.1.3.1   Bridges. If the interconnection is local (within a mile or so), it is physically and economically viable to arrange the LAN in separate segments, connected by bridges. This is the simpler approach and, although the network is physically partitioned, individual machines still appear to be directly connected to one another.

Bridges act at the level of the underlying physical network, for example token ring or Ethernet. They are totally unaware of higher-level protocol information such as the structure of TCP/IP packets or SNA request/response units. Bridges move packets between the two networks they link based purely on the hardware addresses, which appear in the low-level protocol information. Consequently, a bridge linking two token rings is equally able to pass TCP/IP, SNA, and any other packets between the LANs. The only restriction, apart from the distances covered, is that the two networks being bridged together must, obviously, be of the same type.

Figure 9.4 shows two token-ring networks bridged together. Machine A is equally able to communicate with machines B and E, even though one is reached directly while the other is reached via the bridge. As far as A is concerned, all machines appear to be directly connected to the same token ring.

Bridges are frequently *adaptive.* Adaptive bridges learn which machines are on which of the networks they join. Moving packets from one network to the other unnecessarily does not cause errors. A packet on the wrong LAN will simply be ignored. It does, however, mean more traffic on the LANs until the bridge learns the correct location for each

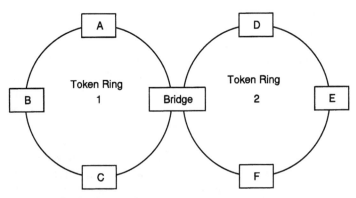

Figure 9.4   Bridged networks.

machine. Internally, adaptive bridges keep simple directories. These map hardware machine addresses to the identifier of the LAN on which that machine is attached. This kind of directory is a simple example of a *routing table*. It maps a network address to a destination network, albeit at a very low level. Management of the tables is completely automatic and is handled locally within each bridge.

Bridged LANs do not require any explicit routing services. The routing which they perform is handled automatically by adaptive bridges. Any machine can communicate with any other simply by putting packets onto the LAN using the correct address, perhaps discovered directly from a naming service. Bridged LANs built this way suffer approximately the same traffic limitations as a single LAN does. The partitioning of the LAN can help if machines on one side of the bridge tend to communicate more with each other than with machines on the other side.

9.1.3.2 Routers. Where bridging is not feasible because of physical limitations on the network or where different kinds of network must be connected together, the solution is to use a router. Unlike bridging, routing takes place at the level of logical network addresses. For example, in TCP/IP, whereas a physical address might be the hardware address of a specific Ethernet card, the logical address it represents is the internet address of the machine containing the card. Similarly, in SNA, the physical address might be the hardware address of a particular token-ring card, whereas the logical address is the SNA network address of the *network addressable unit* (NAU) associated with that card. (For more information on SNA terminology, see, for example, Meijer [1987] and Martin and Chapman [1987] in the Bibliography.)

Since routing works above the level of the network hardware, it can be used to direct packets across dissimilar physical networks. In particular, it is the technique used when remote LANs are linked together by point-to-point or wide area network (WAN) connections. Routers not only move packets between the networks they connect, they also make decisions about which packets must be sent on which links.

Since they work at the level of the logical network, routers must understand the structure of the packets for the particular network protocols in use. A particular router often deals with only one kind of protocol, for example, TCP/IP. However, it is becoming increasingly common for a single router to be able to deal with a variety of protocols, for example, TCP/IP, SNA, and OSI. Some routers are basically workstations with the appropriate adapter cards installed. Others are special-purpose machines whose sole function is routing.

As well as routing between different forms of LAN, routing may include passing packets from a LAN environment to a WAN environment for longer distance communication. This long distance communication usually uses cabling provided by a telephone company and requires low-level serial protocols mandated by them. Frequently, long distance connections are implemented as point-to-point links between two specific machines. At each end of the link, the machines are also connected to one or more LANs. The machines on the link route packets between the LANs. Protocols run over this kind of link include the SNA protocols and SLIP (Serial Link Internet Protocol).

With any topology that involves domains between which only indirect communication is possible, some means to allow the machines in one domain to contact those in the other domain is needed. Packets originating in one domain must be routed to the other domain if it contains the destination machine. The main problem in routing is deciding which domain is the appropriate target for any given packet. This is not too difficult if only directly linked domains are involved. However, it is quite possible that the final destination for a packet is not in the other domain, but that it needs to pass through it to get there.

A route is really a list of addresses defining the path to be used to move packets through the network. If the entire path is defined, the route is called a *complete route*. If it defines only a part of the route, it is termed a *partial route*. It is possible to keep complete routes on all machines that have to communicate. However, in even moderately sized networks, keeping the routing tables up to date rapidly becomes a very difficult and time-consuming task. The usual solution is to keep partial routes on individual machines. The partial routes contain just enough information to allow packets to be sent to the appropriate machine for forwarding.

Figure 9.5 shows a moderately complex network consisting of several interconnected subnetworks. Machines A, D, G, E, and F are connected via Ethernet LANs bridged by H. Machines J, K, and L are on a token ring. Machine I is a router connecting the Ethernet and token ring. Likewise, machines M, N, and O are on a second token ring, while Q and R are on a second Ethernet, with P providing the router function. The two token rings are connected via a wide area network which involves routers L and M. If machine A wishes to communicate with machine R, packets must follow the complete route:

A-H-I-L-M-P-R

It is possible for A to know this complete route. However, it is more usual for A to know only that packets not destined for any of the machines on its Ethernet need to be sent to H for forwarding. Likewise, router I will know to forward packets onto the token ring. Router L will

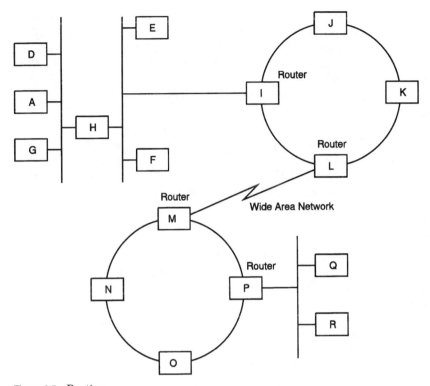

Figure 9.5   Routing.

probably need to know quite a lot of information about the range of addresses served via the second token ring, particularly if more than one wide area connection is available from the first ring. The overall point of this example is that each router needs some routing information to be able to decide whether or not to forward the packets. In most cases, the information required is less than the complete path to the destination.

Routes are held in *routing tables*. These can be thought of as another kind of directory. Whereas in a namespace, the directory maps a name to an address, routing tables map a destination address to a complete or, more commonly, a partial route. Much work has gone into development of automatic methods of keeping such tables up to date.

In some networks, the partial route may consist of a single address. This is the address of the gateway machine to which packets are forwarded for any addresses not in the local domain. For example, in SNA, many networks consist of *subareas*. Subareas are typically associated with mainframe computers or complex network controllers.

Each subarea keeps a routing table which maps subarea addresses. For each destination subarea address, the table holds one or more addresses of subareas to which packets can be forwarded. If there are multiple entries in the table, it simply means that there are multiple routes available. These routes may have different transmission characteristics allowing different qualities of service to be used for different kinds of data.

Structuring of network addresses is sometimes used to ease the problem of determining which gateway to use. Internet addresses show this kind of structuring. Different parts of the address represent groupings of machines. There are even different ways of structuring internet addresses depending on the size of the organization to which an address range is allocated (see, for example, Comer [1991] in the Bibliography).

9.1.3.3 Gateways. The term *gateway* is often used to describe any mechanism for connecting two networks together. Indeed, in TCP/IP internet networks, the term is commonly used to describe any entity performing routing. However, there is a more specific meaning associated with the term, and one which distinguishes gateways from bridges. In this version, a gateway is an entity which performs some specific conversion at a level above that of the network itself. Gateways understand something about the content of the data flowing on the network. A mail gateway, for example, might convert between specific electronic mail formats. A file transfer gateway might convert between the TCP/IP FTP protocol and the OSI FTAM protocol, allowing files to be transferred even though the machines involved do not share a common type of network.

9.1.3.4 Logical and physical addresses. There is one further kind of directory intimately connected with addressing and routing. A routing table for TCP/IP contains internet addresses. However, at the lowest level, physical addresses are needed to allow packets to be delivered over real networks. The actual physical addresses involved depend upon the network to which the device is attached. A further kind of address mapping is needed to allow internet addresses to be transformed into the physical addresses associated with, say, a token-ring or Ethernet card. This is an address-to-address mapping and constitutes a third kind of directory associated with networking.

One problem with physical addresses is that they can change. An Ethernet or token-ring adapter card for a PC or workstation has a preset, unique physical address. Changing the card because of a hardware failure changes the physical address of the machine, but not its network address.

One TCP/IP mechanism for resolving the mapping between internet and physical addresses is called Address Resolution Protocol (ARP). It requires broadcast capability, so it is useful only on LANs. It allows host machines to discover the hardware address associated with a particular internet address. The host that needs to discover the physical address associated with a particular internet address broadcasts its request to the LAN. All machines attached to the LAN receive the request and can reply. The host with that internet address responds with its hardware address. While this form of address resolution is only appropriate for LANs because of the need for broadcast, LANs are the very environment in which these logical-to-physical address mappings are most difficult to maintain because of the sheer number of machines that can be involved.

9.1.3.5 Cross-network routing. The advent of middleware has been driven in part by the need to hide incompatible networks from applications. One aim of middleware is to allow distributed applications to be written independently of the network software underlying the communications. Conventional network applications are limited to a single type of network. Applications written to use TCP/IP interfaces cannot communicate with those written for SNA. Neither sort can use services from applications written to use OSI protocols. Ad hoc solutions are possible to allow such applications to be joined together. These usually involve some form of protocol translation, purposely written for a particular customer situation. Network-independent middleware, such as MQSeries, provides architected facilities and standard APIs to eliminate the need for such special-purpose translators. However, the ability to link different kinds of networks together brings a new routing challenge. While each network can be left to route packets between its own domains, where two different networks meet, a special kind of routing is required. Where a TCP/IP network meets an SNA network, for example, neither network on its own can provide the routing. The middleware layer must do the job. We can term this kind of routing *cross-network routing*. It is not fundamentally different from the routing which takes place within a particular type of network. Indeed the Internet, a linked set of networks based on TCP/IP, already provides this function. However, it supports only one kind of network, namely TCP/IP. Middleware routing must be able to move information between a wide variety of types of network. MQSeries, for example can route messages between a wide variety of networks. Messages originating on an SNA network can be routed to TCP/IP, NetBIOS, DECNet, and so on. And, of course, the inverse routings can also be performed. The concepts underlying this implementation of cross-network routing are covered in Sec. 11.2.4.

9.2 Providing Secure Applications

Security is an important aspect of most commercial computer systems. Applications may manipulate sensitive or financial data of great value. There may also be legal requirements to fulfill for handling certain kinds of data. It may be necessary to restrict access to certain parts of the data to certain groups of users. It may also be important to protect the data against unauthorized access. Services may include sensitive resources.

There are two basic aspects of the protection of data in a single computer system. The first, known as *authentication* deals with being able to be sure that particular users are who they claim to be. The usual solution is based on each user having a unique identifier and having to provide a password. This password is a secret shared between the system and the user.

Once the identity of a user has been proven via authentication, the user's identifier can be used to define what capabilities are available to him or her to operate on the resources of the system. The aspect of security which deals with specification of these capabilities and their enforcement is known as *authorization.*

Authentication and authorization are as equally applicable to non-distributed systems as to distributed ones. There are, however, two additional aspects of security specifically related to distributed systems. The first of these deals with the issue of ensuring that data is not tampered with in transit over the network that connects machines. This aspect is usually called *data integrity*. The second is closely related and deals with ensuring that data is transmitted precisely once, even in the presence of arbitrary network or machine failures. This aspect is known as *assured delivery*.

In the following sections we will look at each of these aspects in more detail.

9.2.1 Authentication

Network traffic associated with a distributed application may need to travel over networks not under the physical control of the organization running them. Long-haul networks frequently include links provided by general network carriers, such as telephone companies. Whenever the network is not entirely under the physical control of the organization operating the applications, there is a danger of unauthorized connection of equipment. Application elements might be run on such equipment and used to attempt to access sensitive or confidential data for malicious purposes.

Distributed applications need a means to prove that traffic being received is from an application element authorized to communicate with them. If a database server, for example, cannot ascertain the validity of the client that is requesting data, it might be about to send confidential data from the database to a competitor rather than to an employee of the company. Similarly, an element requesting work from a server must be able to trust the application element providing the service. Otherwise, for example, that confidential document the user thought was on its way to be printed may instead be transmitting to a competitor.

Authentication is the process by which an application element proves it is who it claims to be. Most authentication schemes are based on the notion of a shared secret. This is information which should be known only to the user and to the system—typically a password. The version held by the system must be stored in a form which cannot be casually viewed. It is normal to encrypt the stored form of the password to prevent it from being stolen. Given that the password is appropriately protected, the ability of users to supply this secret information, when logging on, is taken as proof that they are who they claim to be.

Distributed systems are no different from nondistributed systems in this respect. However, there are additional difficulties which arise simply from the distribution itself. One major problem is that the stored form of the password for a given user will often not reside on the machine on which logon is being attempted. Consequently, at some point the password must be transmitted over the network to allow the version supplied by the user to be compared with the stored form. If the password is sent in the clear—that is, without being encrypted in some way—there is a danger of it being stolen by a malicious application eavesdropping on traffic on the network. On the other hand, any encryption scheme relies on the application elements each being able to decrypt some or all of the transmitted data. Encryption and decryption involve use of keys. These are themselves secret information which must be available to the partners involved in secure transmission of data.

There are two major approaches to encryption of data being sent over insecure networks. They differ in the way encryption keys are made available to the partners. Although we are looking at techniques for providing secure transmission of authentication information, these techniques are equally appropriate for encryption of any data which must be transmitted securely.

9.2.1.1 Private key encryption. In private key encryption, the same encryption key is made available to both partners wishing to communicate. Both these partners are treated as trusted, in terms of the software as well as the physical security of the hardware which contains the encryption key. Both systems must be physically secure, since each contains the encryption key. In addition, every component of the software in each machine must be trusted not to reveal the keys. This may not be easy to arrange if the machines have to run software from vendors. Organizations requiring the highest security in their systems may sometimes require the ability to view the source code of applications from vendors before sanctioning their use.

In private key encryption, the keys are distributed to the partners by techniques completely independent of the network. The keys are loaded into the machines and are stored there until replaced by newer ones. Frequently, the storage used to hold the keys is specially constructed to be resistant to attempts to read the keys out. For example, there may be a cage around the storage module preventing probes being placed on any of the chips or circuitry concerned with key storage. Any attempt to remove this kind of cage or to insert a probe results in the memory being reset, preventing the keys from being stolen.

Obviously, in this kind of encryption, all machines performing encrypted communications must be in physically secure locations. The hardware cost of such systems and their physical protection is not inconsiderable. These sorts of systems are typically used to protect data of considerable sensitivity or value. They are typical of systems used in the financial industry. Indeed, many financial institutions consider that private key systems are the only ones sufficiently secure for their needs.

9.2.1.2 Public key encryption.

While private key encryption is undoubtedly secure, it is, after all, essentially an electronic version of codebook techniques used by the intelligence services for secure communications for decades. It is not appropriate for general authentication in normal distributed applications. This is partly due to the cost of hardware and of administering the keys. Also, the level of physical security needed at every machine is not appropriate for most organizations.

The major alternative scheme is known as *public key encryption*. In this approach, encryption keys are distributed to application elements via the network itself. The keys are held in one or more key stores residing in trusted security servers on the network. These are the only machines that must be physically protected. Keys are transmitted over the network from the security servers to application elements. However, keys are always encrypted more than once. Data flows are such that neither application element ever sees the other element's password or encryption key in the clear. Yet, they do have enough information to prove that the other element was indeed trusted by the security server. This scheme is known as *trusted third-party encryption* and is typified by the *Kerberos* authentication mechanism used in a number of products, including DCE.

Although we have discussed authentication in terms of users proving their identity to the system, passwords and identifiers are equally applicable to application elements. A server, for example, may have to operate with more security authority than the user for which it is doing work. Just as users have to prove their identities to the systems, so, too, do servers. To encompass the notion that both users and applications have identifiers and passwords, the term *principal* is often used to mean any entity which can be given authority to manipulate protected resources.

9.2.2 Authorization

Authorization defines the capabilities which a particular principal has to manipulate specific resources in a computer system. Most operating systems provide mechanisms for restricting access to their own resources. For example, UNIX systems have a very basic mechanism for defining which users can read, modify, or execute given files. While the UNIX approach is fairly basic, some security schemes are very sophisticated and flexible, covering many more kinds of resources. For example, IBM's Resource Access Control Facility is an extremely powerful authorization system appropriate for mainframe-class machines and resources.

Definitions of capabilities typically take the following kinds of form:

Principal A can view resource Y.

Principal B can view and update resource Y.

Principal C can delete resource Y.

Improved efficiency of systems management comes from grouping principals together and applying authorizations to the group rather than to the individual. For example:

Principal A belongs to group G.

Principal B belongs to group G.

Principal C belongs to group G.

Group G can view and update resource Z.

Group G can view resource T.

Group G can view update and delete resource U.

Any new user joining the organization can have all appropriate authorizations enabled simply by adding them to the group. Definitions such as these have come to be called *access control lists*. Indeed, an increasing number of UNIX systems now include this capability for protection of files, greatly improving flexibility of authorization over the native UNIX file permissions. International standards are also beginning to appear. In distributed systems, DCE has extended the notion of access control lists to cover authorization to resources which can be accessed remotely. Currently, the set of resources protected by any of these schemes is restricted to those which form part of the underlying system. Examples include files, devices, and communications links. Protection for general application resources, other than those which are implemented as underlying system components, is not readily available. DCE does provide mechanisms for extending its access control lists to other objects, but this requires development of a significant piece of application-specific code. Over time, we can expect increasing use of facilities like this to improve the authorization capabilities available to applications.

Of course, defining the authorizations that principals have is only part of the story. Any underlying authorization system must allow applications to test the authorizations on behalf of the user for which they are executing. Remember that servers often operate with more authority than the user or application making the request. In this case, although the server may have access to some specific data, the user may not be entitled to access it. Security systems provide application programming interface calls, which allow servers to test the authorization for a particular principal to a particular resource. The implementation of the function provided by this interface involves querying the access control list to check the authorization.

9.2.3 Data integrity

One of the prime functions of a reliable network is the transmission of data unchanged from its source to its target. Two kinds of modification may occur to data in transit over a network. First, data may be modified by errors in the network itself. Networks are frequently much less reliable than any other part of a given computer system. This is hardly surprising given the harsh electrical environment in which many of them operate. This is particularly true of long-haul networks running over standard telephone lines. Second, the data might be intercepted and modified deliberately with malicious intent.

Low-level communications software is written to take network errors into account. The network layer detects alteration of bits in the data. Typically, this detection is carried out using *checksums*. Checksums are simple computations based on the data being sent. Usually, the data to be transmitted is treated as a sequence of integers. These are added together and a simple overflow rule is applied if the total overflows the field in which the checksum is being computed. The checksum is transmitted along with the data. The checksum is recomputed at the receiving machine and, if it matches that sent with the data, the assumption is made that the data was transmitted without errors. If the checksum does not match, the network will retransmit the data, assuming that an error has occurred.

While normal checksums are used in all networks to provide protection against errors, they do not protect against deliberate modification of the data. A malicious application could intercept data on the network, modify its contents, and recompute a correct checksum. The recipient would believe that the data had been transmitted successfully. A more sophisticated approach is required to protect against deliberate modification. Obviously, some kind of encryption must be involved. One approach is to encrypt the checksum with a key available only to the authenticated application elements participating in the exchange. It will be essentially impossible for a malicious application to reconstruct the checksum correctly, allowing the changes to be detected. Any traffic with an invalid checksum can be discarded. Either there was a network error or some malicious attempt was made to alter the data.

An even more secure approach to data transmission over a network involves preventing it even being read by any application which is not properly authenticated. To do this, not only the checksums but also all the data must be encrypted. Once again, only applications which have access to the appropriate encryption keys can decrypt the data. Eavesdropping is virtually impossible.

There is one practical difficulty in data encryption. The most secure implementation of encryption commonly available uses the so-called *Data Encryption Standard,* or DES. The research associated with this encryption was funded by the United States government. To use it for data encryption, organizations outside the United States must apply for specific licenses. U.S.-based vendors are prevented by law from supplying DES-based software that can be used for data encryption, to non-U.S. customers who do not possess such a license. The licenses tend to be granted only to major financial institutions. The consequence is that general encryption of data is not readily available outside the United States. This situation is becoming more critical. As the demand for distributed systems grows, more and more organizations will need the security of being able to encrypt their data, even though they do not qualify for the appropriate license. It is likely that software houses outside the United States will be only too eager to provide solutions which U.S.-based vendors are prevented from supplying by law. The current situation is made the more bizarre because the algorithms used by DES are in the public domain, having been published in computer science journals.

9.2.4 Security services and MQSeries

MQSeries products utilize the security mechanisms of the platforms on which they run. Their local resources, such as queues, tend to be protected by the same techniques which the platform uses for general resource protection. Data in transit may be protected by mechanisms inherent in the underlying network if they exist. In addition, MQSeries provides the ability for a customer's own encryption schemes to be installed and used. MQSeries messages carry information about the user running the application which caused them to be sent. This contextual information is available to applications to allow appropriate authorization checks to be made against resources being accessed.

9.2.5 Assured delivery

Although basic network support typically ensures reliable delivery of packets between machines, over an unreliable communications medium it is not usually capable of assured delivery. To guarantee that data will be sent exactly once and that communications will restart and recover properly after a major machine or network failure, more than simply reliable packet delivery is required. In particular, both partners in an assured delivery system must remember which data has been successfully transmitted and hardened. Hardening is the process of writing data to a storage medium, such as a disk, and ensuring that it is recoverable even if there is a hardware failure. We will cover some of the details of how this is achieved in practice in Sec. 9.4. The point to bear in mind here is that assured delivery guarantees that a specific interchange between two communicating applications takes place exactly once, despite almost arbitrary failures of machine or network.

Assured delivery is a feature of some middleware layers, such as MQSeries. The components of MQSeries which move messages between queue managers do so in a manner which allows them to provide assured delivery. The techniques used involve network connections, which can be restarted, and hardening the data put onto queues by the queue manager.

Assured delivery in middleware is by no means universal, however. For example, RPC techniques, including DCE, do not provide it. To get assured delivery over a DCE connection usually means running a full DCE-based transaction processing system, such as Transarc Encina or CICS/6000.

9.3 Data Definition and Conversion

Distributed applications are likely to operate in heterogeneous environments. Machines participating in distributed applications often have very different hardware architectures. These differences are pervasive, extending down to the representations of data used. Commonly encountered differences include:

- *Byte ordering:* Personal computers based on Intel microprocessors use a different ordering for the bits which represent integer values from those used by most other systems.

- *EBCDIC and ASCII:* Many mainframe and midrange business systems use encodings for readable characters based on the EBCDIC proprietary standard rather than the ASCII encoding defined by the American National Standards Institute.

- *National Languages:* The original ASCII encoding standard really supported only United States English. Various extensions have allowed it to support a variety of European languages based on character sets similar to that of English. These are the so-called Latin-1 languages. However, there are many languages, particularly those of Far Eastern countries, which cannot be represented by simple, eight-bit ASCII encodings. The lack of standards for these encodings has encouraged a variety of incompatible implementations to appear. Fortunately, standards work is under way and, in the future, consistent encodings for all world languages should be available.

- *Floating-point numbers:* There are at least two commonly used floating-point number representations. IBM mainframe machines use a proprietary representation developed many years ago. Standards work at the IEEE subsequently resulted in a very similar, but subtly different, representation which most manufacturers have now adopted.

Differences such as these will continue to affect transmission of data between machines with unlike architectures. At some time in the future, all machines may finally use a common set of representations. Indeed, trends towards this can already be seen in the industry. However, for the foreseeable future, there will be machines with differing data representations which need to communicate with one another. Consequently, middleware layers have to deal with the problem of converting one machine's representation into that of another machine.

There are two main approaches to data conversion. They differ in where conversion is performed and the precise nature of the data actually transmitted.

9.3.1 Conversion via canonical forms

A *canonical form* of data is a definition which is common and independent of any specific machine architecture. Conversion schemes based on canonical forms define the precise format of each data type as it appears on the network. Each machine must convert the data to this format before putting it on the network. Machines receiving the data convert from the canonical form to their own representations before using it.

An advantage of canonical forms is that each machine need provide only a single pair of conversions. A machine only ever converts between its own format and the canonical form. A significant disadvantage is that communication between identical machines whose representations differ from the canonical form requires two conversions. In this case, conversion is entirely unnecessary, because both machines use the same representation.

9.3.2 Conversion via receiver makes it right

The main alternative to canonical form conversion goes under the rather unwieldy title of *receiver makes it right*. In this scheme, a sender puts data onto the network in the sender's own representation. The sender also includes enough information so that the receiver of the data can know what representations were used. It is the receiver's responsibility to convert the data into the receiver's own representation. The advantage with this scheme is that no conversion is necessary if both machines use the same representation for a particular data type. The disadvantage is that every machine must be able to perform every conversion. Fortunately, the continued rationalization of data representations means that there are now relatively few different representations. The appearance of entirely new ones is also unlikely.

9.3.3 Self-defining data streams

So far we have looked only at techniques for converting known fields within some data being transmitted between partners in an application. We have not addressed the problem of determining how to represent the structure of the data. Application data can consist of arbitrary combinations of integers, character strings, floating-point numbers, and so on. Without knowing which parts of the transmitted data are in which format we cannot apply the appropriate conversions. The definition of the structure of the data being transmitted is data about data, often called *meta data*.

One advantage of the remote procedure call model of distributed applications is that the type and format of every item flowing is known explicitly at client and server. The interface definition, a fundamental requirement of RPC programming, makes the meta data explicit. Consequently, there is no need to pass meta data with the data itself. Both partners know exactly what to expect in any transmitted data.

In more general cases, such as message queuing, where the structure of the data being passed does not have to be made explicit to the middleware layer, additional meta data must be supplied. Standards do now exist for defining this meta data. The most important is probably the International Standards Organization's Abstract Syntax Notation Version One. Abbreviated ASN.1 and pronounced "ASN dot 1," this notation is used for the definition of data for application programs in the presentation layer of the OSI model of distributed applications. ASN.1 has also been adopted by a number of other bodies for data representation definitions.

ASN.1 is a formal language for definition of abstract data types. It provides a human-readable form of definition. Compilers for this form of definition generate *concrete data descriptions* of the same information, for use in communication protocols. A concrete data description is a compact description of the data which is being sent across the network. It travels with the data so that the receiving system can determine the format of the information it is being sent. With ASN.1, data in transmission is in a canonical form. The concrete data description defines what is being transmitted. Basic Encoding Rules, another output from the ASN.1 compiler, describe how each item of information should be converted between the canonical form and the local system's data representation. These rules are normally emitted from the ASN.1 compiler as source code for an encoder/decoder module which can be used in the application programs at each end of the communications link.

9.4   Resource Recovery

In many types of application, the data being manipulated is of great importance. Nowhere is this more evident than in the transfer of funds between accounts. A business transaction may involve moving data that represents many thousands of dollars. Clearly, care in handling such data is vital. Organizations which frequently do so place stringent requirements of data reliability on their computing systems.

The quality of the basic hardware components of computer systems has increased steadily, especially in recent years. However, failures do still occur. Processors may fail, disk drives may fail, and networks may fail. Some kinds of failure have nothing to do with the computing machinery itself. Interruptions to main power supplies, while rare, do still occur. Although steps can be taken to protect installations from various kinds of hardware failure, a vital role in the protection of data from loss is played by system software which supports applications. This system software gives applications the ability to deal with resources as if they are 100 percent reliable, while taking account of the fact that they are not. We can term resources protected in this way *recoverable resources.*

Recoverable resources are not at all uncommon. Typical relational database managers provide recoverable resources in the form of the tables within the database being managed. The database manager provides a recoverable resource, namely the database. This recoverable resource is based on the file system software and disk drive hardware of the underlying system. These components are not usually recoverable in themselves. A head crash on a disk drive invariably results in loss of data on that drive. The challenge for resource managers is to provide highly recoverable resources based on components which may fail.

There is a distinction between the levels of reliability normally considered adequate for general-purpose file systems and those associated with business-critical applications. The additional level of reliability associated with the database systems is not achieved without cost, of course. Not only is there the cost of the software to be installed and its administration, there is a performance penalty to be paid. The database manager is carrying out a significant amount of work to protect the data, in addition to the other functions it provides. The cost involved dictates those applications for which such protection is appropriate. Normal file-based applications, such as the word processor being used to create this text, do not warrant these costs. Business-critical commercial applications certainly do.

The important point about recoverable resources is that, in the event of failure of the underlying system, they can be restored to a known state. During a failure, the typical sequence of actions is as follows:

1. The system fails.

2. The underlying fault is rectified. For example, the failing hardware is replaced.

3. The underlying system is restarted.

4. The resource manager is restarted and performs resource recovery. The resource is restored to its state just prior to the failure.

5. The resource manager allows work to restart.

Two things should be noted about this set of operations. First, the resource manager goes through a distinct recovery phase in which it resolves any inconsistencies in the recoverable resources. Second, none of the services can be restarted until the problem which caused the failure has been rectified. If a hardware component must be replaced, there could be a considerable delay. Although the system will eventually be recovered, the loss of service, sometimes called an *outage,* may be unacceptable. The usual way to overcome the delay is to have a separate system available to take over the work load of the one which failed. A system configured to do this is often called a *hot standby*.

So far we have considered how a failure may affect an application executing on a single machine. However, a distributed application may have elements running on a number of different machines. Failure of a component of any one of these machines may cause loss of data unless special precautions are taken. In addition, data integrity frequently demands that update on more than one machine occur simultaneously, at least from the standpoint of the application. An example will help illustrate the point.

Suppose an application has to handle a financial transaction which includes the transfer of money from one account to another. In addition, suppose that the accounts involved are managed on two different machines connected via some network. For the transaction to be processed correctly, one account must be debited and the other credited by the same amount. As long as everything proceeds correctly, this is, of course, trivial to arrange. However, once we introduce the notion that the hardware components involved can fail, we can immediately see data integrity problems. If the debit action works but the credit fails, the bank will have improperly obtained funds. It will also have gained at least one irate customer! Conversely, if the credit works but the debit fails, the customers may be delighted, but the bank will have lost funds.

It is clear from the example that proper operation of the transfer requires that both the operations must work. Conversely, if any failure occurs, the resources must be returned to the state they were in before the transfer began. It must be as if nothing had happened. When operations are dependent on one another, as they are in this example, we group them together and refer to the group as a *logical unit of work*. A logical unit of work is a set of operations which must all either work or not work. Importantly, if any of them fails, we require that the state of the system be returned to that which was in effect before the logical unit of work began. In our example, the logical unit of work consists of the credit and debit operations. In general, any number of operations may be involved. If some of these operations run on different machines, or even involve multiple resource managers on the same machine, the group is called a *distributed logical unit of work*.

The ability to support logical units of work is not a property of all applications. Logical units of work are not normally supported directly by operating systems or networking software. For example, many failures cause the loss of a connection between client and server when using a traditional remote procedure call. Recovery from such a loss is entirely the application's responsibility. The underlying networking support provides no assistance.

9.4.1   Local logical units of work

Local logical units of work are almost always provided by a database manager. Applications define a unit of work to the database by using verbs such as *commit* after a sequence of operations has been performed. Up until this point, changes made to the data can be regarded as pending. The database manager guarantees to make them permanent once the commit verb has been issued by the application. Similarly, the application can issue the *backout* verb to back out any pending updates. Application failure before a commit is equivalent to a backout.

Resources managed by systems which support logical units of work are recoverable. They can be reset to a known, consistent state after any failure. Indeed, we can think of an application running against a set of recoverable resources in terms of a series of well-defined state transitions. Each transition represents a logical unit of work. The units of work move the application data between well-defined states. Because units of work are indivisible, the data is always in one of these states and never in some intermediate state. Because the states are well defined, the application code can recognize them and take appropriate action. The application will never be presented with a set of resources in a state it is not prepared for.

It is clear that the database manager needs to record some state information for each logical unit of work it is currently processing. If the machine should fail with a unit of work in progress, the database manager must be able to return the database to the state it was in before the failure, during the resource recovery phase. The information needed to do this is stored in a *log* during the normal operation of the system. When recovering from a previous failure, the database manager reads the log in order to determine how to recover each unit of work that was in progress when the failure occurred.

Logging is a vital part of the operation of any recoverable resource manager or transaction processing system. At its simplest, a log is merely a file written sequentially. Each entry in the file, known as a *log record,* describes some action taken by the resource manager. Frequently, these records also contain some or all of the data involved in the operation, providing a copy which can be used to re-create it if necessary. Typically, each resource manager vendor provides a log as part of the resource manager itself. Transaction managers also use logs and may provide their own or use one provided by a resource manager.

Because the log is fundamental to the proper operation of the system, great care is taken when designing and implementing it. It is a key element in robust, recoverable systems. It can also have a major effect on performance. Since information in the log is required for recovery, it needs to have been written to disk before the operation itself proceeds. Most normal file systems write data to buffers in main storage. These buffers are flushed to disk only when they fill, or after some predetermined interval. The buffering is vital to maintain overall system performance since processors and main memory are orders of magnitude faster than disk drives. The approach used by file systems is not sufficient for handling log data. We cannot afford to lose log records in the event of a power failure. Fortunately, careful analysis reveals that not all log data needs to be written to disk immediately. There are, however, key events which do require data to be written to the disk before an operation can proceed. Log data can be buffered in main memory, just like normal data, and can be written to disk when the buffers fill or when one of these special events occurs. The act of flushing the log records to disk because of one of the special events is known as *forcing* the log.

The log itself must be available in order for recovery to be performed. Consequently, it must be protected from data loss due to failure of the underlying disk system. One common way to do this is to have two copies of the log being written. This is usually referred to as *dual logging*. Naturally, the two copies of the log must be on different physical devices, so that an error on a single disk cannot destroy them both. Some operating systems facilitate dual logging by providing a facility called *mirroring*. A mirrored file or file system is automatically written as two or more copies when an application writes to it once. Regardless of the implementation, dual logging incurs a time penalty when the log's buffers have to be written to disk. Multiple write operations are required: one for each physical device containing a copy of the log. Allowing the system to proceed after only one copy has been written puts the integrity of the log at risk. For example, the disk to which the log has been written might fail before the second copy can be written. Consequently, the system must wait until both copies of the log records have been written before allowing the application to continue.

9.4.2 Distributed logical units of work

So far we have looked only at the case in which all the resources being used by an application are managed by a single resource manager. Once more than one resource manager is involved, and especially when more than one computer system is involved, support for distributed logical units of work is required. Most database managers can provide distributed units of work for their own databases. This means that databases on more than one machine can be updated and can participate in a unit of work. However, if more than one kind of database manager is involved, a transaction processing system is often needed to provide the services required for distributed logical units of work (see, for example, Gray and Reuter [1993] in the Bibliography).

Distributed logical units of work require a special protocol to be followed. It is known as the *two-phase commit* protocol and defines how resource managers and the logical unit of work coordinator operate together to provide distributed, recoverable resources. The coordinator of the logical unit of work is the entity which receives the application's request to commit or back out. It has the responsibility for ensuring that appropriate requests are made to the resource managers to fulfill the request.

9.4.2.1 Two-phase commit. For the purposes of this discussion, we'll consider the common case in which the logical unit of work coordinator is a transaction manager, and the resource managers are databases. The resource managers are each capable of committing and rolling back the data which they control. They have their own logging facilities and, individually, provide recoverable resources. The challenge for the logical unit of work coordinator is to make these separate, individually recoverable resources appear as a single recoverable entity as far as an application is concerned.

The two-phase commit protocol is driven by the coordinator—in our case, the transaction manager. It proceeds as follows. The application issues a request to commit the work it has performed. This request goes to the transaction manager, whose programming interface must supply the appropriate verbs. Contrast this with the case in which only a single resource manager is involved. In that case, the commit request goes directly to the resource manager. On receipt of the commit request, the transaction manager requests each participating resource manager to prepare to commit the work it has performed. This is the first phase of the two-phase commit, known as the *prepare phase.* In preparing to commit, each resource manager must take the actions appropriate to be able to commit during the second phase. This generally requires the resource manager to record state information on stable storage, usually by writing log records. The key point about the prepare phase is that it is the last point at which a resource manager is allowed to reject the operation. Once a resource manager has replied to the transaction manager indicating that the prepare is successful, the commit which follows must be successful. In effect, by declaring that it has successfully prepared to commit, the resource manager is guaranteeing to the transaction manager that it will be able to complete the commit if asked to do so.

The transaction manager collects the prepare replies from the resource managers. If all of the resource managers prepared successfully, the transaction manager knows that the commit will be successful. It sends a commit request to each resource manager. All resource modifications which were part of the logical unit of work get committed together. This constitutes the second phase of the two-phase commit protocol, and is called the *commit phase*.

If, during the prepare phase, one or more resource managers reply indicating that they cannot commit, the transaction manager will send backout requests, rather than commit requests, during phase two of the protocol. In this case, all resource modifications are rolled back together.

If a resource manager fails during the two-phase commit protocol, or if connection with the transaction manager is lost, the resource manager affected will not know the outcome of the logical unit of work. It discovers the result by contacting the transaction manager during its own restart processing. The transaction manager logs important state information, including the outcome of all two-phase commit decisions. Consequently it has the required information.

Of course, if the transaction manager loses contact with a resource manager prior to or during the prepare phase, it will not have received a reply to the prepare request. It will be unsure about that resource manager's willingness to commit. Different transaction manager implementations take different approaches to this problem. However, one common technique is to assume that any missing replies indicate that the resource manager affected is unable to commit. The logical unit of work is rolled back. This version of the protocol is known as *two-phase commit with presumed abort*.

In our example, we used a transaction manager as the transaction coordinator. Coordination of distributed logical units of work is a major role for transaction managers, such as IBM's CICS, Transarc's Encina, and UNIX System Lab's Tuxedo products. Transaction managers provide many other functions in addition to transaction coordination. Individual products vary, but these functions may include scheduling of work, distribution of work among multiple machines, and provision of their own recoverable resources. Some products, notably CICS, provide an operating-system-independent environment for applications. CICS programs are highly portable, even between machines with radically different architectures and operating systems. Application portability has been a significant factor in CICS' enormous popularity in commercial applications.

The two-phase commit protocol requires that the coordinator—in our case, the transaction manager—and the resource managers have an agreed way of communicating with one another. The transaction manager, for example, has to be able to request each resource manager to commit. The transaction manager and resource managers must agree on the mechanism used.

In the open systems community there is now a standard which covers the interaction between the various parts of a transaction processing system. The standard is controlled by X/Open, one of the major open systems standards bodies. The standard is based on the model of distributed logical units of work which we have just covered. The interface by which transaction managers and resource managers communicate is known as the XA interface. It is a series of function calls which allow transaction managers and resource managers to communicate. It is based on concepts from the CICS Resource Manager Interface (RMI), which performs the same function for that particular transaction manager and the database managers that support it.

9.4.2.2 Queue managers as resource managers. Database managers are not the only kinds of resource manager, though they are the most common today. Other examples of resource managers include transactional file systems. These store data in entities which behave more like files than full databases. Usually, they include keyed access to data which is typically stored as sets of fixed-length records. Because the file system can participate in logical units of work, it can be kept consistent with other application resources, such as databases.

Databases and transactional file systems provide recoverable resources whose primary function is storage. However, queue managers can provide recoverable resources whose primary function is delivery. This is true of the MQSeries queue managers. Such queue managers can participate in logical units of work. Message operations can be committed and rolled back in just the same way as database operations. The queue manager logs queue operations just as a database manager logs database operations. As a consequence, distributed logical units of work can include not only database updates but also message operations. If the unit of work rolls back, the database is returned to its state before the operation started, and any messages output by the application are removed from the queue before they can be sent. With a database, updates made by one application are not apparent to other applications until they have been committed. The same is true for message queues. Messages placed on queues are not visible to other applications until a commit has been processed. Likewise, messages removed from queues will reappear if a backout occurs.

The following example illustrates a unit of work including both database and queue operations.

1. Start logical unit of work.

2. Update database.

3. Put message.

4. Application terminates abnormally.

5. Recovery processing starts.

6. The transaction manager and database manager restore the previous view of database.

7. The transaction manager and queue manager restore the previous view of the queues.

The resources are in the same state at the end of this sequence as they were in before the unit of work started.

9.5 Systems Management

The topic of systems management covers nearly all of the day-to-day tasks which keep a computer system operating. These vary from the definition of new users to the archiving of important data. None of the operations involved is new. Organizations providing computing services have always had to perform them. The need to perform such tasks has encouraged the idea of centralized service provision. In this approach, all the major machines involved in providing an organization's computing systems are gathered together in a single site, or a small number of sites. Each site, commonly known as a *data center,* has a significant amount of equipment. It also has an operations team, responsible for managing and maintaining the service. By gathering the equipment to be managed together, the operations staff are able to cater for many systems. Consequently, the staffing costs are spread over multiple machines, making the systems management more affordable.

The data center approach is well suited to a centralized environment in which a small number of very powerful, large machines provide the computing service. It can be an extremely cost effective way of providing a computing service to a large number of users. In contrast, a distributed environment, by its very nature, often requires that machines be geographically dispersed. Typically, they are nearer to the end user than is the case with the data center approach. This distributed network of smaller machines provides a number of challenges for systems management.

It is normally impractical to have systems management staff located at each site which has a machine. The cost alone usually precludes it. When machines are grouped in a data center, each member of the operations staff participates in looking after more than one system. If the machines are distributed, each will require one person. The result is a significant increase in the number of operations staff. The only way to overcome this problem is to allow operations staff remote access to the machines. This way, although the machines are geographically dispersed, the staff are not. They can continue to manage multiple systems. However, the loss of physical access to the machines, which this approach involves, presents the most significant challenges to systems management. How do you back up a disk drive on a machine when you can't physically put a tape into its tape drive?

Solutions to the problem require sophisticated, distributed systems management software. These solutions may also have implications for the amount of traffic on the network connecting the machines.

In the rest of this section we will examine the issues involved with various categories of distributed systems management. We will divide the topic into a number of sections, each of which will deal with one particular aspect of systems management. In most cases, the topics covered represent what is being done on well-managed sites today. There are few new operations mandated by distribution alone. However, the need to perform these tasks in a distributed environment, without physical access to the machines involved, serves to make the operations more difficult.

9.5.1   Host management

The term *host management* describes the operations required to maintain individual computers within the distributed system. All machines, regardless of the size and power, require some regular operations to be carried out. These operations do not include anything to do with the data or the operation of applications on the machine. We will cover those later. Instead, it covers management of the hardware and software configuration of the machine and issues to do with how well it is operating. There are a number of aspects of host management which we will cover in the following sections.

9.5.1.1   Installation management.   New applications, operating system updates, and networking software changes all need to be applied to the machines on the network from time to time. In a distributed environment the installation process must be capable of being carried out remotely. This normally means that the remote systems must be able to download the new software over the network from some other machine. Machines which provide this service are sometimes known as *install servers*. Since applications and operating systems are frequently tens or hundreds of megabytes in size, the intervening network must be capable of supplying significant bandwidth to enable such operations to be carried out remotely. Local area networks rarely have problems in supporting this kind of operation. This is not necessarily the case for long-haul networks.

There are some especially difficult tasks when dealing with installation of distributed systems. Sometimes, it may be necessary, for technical reasons, that some fix or new software level be installed simultaneously on all the machines on the network. If this installation is to include users' desktop machines, it may be necessary to take special steps to ensure that all the affected machines are left running during the overnight or weekend period in which the installation is taking place. Quite apart from the technical difficulties, organizational impacts such as this require careful planning and serve to make the task more difficult than when all machines are within physical reach. Also, of course, network failures during such an operation could have serious consequences if the only way to recover is to restart the remote machines.

9.5.1.2 License management. Copies of software in use in an organization need to be properly licensed. Failure to have an adequate number of licenses can render an organization open to prosecution. Before distributed systems, and especially local area networks, were so prevalent, use of a product was simply controlled by the number of physical copies purchased. Each had to be installed separately. With the advent of file servers on networks, a considerable maintenance benefit can be obtained by installing the software once and allowing machines to use it via the network. The clear savings in installation time and disk space make this a very attractive alternative for customers. Software which is used this way is sold using multiuser licensing. The customer specifies the maximum number of concurrent users and pays the appropriate fee. Unfortunately, the only safe way to ensure that enough licenses are purchased is to specify one per attached workstation. This may far exceed the actual concurrent use. Recently, moves have been made toward so-called *user-based* pricing. Here the software can police its own use and prevent more than a specified number of copies of itself running concurrently. Attempts by users to run more copies than the license permits fail. With this technology, customers can pay for a realistic number of licenses and be sure that it will never be exceeded. An increasing number of application programs can now offer this feature. In addition, operating systems are starting to offer the technology for applications to use. A good example is the network license server in IBM's AIX operating system.

9.5.1.3 Performance management. The performance of any computer system is subject to variation over time. The reasons for this are many and varied, but include

The actual application load on the system

The number of attached users

The organization of data on fixed disks

The number of recoverable errors occurring

The aim of performance management is to spot trends in the performance of systems so that appropriate action can be taken before problems manifest themselves. The actions involved depend on the particular cause of the decreasing performance. If the cause is poor organization of data on disks, the remedy is to archive, initialize, and restore the disks involved. On the other hand, if the cause is the popularity of a new, memory-intensive application, the remedy is to install more memory.

In a distributed environment, the collection of appropriate performance data must be possible from remote machines. The remedies must also be capable of being implemented remotely. Frequently, the remedies are standard systems management tasks which have to be carried out regularly in any case. Performance monitoring can help identify exceptions which need more urgent attention. Fortunately, distributed systems monitoring products are beginning to appear. Typically, one workstation acts as a monitor, collecting performance data from a large number of other machines. Though the monitoring workstation may be dedicated to that task and may use highly sophisticated graphical representations of the data being collected, the load placed on the machines being monitored tends to be fairly light, not adversely affecting their ability to perform useful work. One drawback with most currently available systems is that the tools tend to be restricted to one kind of machine group. For example, one product may be able to monitor open systems machines, while a different one monitors PC-class machines and yet another monitors Macintoshes. This can be a drawback in a heterogeneous environment, where operations staff will need to be conversant with multiple tools.

9.5.1.4 Problem determination. Systems used in production environments need to provide good problem determination facilities to allow faults to be traced and rectified quickly. The whole aim is to minimize the time during which the system is unavailable. Typical facilities provided by operating systems include error logs, in which system and application errors can be recorded, and trace systems, which can be turned on to yield more information as the error is re-created.

If a remote system detects a problem, it must be able to warn the operations staff. Most techniques today rely on the notion of *alerts*. These are messages which can be sent from a machine as it detects a problem to some designated remote control center. Alerts are usually a feature of the network management system being used. Of course, it is possible for a machine to fail without having the opportunity to issue an alert. In this case, it is necessary for the network management system to poll the machines on a regular basis so that it can detect the failure. The effect is that failures which do not result in an alert take a little longer to detect.

Once a failure has been detected, some actions are obviously necessary. Some operating systems can be configured to restart automatically after a system failure. If this is successful, operations staff can access the error records stored by the system and analyze the failure. If the system cannot be restarted, the situation is more difficult. A local member of staff may have to examine the machine on behalf of the operations staff. Some machines carry external indicators which warn of problems encountered during restart. Values in these indicators might need to be reported to the operations staff. The important point to note is that, while much recovery and repair work can be carried out remotely, if the system is not running well enough to allow suitable network communications, there is no alternative to using local staff. While they do not have to be highly trained, they must know enough about the machine to assist the operations staff in initial problem determination. The level of skill required varies from system to system. In the best cases, knowledge of how to obtain external indications of problems, how to switch the machine on and off, and how to tell if it is running is required. Even these apparently simple tasks can involve some degree of skill. Some systems, for example, require separate components to be switched on and off in a specific sequence.

Factors such as the appropriate education for local staff must be taken into account when assessing the cost of systems management for a distributed system.

9.5.1.5 Inventory management. The goal of inventory management is the maintenance of a database showing the precise configuration of the hardware and installed software on each machine in the distributed system. The reason for doing this is rather different from that which motivates license management. An inventory allows operations staff to assess the impact of a forthcoming change on the machines in the distributed system. For example, a new level of some networking software might require a particular level of operating system for one specific class of machines on the network. The inventory allows machines needing an operating system upgrade to be identified and the necessary work to be performed in advance of the network upgrade.

The inventory allows an organization to monitor the hardware it has installed. It is also a method of verifying that the number of installed copies of licensed software does not exceed the number of licenses which the organization has purchased.

Keeping a hardware inventory up to date is a particular problem when there are PC and workstation machines in the network. It is very simple for end users to change the configuration of these machines, adding cards or swapping devices around. The inventory can quickly become inaccurate if such changes are not always recorded. Providing a mechanism to ensure an accurate inventory on these machines is a significant challenge. Because the numbers of machines can be very large, the problem can become intractable.

9.5.2 Network management

The term *network management* describes the operations required to maintain the networks which connect the components of a distributed system. Increasingly, network management products include not only functions for managing network components but also functions for managing hosts. In the main, the host management functions are extensions to the functions provided to manage the network itself. However, it appears that newer products are being positioned as the focal point for all the functions of distributed systems management. Given the central role played by the networks involved in any distributed system, this is hardly surprising.

Since we have already discussed host management, this section will cover only those aspects of network management which concern the network itself.

9.5.2.1 Performance management.
Just as in host management, the aim of performance management is to spot trends in the performance of systems so that appropriate action can be taken before problems manifest themselves. As with hosts, changes in performance of the network can be caused by a variety of factors. Analysis of the performance of various components of the network allows trends to be spotted and remedies to be implemented.

Although some elements of the network (such as communications controllers, routers, bridges, and, indeed, the network medium itself) may be special-purpose components, others may be hosts containing appropriate network adapters and running suitable software. Internet routers are a good example. They can be purchased as special-purpose units. Alternatively, a workstation with appropriate adapters and software can be configured to do the job. In addition, any host attached to the network can be viewed as part of the network, at least for some network management tasks. If a specific host is heavily loaded with application work, for example, its ability to handle network traffic may be impaired and a bottleneck might be created, affecting other machines in the network. In this case, the solution to the network performance problem might involve upgrading the host or reorganizing the network to remove the bottleneck.

Just as with host performance management, performance data must be collected from the network components to allow the network's performance to be monitored and analyzed. Most network protocols include provision for sending management data such as this between elements of the network. This allows performance data to be collected at the appropriate site for analysis by the systems administration staff.

9.5.2.2 Problem determination. Identification and rectification of faults on the network are similar to host problem determination in many respects. The same kinds of difficulties are caused by the operations staff not being local to the equipment which has developed the fault. As we have seen, some of the network components are actually hosts. However, some of the equipment is very specific to the network, particularly if it involves long-haul connections managed by an external organization. Faults developing in such connections cannot be resolved by an organization's operations staff. They need to be rectified by the carrier providing the network service carrier.

Other problems on the network are reported and treated much like those on host machines. Often, equipment detecting a fault sends an alert to the systems administration staff controlling operation of the network. Some alerts may be detected and handled automatically by the network management system. Others require the intervention of operations staff. As with host problem determination, if the fault is so severe that all contact is lost between the operations staff and the portion of the network involved, remote resolution of the problem is clearly impossible. Local assistance will be required.

9.5.2.3 Difficulties associated with heterogeneous networks. Whereas specific network protocols provide mechanisms for reporting problems, the design of these features differs among different networks. Many organizations have a variety of networks installed. For example, mainframe hosts might be connected via SNA, workstations via TCP/IP, PCs via NetBIOS, and Macintoshes via Appletalk. All these networks can be connected together using the appropriate hardware and software, and information can flow among the machines involved. Managing the networks is, however, complex. There is usually no single point of control from which the whole network management task can be accomplished.

Future network management products will address this heterogeneity of networks more thoroughly than those which exist today, increasing the level of integration and allowing management of a wider variety of network types from within a single environment.

9.5.3 Distributed data management

Data management includes all the tasks required to ensure that the data needed by the applications in the distributed system is always available. This data includes files (in local and distributed file systems) and databases.

9.5.3.1 Remote archiving. One major element of data management is *archiving*. Data which is impossible or expensive to re-create must be backed up to offline storage on a regular basis. While archiving is a relatively simple task when the system involved is local to the operations staff, remote archiving introduces some major complications. The most obvious one is that of handling the media onto which the archive is being made. Typically, the media are high-density tape cartridges. For maximum flexibility, the media should be at the same site as the operations staff. This immediately implies that the data being archived must traverse the network before being stored. The archiving operation may be subject to the same kind of failures as any other distributed operation and must be robust in the event of such failures.

In addition, if the data being archived is sensitive, it will probably need to be encrypted before transmission. There is little point in protecting data transmitted by distributed applications if the archiving operation sends large amounts of data across the network unprotected.

Archiving tends to involve significant amounts of data. Normally it has to be carried out while the data involved is not being operated on. Applications are usually unavailable while archiving is in progress. The consequence of this is that there is often a restriction on the time available in which to carry it out. The time available and the amount of data involved place requirements on the bandwidth of the intervening network. Wide area network connections may not be able to supply sufficient bandwidth without requiring prohibitively expensive media and services.

Remote archiving has three major attractions. First, staff local to the machines carrying the data being archived do not need to be involved in the process. Second, the archive is physically remote from the machine, providing an element of support for disaster recovery. If the machine is destroyed by a fire or other catastrophe, the data is safely archived. Obviously, data at the operations center itself must be protected against similar disasters. However, disaster recovery of such a center is not, in principle, different from recovery of a traditional data center. The third major attraction is that data recovery at the remote site can happen across the network. This should lead to relatively short periods during which the data is unavailable after a failure.

9.5.3.2 Local archiving. An alternative to sending archive data across the network is to run the archiving operation locally and to send the media by some independent means to the operations staff. While minimizing the requirements on the network, this approach has two drawbacks. First, staff local to the machine have to be able to handle the backup media, loading it into the machine each evening and retrieving it each morning. Second, there may be a considerable delay in retrieving data after a failure. If a disk drive fails just after the tape has been sent to the operations center, it may be hours before it can be recovered from the carrier.

9.5.4 User management

There are many tasks associated with the management of users and their accounts on the various machines involved in a distributed system. The ideal environment for end users is one in which the entire distributed system, with its heterogeneous community of machines and networks, appears as a single environment, regardless of how they access it. Today this goal is achievable, but only across machines of the same broad class. The PCs in a network can all be made to look the same to end users. But they cannot be made to look the same as the UNIX machines in the same network without application support. This is one of the major attractions of distributed applications. They allow users to access machines other than the ones on their desks without being aware of it. This seamless access for the end user requires additional work in account management in today's heterogeneous environments.

9.5.4.1 Account management. Account management is a simple task in a local environment or even in a network of similar machines. Most operating systems provide user accounts in roughly the same way. In a heterogeneous distributed environment, however, a number of issues arise.

The first issue is one of account allocation. Although account management is one of the tasks which can be devolved to remote sites or groups of sites, and indeed large organizations need to do this, the distributed system is effectively a single entity. There needs to be a way to prevent allocation of duplicate user account names by different sites. Worse, as multiple kinds of systems may be involved and accounts must be set up on each, a user account naming convention must take into consideration the capabilities of the least flexible system.

The second issue is concerned with maintenance of multiple accounts. Because different classes of machines are usually unable to share accounts, each user may have to have several accounts, one on each class of machine. This may be the point at which the apparently seamless environment breaks down. One aspect of account management, namely the assigning of passwords, is almost always the end user's responsibility. In a heterogeneous environment users may need to do this for a number of different accounts.

9.5.4.2 User assistance. Many large organizations provide some kind of help facilities over and above those in the systems or applications themselves. These often include a *help desk* which users can contact in the event of a problem or in order to find out more information about some aspect of the system. Some installations also include online help. In either case, provision of these facilities is made more complex by the fact that the users are remote from the support staff. It is not possible, for example, for a member of the operations staff to visit the user in the case of an intractable problem. Online help needs to be provided in such a way that it can be viewed, regardless of the particular machine to which the user is directly attached. Full multimedia online help may be very attractive, conceptually, but is of no value to a user with an ASCII, character-based terminal attached to a midrange UNIX machine. The reality of commercial systems is that low-cost, character-based terminals are, and are likely to remain, the most common form of access to computing services for end users.

9.5.4.3 Accounting management. Some organizations charge for the provision of the computing service to end users. Where the service is provided by an organization for its own internal use, this may be a notional charge. Even so, it is an important way in which use of resources can be monitored. It provides a basis for future purchasing decisions by allowing usage trends to be seen, for example.

Accounting for use of a single system by a given user is relatively straightforward. Accounting measures include the length of time the user was actually logged on and the amount of processor time used by processes belonging to that user. In a distributed system, the correct measures are quite different. A user may log on to a particular machine on the network, and then use applications which spend most of their time executing on other machines. Typical client-server environments make little or no provision for apportioning the time used in executing a service to the clients making requests for that service. Whereas operating systems almost always provide this information, distributed systems middleware has yet to get to a similar level of sophistication.

9.5.4.4 Authorization. Most practical computer systems require some level of protection to be applied to the various resources being used. Users are authorized to the resources they need while others are protected from accidental or malicious damage. Whatever the level of protection required, users need to be authorized to access the appropriate resources. In a local system, this is relatively straightforward. The simplest situations involve a single security manager governing all the resources. This would be the case, for example, where the resources involved were tables and databases on a single database manager. In this case, all authorizations could be carried out using the database manager's security system.

Where multiple resource managers and multiple systems are involved, life is considerably more complex. A single distributed application may be able to access data from different resource managers on different machines. The end user may have to have an account on each machine to allow the authorizations to be set. The application must be able to operate under the user's accounts on each machine involved so that authorizations can be checked. Finally, the authorizations themselves cannot be managed through a single security manager, since each individual system, and even each resource manager, may have its own. This diversity means an increased work load for the systems administration staff and an increased requirement for education.

9.5.5 Change control

However much support staff and systems programmers dislike it, change control is a vital component of any moderately complex installation. The goal of change control is to prevent updates being made to the system which might cause loss of service. Almost any proposed change, from installation of a new application to the altering of some system parameter in search of more performance, could cause some or all of the system to fail. One of the perceived advantages of distributed systems is the apparent ability to continue processing work load even when some part of the system is unavailable. While this is undoubtedly true for some types of middleware, with MQSeries being the obvious example, the fully synchronous nature of many client-server applications and the sheer complexity of the environment make it difficult to achieve in practice with traditional approaches.

The challenge for change control is the identification of potential interaction between changes which seem harmless on their own. In a distributed environment, the scope for interaction is increased. A new level of operating system in one machine might introduce a problem into its networking software which becomes apparent only when it attempts to exchange information with another vendor's machine. One major problem in heterogeneous distributed systems is that it is so difficult to test changes before they are applied. Without having a test environment that includes every kind of hardware and software component in the actual system, it is virtually impossible to be sure that a change to be made will not cause problems when it is made on the live production system.

9.5.6 Systems management and MQSeries

Products specifically intended to operate in distributed environments obviously need to provide systems management facilities which can be run remotely. MQSeries is no exception. Indeed, newer versions of

MQSeries products provide two methods of performing systems management: one based on administration message queues and the other based on portable scripts containing systems administration commands. We'll look at these two techniques in the sections which follow. Every administration operation required by MQSeries can be carried out using these techniques, apart from the initial creation of the queue manager itself.

9.5.6.1 Managing MQSeries via queues.

Each MQSeries queue manager has some standard queues, built when the queue manager is initially created. One of these queues, called SYSTEM.ADMIN.QUEUE, is specifically for systems administration. Messages arriving on this queue are processed by an application called the *command server*. This program, shipped with MQSeries, interprets the messages as requests to perform systems administration operations on the queue manager. A standard set of administration messages is defined, which the command server can process. Messages can be placed on the queues by the systems administration application. This interactive application is used by operations staff to administer the network of MQSeries queue managers. It is based on modern GUIs, such as Motif or Presentation Manager, and provides a user-friendly administration environment. Operations are selected, the target queue manager chosen and appropriate values supplied. Then the application builds the necessary message and sends it to the target queue manager's administration queue. Results of the operation are returned, also via queues.

By using a queue as the mechanism for sending administration requests to the queue manager, all the benefits of messaging and queuing are gained for the systems administration tasks. Remote systems can be managed this way, allowing a single point of control for the distributed system if desired. Time-independent processing of the requests is also achieved. If requests cannot be sent immediately to the target system, they are queued and delivered as soon as the required communications links become available. The message-based administration interfaces are fully documented and can be used by other applications.

9.5.6.2 MQSC and portable scripts.

While interactive administration applications are sufficient for many operations, they are not appropriate where bulk updates must be made to many resources on many queue managers. For this kind of work there is no substitute for being able to prepare a batch of requests in advance, using a text editor. The Message Queuing Script Commands (MQSC) are designed to fulfill this need. These commands can be put together in normal text files. They are interpreted by the MQSC processor, shipped with the newer MQSeries products. Because the commands have a syntax defined by MQSeries rather than by the underlying operating system, scripts written using MQSC are portable between queue managers, even when they are running on different classes of machine. Scripts written for a queue manager running on IBM's MVS operating system can be interpreted and executed on an RS/6000 running AIX, for example. This portability allows scripts to be written in one location and then shipped to another location for processing, regardless of differences in the type of machine.

9.5.7 Future systems management models

One thing which is clear from any examination of the current state of distributed systems management is that there is no integrated solution capable of performing the required functions in a consistent way across the wide variety of machines found commonly in modern, heterogeneous environments. The functions do exist, but they are insufficiently broad in the scope of the machines to which they can be applied. Middleware products are beginning to address some of the issues, but there is still a long way to go. DCE, described in more detail later, provides a basic, synchronous client-server architecture which forms the basis for a systems management framework known as Distributed Management Environment (DME). Like DCE, DME is a product from the Open Software Foundation. It has found wide support in the industry and among customers. For that reason alone, it is likely to provide the basis for future systems management products which do span a broad enough spectrum of machines to satisfy many customer's needs.

DME views a distributed system as being of composed of objects. These might be physical objects, such as computers and network controllers, but equally might be more ethereal entities, such as applications. The objects provide management interfaces which allow them to be administered. DME provides the linkages between management applications and the objects to be managed. The linkages are based on existing network management standards and new distributed object standards, particularly the Object Management Group's Common Object Request Broker Architecture, known as CORBA.

Systems management is probably the single most difficult aspect of the implementation of a heterogeneous distributed system. It is hard because it is the one place in the overall system where the underlying heterogeneity is exposed with a vengeance. Industrywide initiatives such as DME are important because they promise to provide environments in which that heterogeneity can be hidden once more. In the meantime, individual products can reduce the impact of their own systems management tasks by providing common interfaces regardless of the underlying machine architecture. As we have seen, this approach has been adopted by MQSeries.

9.6   OSF Distributed Computing Environment

The Open Software Foundation (OSF) was formed a number of years ago by a group of computer systems vendors which included IBM, Digital Equipment Corporation, and Hewlett-Packard, among others. Describing itself as a not-for-profit organization, its prime goal is to make available technology to assist in the development of open systems. It does this by acquiring technologies from the industry and integrating them into packages. These packages are then made available in source code form. Interested parties can license the source code, port it to the platforms in which they are interested, and sell the resulting products.

The OSF has a number of packages available. These include OSF/1, a portable implementation of UNIX which adheres to many additional international standards, and Motif, probably the most widely used graphical user interface on open systems today, as well as the Distributed Computing Environment.

DCE is a middleware layer. It depends on an underlying network. Currently, this is almost invariably TCP/IP, although in principle any network could be used. DCE provides an abstraction of the network. It is based on the traditional, synchronous client-server model of distributed applications, in which client applications request services from servers which may be on the same machine as the client or remote from it.

DCE provides some of the distributed communication services we have looked at in this chapter. Specifically, in addition to providing a model for distributed computing and a programming abstraction, in its remote procedure call, DCE provides the ability to locate services and the support required to enable secure distributed applications to be built. In this respect, it has been far ahead of any other approach, in the open systems arena.

9.6.1   Distributed application model

DCE is based upon the traditional, synchronous, client-server model. A client requests a service from a server and waits until the results of the request are available. As might be expected, given this model, DCE communications is based on remote procedure call (RPC). To the client, a request for service looks like a function call and has very similar semantics.

The organizational aspect of the DCE distributed application model is based on the concept of a DCE *cell*. A cell is an administrative domain for DCE. Typically, it represents a collection of machines administered as a single entity. For example, a user has a logon identification and password for the cell. This allows the same level of access, regardless of which specific machine the user is directly connected to.

Figure 9.6 shows a pair of interconnected cells. A cell is usually a set of machines connected by local area networks. It is normal that communication within a cell is faster than between cells. In Fig. 9.6, cell A consists of a single token ring connecting machines A, B, C, and D. Cell B contains an Ethernet connecting machines E, F, G, and H as well as a token ring connecting machines J, K, and L. Machine I is acting as a router linking the two LANs. Machines D and E are linked via some point-to-point wide area connection. Each is also a router. TCP/IP is the underlying network protocol for the whole network.

Cells can vary widely in size. The boundary of a cell is more likely to be determined by organizational considerations than purely technical ones. Since the cell is the unit of administration, cell boundaries must be chosen to match available systems administration facilities.

The DCE RPC includes the ability to perform data conversion. Because the RPC mechanism knows the precise form of each parameter being passed between client and server, it can map the machine-dependent representations in an appropriate way. DCE uses the *receiver makes it right* approach discussed earlier.

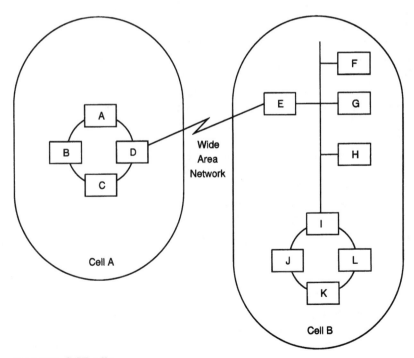

Figure 9.6 DCE cells.

9.6.2   Directory service

The DCE directory service provides the function by which clients locate the services they wish to use. Clients refer to services by name. The directory provides a name-to-address mapping returning the addressing information necessary for the client to contact the server. The DCE directory service is often referred to as the name server. The directories themselves constitute the DCE namespace.

The DCE directory can actually provide a much more general directory service than just the name-to-address mapping required to locate services. It provides a very general programming interface to allow applications to use its facilities to store and retrieve all kinds of information. This kind of access uses application-defined structures within the directory.

The DCE directory plays a vital role in a DCE-based distributed system. Without it, clients are unable to locate servers. Consequently, the directory is a potential single point of failure. It is also a potentially heavily used resource. To overcome these potential problems, the directory is implemented so that it can be partitioned and run on several different machines, while appearing as a single entity to applications. The partitioning can be used to replicate some or all of the namespace, so that if the machine running one replica becomes unavailable, requests can be routed to another copy. In addition, the directory can be split by content. Different parts of the directory run on different machines, spreading the processing load associated with requests to the directory.

The directory is divided internally into two parts, though the division is not apparent to applications using the service. One part of the implementation is optimized to operate within a cell. Not surprisingly, this component is known as the *cell directory service* (*CDS*). The CDS processes directory requests, which can be handled entirely within the cell. The second part of the directory is known as the *global directory service* (*GDS*). It works together with the CDS to process requests which span cells. Such requests involve retrieving information from the CDS of a remote cell. The global directory service can operate in a manner compatible with the OSI X.500 international directory standard, increasing the possibilities of interoperating with other directory implementations.

9.6.3   Security service

DCE provides both authentication and authorization services as part of its security service. Authentication requires that users log on to the cell. This is much the same as logging on to a local area network of interconnected PC-class machines. The user is a principal, as far as the

DCE security service is concerned. Each principal has certain privileges to the resources in the cell. DCE provides *access control lists* (ACLs) and suitable management facilities. However, the ACLs provided affect only the DCE resources. For example, users can be authorized to perform specific actions against particular subsets of the information in the CDS. Facilities are available, within the comprehensive set of programming interfaces provided by DCE, which allow application-defined ACLs to be used and managed. This does, however, require programming effort.

DCE also provides the kinds of protection we discussed earlier, for data in transit across a network. Specification of the level of protection required is made when setting up a connection between client and server. The processing necessary for any encryption is then carried out below the RPC interface, making it entirely transparent to the application.

DCE is unable to provide assured delivery. Loss of a network connection during an RPC call leaves both client and server in an indeterminate state. Application code is responsible for providing the means to recover an operation which fails.

9.6.4 Additional DCE services

DCE provides a number of other services and a distributed file system, all of which depend on the basic directory and RPC mechanism. The *distributed time service* (DTS) helps keep the clocks on individual machines in the network synchronized with one another. An external time provider can be used. An example of a time provider is the radio transmissions, linked to atomic clocks available in many countries. The *threads service* allows a single application to be multithreaded even on platforms such as UNIX, many implementations of which do not support the concept. Threads allow an application to continue processing, even when part of it is blocked waiting for some event. This is vital in an RPC-based client-server scheme. Without it, the synchronous nature of the RPC would prevent further work being done in a client while waiting for a reply from a server.

Finally, DCE provides a robust, high-performance distributed file system, known as DFS. Based on Transarc's Andrew File System, it provides the ability for machines to reference files on one or more file servers as if they were local.

9.6.5 DCE component interrelationships

The diagram in Fig. 9.7 shows how the various components of DCE relate to one another. The interrelationships are quite complex, with components relying on one another. Most fundamental are the RPC

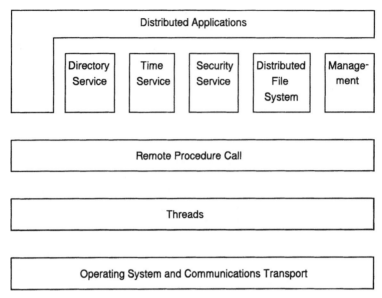

Figure 9.7   DCE components.

and threads services. The directory and security services are tightly coupled with the RPC mechanism. They rely on it, and it uses their services. The time service and the distributed file system also make extensive use of the other services. Similarly, all the systems management applications provided with DCE are fully distributed and make use of DCE facilities.

9.6.6 Summary

DCE provides a number of the basic services required by a distributed system. It provides a name service, though it relies on the underlying network to provide routing. It provides comprehensive security facilities, including protection of data in transit over the network. It also automatically converts between machine-dependent data formats.

However, DCE provides no resource recovery facilities, nor does it assure delivery of requests and data between clients and servers. In addition, it provides systems management only for its own components. These limitations should not be seen as omissions in DCE. The scope of DCE was chosen by the OSF to cover the basic requirements for distributed computing while leaving room for vendors to write applications based on DCE to provide additional functions for those customers requiring it. DCE can be viewed as the lowest common denominator. It includes all the functions which any secure distributed application must have. However it is not, in itself, sufficient as a development base for most large-scale commercial applications. Transaction processing vendors have recognized DCE's potential by implementing their systems on top of it. IBM's CICS products for UNIX systems and Transarc's Encina system are the most notable examples of such systems.

10

Approaching Application Design

There are many methods for designing applications. In this chapter, we will see how applications based on messaging and queuing can be designed using some of these techniques. The aim is not to recommend one specific design technique. This is, after all, a book on messaging and queuing, not on application design. Instead, our objective is to show how to derive an MDP-based implementation, regardless of the design approach chosen.

We start with a brief review of design methods, for readers unfamiliar with the techniques. The review includes a pair of short case studies, to illustrate the methods. Then we look in more detail at application design using the two major approaches known as function-oriented design and object-oriented design. For each of these approaches, we will derive a design from the same specification. We will also show how to construct an MDP-based implementation from the resulting designs.

10.1 A Review of Design Strategies

This section is a short review of the major strategies available to application designers. We will look at the goals of a design method, examine the current state of the literature on the subject, and then describe each of the major classes of approach, function-oriented design and object-oriented design, using simple examples.

10.1.1 Design method goals

The design of a piece of software defines how its external specification is to be implemented. It expresses the intent of the software. A good design partitions the solution into units which are understandable in their own right, without too many complex interdependencies. This is important for a number of reasons. First, it allows implementation of units to be carried out in parallel. This not only provides the basis for reducing the elapsed time required for development, but greatly simplifies the planning involved. Second, a well-structured design can be used in the education of the implementers. This task is too often forgotten during the design phase of a software development project. Finally, the design forms the basis for future maintenance and enhancement of the application. To be useful, the design must be capable of being kept up to date with the implementation it defines. Often the tendency is to discard the design as soon as an implementation is under way. The cost of keeping the design and implementation in step is often considered too much of an overhead as time pressures increase towards the end of the project.

The design methods we will consider provide a framework in which designs can be constructed and maintained. The methods try to yield designs that have the desirable properties which we have just listed. Ultimately, of course, design is a creative process. Consequently, no method alone can produce good design. The effectiveness of a design is ultimately dependent on the skill and experience of the design team. However, design methods do help designers to express their intentions in ways which are likely to communicate well to others.

All computer-based systems are, to a greater or lesser extent, models of part of the real world. This is clearly true for systems whose essence is simulation. Modern flight simulators, for example, are such sophisticated and accurate models of real flying that pilots can carry out much of their training on them. However, even mundane systems, such as payroll, have an element of modeling in them. Tax codes, salaries, and pay slips are very real things to the employees involved.

The way in which computer-based systems parallel the real world leads us to a general view of software as having two kinds of component, namely *entities* and *operations*. Entities are simply the "things" in the system. Operations are what the system does to the entities. Entities are associated with the nouns of human language, while operations are associated with verbs.

Since there are these two classes of component, it should come as no surprise to discover that there are two main kinds of approach to the design of software systems. Each takes one of the components as its starting point, making the other subordinate. Indeed, the only surprise should be that while one approach, function-oriented design, has been in use for many years, the other, object-oriented design, is relatively new and untried.

Function-oriented design considers the operations of a system as its starting point. All other aspects of the design are derived from the operations. Object-oriented design considers the entities as the starting point of the design, deriving other aspects, including the operations, from them. There are techniques which superficially appear to span the two approaches. In the main, they tend to have much greater affinity with one, sitting somewhat uncomfortably with the other.

We will look at both function-oriented design and object-oriented design shortly. Before doing that, however, it is worth looking briefly at the current state of the literature on design.

10.1.2 Design method literature

The two design approaches have caused something of a split among authors of works on design. Though each side does, in general, admit some advantages for the opposing method, there is a clear division. Each method more or less claims that it is the only way to proceed, though the reasons employed are rather different. Proponents of function-oriented design tend to have significant intellectual investment in their methods. Their approach has been refined over many years and has gained academic respectability in addition to being used successfully on a wide variety of real design tasks.

What the proponents of object-oriented design lack in longevity, they more than compensate for with fervor. Unfortunately, there has tended to be a shortage of good examples of object-oriented design not intimately entwined with a particular language. Whereas books on software engineering often cover function-oriented design in depth, object-oriented design tends to be treated rather superficially. The same is true of books on particular object-oriented programming languages, such as C++, Smalltalk, or Eiffel. There also tends to be some confusion between design and implementation, since the main goal of these works is to describe the language. Object-oriented languages may be helpful in implementing object-oriented designs, but they are by no means mandatory. Indeed, in distributed systems the object-oriented features of a language may be of little direct use. Programming language features are almost invariably restricted to a single process on a single machine.

Recently, however, a number of texts have been published dealing more directly with object-oriented design methods. Though all authors claim unique aspects for their own methods, there is a considerable amount of agreement between them in many important areas (see, for example, Booch [1994], Jacobson et al. [1992], Rumbaugh et al. [1991], Coad and Yourdon [1991 a and b] in the Bibliography).

This is not the place to pursue the arguments in favor of or against each approach, though we will mention some of the salient points in the following sections. Instead, we simply recognize that both approaches have merit and that many organizations will choose to use one or the other as the basis for their development activities. In either case, the resulting form of design can be implemented using MDP. Indeed, other design approaches can also yield systems implementable with MDP. Even if the precise system you use is not identical to those discussed here, the methods used to derive the MDP implementation from the design should apply. They should allow you to implement systems using MDP from designs derived using your favorite method.

10.1.3 Function-oriented design

Function-oriented design takes as its starting point the function or functions provided by the system in question. To illustrate the approach, we will develop a simple design for a machine which issues tickets for use on public transport. Before we can design the machine, we need to have a specification for the facilities it offers, as follows:

- The machine issues tickets for use on public transport. Customers place coins into the machine and then specify the required journey via appropriate buttons on the machine. The machine issues the ticket. Change is given as required. Ticket requests are rejected if the value of the coins entered is too low or if the required amount of change cannot be given because the internal store of change does not contain the required coins. Customers can elect to abort the operation at any time prior to the machine beginning to issue the ticket.

From this specification, we will analyze the required system and then construct a data-flow diagram. This will represent the top-level design for the machine. Finally, we will see how structure diagrams can be derived for some of the functions involved.

10.1.3.1 Essential systems analysis. There are many ways of analyzing systems from their specifications. We are going to use an approach called Essential Systems Analysis (see McMenamin and Palmer [1984] and Stevens [1991] in the Bibliography). This technique is based on previous approaches to structured design. It is a refinement and overcomes a number of problems that existed with the earlier methods.

The key to the analysis is in extracting the *essence* of the system being designed. The essence of a system is the activities and memory required for it to function. Not surprisingly, these are termed *essential activities* and *essential memory*. Essential activities are those functions at the heart of the system. These functions are the reason for the system's existence. Essential memory is the data which must be stored to allow the essential functions to operate.

The essence does not include all the activities and memory of the final system, only those essential for it to function. For example, activities such as administration are generally not considered part of the system's essence. This is because they are required only to overcome the fact that the underlying technology is not perfect. The concept of essence allows the design problem to be bounded. It allows the designer to concentrate on the most important aspects of the design without becoming too involved with particular details. Overlooking some aspect of the design, which later turns out to be part of the essence, might invalidate a lot of work. Just as with any design method, there are key stages at which decisions have to be made. Determining the essence of the system is such a stage in this design approach.

One simple approach to extracting the essence of a system is to look at the functions described in the specification. For our ticket machine, the essential activities are collecting coins and issuing tickets. Cancellation of a request is also an essential activity. In this case, all of these activities relate directly to what we might call the system's user interface and, hence, to the customer's view of what is happening in the system. This kind of external view of the system is often a good starting point for determining essential system activities. It might not be complete, however, as we shall see shortly.

Having identified the essential activities, we can turn our attention to the essential memory. We have to determine the information the system needs in order to be able to operate. Some information is provided in the events that trigger each activity. For example, when the customer requests a ticket for a particular destination, the request itself will contain information about the destination. However, information on the request is not sufficient in itself. Extra data is needed. In this case, a table relating destinations and ticket cost is required. Since coin entry and destination selection are independent activities, we will also have to remember the current value of coins entered. Finally, we also need to remember the current state of the change available in the machine. These are the three basic pieces of essential memory.

The next step in the design is to create a data-flow diagram based on the essential activities and essential memory we have just derived.

10.1.3.2 Data-flow diagrams. A data-flow diagram is a relatively formal way of representing the functions performed by a system and the data used in those functions (see, for example, Yourdon [1989] in the Bibliography). As with the analysis step, there are many variations of data-flow diagrams and some disagreement between proponents as to exactly what information should be represented. As before, we will sidestep these debates and choose one particular approach, in this case using notation similar to that of Gane and Sarson (see Gane and Sarson [1979] and Stevens [1991] in the Bibliography). Although we will defer discussion of the relationship between function-oriented design and MDP until Sec. 10.2, it is worth noting here that the data flows in these diagrams identify communications between components which could possibly be implemented using MDP.

Figure 10.1 shows the data-flow diagram derived from the essential activities and essential memory we discussed in the previous section. Activities appear in the larger boxes, divided horizontally, while memory appears in the smaller boxes divided vertically. Arrows indicate the flow of data between essential memory and the essential activities. Arrows also represent information arriving at and departing from the system.

Three events, namely, a cancellation request, the arrival of a coin, and a ticket request, cause some processing to occur. As a consequence, this form of data-flow diagram is designated an *event-partitioned data-flow diagram*. In this particular case, the events are quite separate. Each involves only a single activity, though the essential memory is shared.

Coin arrival triggers the *Evaluate Coin Value* activity. In practical machines, this is probably some combination of hardware and software used to distinguish between the various valid coins accepted by the machine. Notice that, at this stage in the design, we need not worry

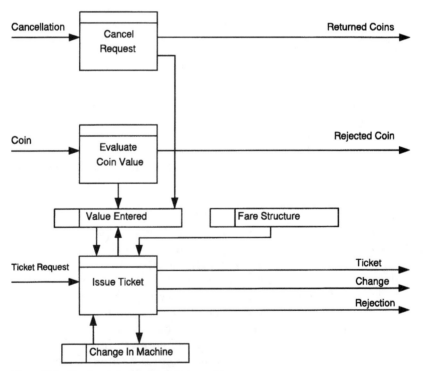

Figure 10.1   Top-level design for ticket machine.

about how this is carried out. We simply require that activity take a coin as input and produce either a rejected coin or an update to the current total value of coins entered. This running total is one of the pieces of essential memory we discovered previously.

A ticket request triggers the *Issue Ticket* activity. The destination is information associated with the request. The activity needs to determine the cost of traveling to the destination requested, compare this with the current total of coins entered, and issue a ticket and change, or reject the request, depending on the outcome. It also needs to check the availability of any change it may need to issue. In the specification, one of the criteria for rejecting a ticket request was the lack of suitable change. This particular activity needs to examine all the essential memory we identified for the machine. It also will need to update the value of coins entered and possibly the information about the change in the machine if a ticket is issued.

A cancellation request triggers the *Cancel Request* activity. This activity returns to the customer any coins entered and resets to zero the value of coins entered, thus to be ready for another customer.

The data-flow diagram is a fairly complete representation of a design which meets the specification. We have not attempted to define how the essential activities themselves operate. We will look at that level of design in the next section. Before doing that, however, we must point out some potential shortcomings in the design. These have been left in deliberately. They serve to illustrate that no design method, however sophisticated, removes the need for designers to make decisions and judgments.

First, look again at the design of the *Cancel Request* activity. We have a piece of essential memory containing the value of coins entered, but we have said nothing about the fate of the coins themselves. Returned coins must come from somewhere. We might choose to retain the coins entered by the customer, returning them if the request is canceled. Alternatively, we could choose to return coins using the same techniques by which we give change. For the purposes of this chapter, the issue is not which to choose. Instead, the issue is whether or not the mechanics of coin return constitute an essential activity for this design. We might argue that such considerations are at a more detailed level than we are addressing. However, if this is the case, we had better be sure that nothing in the design, as it currently exists, depends on the mechanism by which coins are returned. In addition, the method of coin return is visible to the customer. One method returns exactly the coins entered, whereas the other probably does not. The design method we are using gives us no real help in decisions of this kind. It all comes down to our interpretation of the term "essence." As designers, we are responsible for such decisions, which is as it should be. The best help that design methods can give is to highlight where such decisions need to be made and to give us a framework in which to think about them.

Another issue of essence is highlighted by the task of replenishing the machine with change, tickets, printer consumables, and so on, and retrieving the money collected. At first sight, it is reasonable to treat these as ancillary activities, especially considering the form of the specification we were given. However, though ancillary as far as the ticket-buying customer is concerned, these tasks are certainly essential from the standpoint of the organization installing the system. To them, the essence of a ticket machine is the generation of revenue. For them, adequate protection of the collected coins against theft will be essential, for example. In this case, the considerations were not in the specification. Once again, no design method, however sophisticated, can protect us from incomplete specification. Some deficiencies will become apparent during the design phase, but others might not. Good design still relies on good requirements gathering and clear specification. Space does not permit further investigation into requirements gathering and specification. However, for readers requiring more information, most standard texts on software engineering cover these topics in some detail (see, for example, Sommerville [1989] in the Bibliography).

10.1.3.3 Structure diagrams. The final stage in the design is to expand each of the essential activities into a form which can be implemented in some programming language. One graphical representation which can help in this task is known as a *structure diagram*. Basically, a structure diagram is a hierarchical view of the modules involved in the design and includes a representation of the interactions between them. Figure 10.2 shows a structure diagram for the *Issue Ticket* activity from the data-flow diagram in Fig. 10.1. The *Issue Ticket* box represents the entire function. The boxes below this one represent lower-level functions which together constitute the *Issue Ticket* function. Often, the hierarchical breakdown of a function into its constituents in a structure diagram will define the functions in the individual subroutines of the programming language being used. In addition to the hierarchy itself, some additional notation can be used on these diagrams. Figure 10.2 shows one additional kind of symbol. The diamond shapes on the *Issue Ticket* and *Check Request* boxes are conditional execution markers. In the case of the *Issue Ticket* box, the symbol indicates that either *Process Request* or *Reject Request* will be invoked, but not both. In the case of *Check Request,* it means that *Check Change* might be called. These symbols on their own are interesting but not terribly informative. We really need to know under what conditions the options in the condition will be exercised. It is usually difficult to get that detail directly onto the diagram. The tendency is to label each such symbol and then to provide a table which includes the relevant information along with the diagram.

Other information is commonly included in the annotation of a structure diagram. Most important is the data flowing between the various boxes. If the boxes eventually map to subroutines, the data flows will be parameters and return values. Again, due to space constraints, the data is usually tabulated and the relevant connections are labeled. Rather than resort to labels and tables, we will describe the structure diagram here in the text so that we can supply explanations. Remember, though, that this is not the normal way to annotate such diagrams.

Issue Ticket uses three functions. First, it performs the checks necessary to determine whether the request can be processed successfully. To do this, it invokes the *Check Request* function, passing in the requested destination and receiving back an indication of the validity of the request, the fare, and the change to be given. If the request is valid, *Issue Ticket* invokes *Process Request* to perform the request. It passes in the destination, the fare, and the change to be given. If the request is not valid, *Issue Ticket* instead invokes *Reject Request.*

Check Request uses two functions. First, it checks to see if enough money has been placed in the machine to satisfy the request. To do this, it invokes *Check Coin Value,* passing in the destination. It receives back an indication of whether the request can be satisfied, the fare itself, and the value of the coins currently in the machine. We do not have a definition of how *Check Coin Value* works. However, it is clear that it will need to access the *Fare Structure* and *Value Entered* essential memory to com-

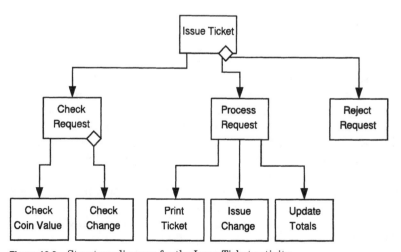

Figure 10.2   Structure diagram for the Issue Ticket activity.

plete its processing. *Check Request* subsequently invokes *Check Change,* unless *Check Coin Value* has already rejected the request. *Check Change* ensures that change can be given. *Check Request* passes it the fare and the value of coins in the machine and receives back an indication of the validity of the request. Again, we do not have a definition of how *Check Change* works. However, it is clear that *Check Change* will need to access the essential memory which holds the information about the change available in the machine.

If the request is valid, *Issue Ticket* invokes *Process Request,* passing it the destination, the fare, and the change to be issued. *Process Request,* in turn, invokes *Print Ticket,* passing it the destination and fare. Then it invokes *Issue change,* passing it the change to be issued. Finally, it invokes *Update Totals,* passing it the fare and the change to be given. This function updates the essential memory to reflect the fact that the ticket has been issued along with some change. Once again, we have not provided details of the design of these functions, but it is clear that *Update Totals* needs access to the *Value Entered* and *Change in Machine* essential memory.

The hierarchical nature of structure diagrams means that it is relatively easy to design functions top down, with lower-level boxes representing additional levels of detail and refinement. As the level of detail becomes greater, it is possible to write pseudo code or even program code directly from them.

We have now seen the major elements of function-oriented design methods. It is time to turn our attention to object-oriented design. Later, we will return to the subject of function-oriented design to see how message-driven implementations are derived.

10.1.4 Object-oriented design

Object-oriented design takes as its starting point the entities in the system being designed, rather than the functions. These entities are mapped to *objects,* a term which, in this form of design, has a very specific meaning. An object comprises data and a set of functions for manipulating that data. The data in an object is never accessed other than by functions of the object itself. This leads to a high degree of modularity. The data in an object is encapsulated. The functions of an object are known as *methods.* Objects interact with one another by sending messages which cause methods to be invoked. This concept fits very nicely with the general idea of objects being distributed over a set of machines, executing independently and sending messages to one another. However, some object-oriented languages, notably C++, collapse the notion of passing a message to that of a simple function call, immediately restricting invocation to be within a single operating system process. Although this is an extreme case, we should not, in general, expect object-oriented languages to provide us the mechanisms required to distribute applications across multiple machines or even multiple processes. This is not to say that such languages are not valuable in their own right, merely that they, like almost all computer languages, are designed for use in a local environment.

Recent developments in object-oriented technology and, in particular, the Common Object Request Broker Architecture (CORBA), do provide mechanisms for remote method invocation. Unfortunately, current implementations provide techniques based only on remote procedure call. No solutions based on message passing have yet been devised. However, as we shall see, this does not prevent us from using object-oriented techniques within single programs and using message passing to communicate between them.

To illustrate object-oriented design, we will work through a simple example design for a word processor. As before, we need to start with a specification, as follows:

- The word processor manipulates documents. A document has a title. Documents are composed of sections, and each section contains one or more paragraphs. Documents have an associated style, which describes elements such as the typefaces in use and the page layout to be used when printing the document. Different styles can be used with the document to alter its appearance without altering its content. This version of the word processor is not WYSIWIG; however, it must be possible to preview the final appearance of the document on screen to check it before printing.

From this specification, we will analyze the required system and then construct a top-level design based on object-oriented design. Finally, we will see how structure diagrams can be derived for some of the object methods involved.

10.1.4.1 Selecting the objects. As with function-oriented design, we depend heavily on the specification when deriving a design. All the comments about specifications, which we made when looking at function-oriented design, apply equally to a design based on objects. Whereas in function-oriented design, we started by trying to define the essence of the system, with object-oriented design we try and identify the objects. A good starting point is to look at the external specification and to note the nouns which occur there. For our word processor, we end up with the following list:

Document

Title

Section

Paragraph

Style

Typeface

Page

The specification also contains information about functions, for example, *Preview*. We cannot capture those just yet, but we will return to them shortly.

We have a list of nouns that appear in the specification, but are these all objects? Look for a moment at *Title*. The title of the document could be represented either as an object in its own right or as data associated with the document. We would call it an attribute of the *Document* object. Which is the correct way to represent *Title*? Once again, we have reached the designer's dilemma. In function-oriented design, we had to decide what aspects of the system were essential and what were peripheral. In object-oriented design, we have to decide which entities constitute real objects and which are merely attributes of other objects. Fortunately, guidance is at hand. A noun in the list probably represents an object if there are specific operations which apply to it. These operations must be more than simply inquiring or setting the value associated with the noun. In the case of *title*, it is pretty clear that there are no such operations and that we can represent it as an attribute of the document. In a similar manner, *Typeface*, although it probably represents a set of font, style, and size values, is probably an attribute of *Style*. Removing *Typeface* and *Title* gives us a reasonable starting point. We can always create a new object if necessary. Erring on the side of fewer objects is generally recommended when the situation is not clear-cut.

Let us now turn our attention to *Style*. We might reasonably consider that all stylistic elements are attributes of the document itself. Two things mitigate against this. First, style may differ throughout a document on a section or paragraph basis. So style is not simply an attribute of the document. We will see how to handle this shortly. More compellingly, we are told in the specification that a document can be printed using different styles to alter its appearance without altering its content. This strongly suggests that styles are entities in their own right and need a separate existence. Looking at our previous statement, that objects have more function than simply inquiring or setting the values of their attributes, we might ask whether this is true of a style? It is highly likely that we would want to load a style into a document or save the current document style. This means that not only does a style have additional functions which imply that it may be an object, but that it also has a separate external representation. This is a particularly powerful indication that *Style* is indeed an object.

One problematic noun remains. Should *Page* be an object? It appeared on our list of nouns because of the phrase "page layout" in the specification. The layout of a page could include the physical dimensions of the various areas, such as margins, gutters, line length, and so on, as well as attributes like the number of columns being used. These values are all attributes of the document style and do not require a separate object. However, pages have other attributes which change during printing. Pages may be numbered. The physical layout of odd- and even-numbered pages may need to be different to reflect the need for space used in binding them together to produce the final document. It seems as though we might need some object to represent this complexity, and we might be tempted to define something called *Current Page* for use during printing, in the expectation that its true function will emerge. However, this is probably the wrong thing to do at this stage. We should take a parsimonious approach to the definition of objects, defining new ones only when the need is absolutely clear. Currently, the need for some kind of *Page* object is half suspected. There is no reason to create it yet. Creating new objects later in the design is relatively easy. Creating too many too soon is a common mistake. It often leads to poorly specified objects and to a great deal of confusion.

Our initial list of nouns has now shrunk to become

Document

Section

Paragraph

Style

Before expanding on the definition of these, it is time to distinguish the terms *object* and *class*. So far, we have used the term *object* without thinking about the difference between a kind of object, say a *Paragraph*, and an instance of an object, for example, the paragraph you are currently reading. A *class* is an abstract description of an object. It defines the data which every object in the class has, as well as defining the methods which can be used with the objects. An object is a particular example of the class to which it belongs. The distinction is just the same as between a type and a particular variable in normal programming languages. Any variable of type *integer* will have particular properties, such as the maximum and minimum values which can be represented. These properties are shared by all variables of the type, though they can each represent different actual values. The variables are instances of the type.

What we are really seeking from our list of nouns is a set of classes which we can use to describe our design. Each noun in the list becomes a class. Each class has methods associated with it. Our first diagram for the design, in Fig. 10.3, shows the classes we have discovered and the methods which each provides. Each of the nouns from our list appears. We have added two further classes, which will be discussed shortly. Each box in the diagram represents a class. Within the box is a list of some of the methods which the class will probably supply. For example, the *Document* class is shown as providing the following methods:

New   Create a new, empty document

Open   Load a document from a saved, external representation, typically a disk file

Save   Store a document as an external representation, typically a disk file

Print   Print the document

Preview   Display a representation of how the document will look if printed

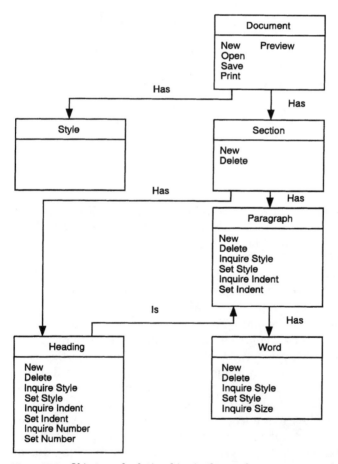

Figure 10.3 Objects and relationships in the word processor example.

It should come as no surprise that these methods are similar to the functions available on the *File* or *Project* pull-down menus on word processors with graphical user interfaces. After all, our design should be providing function which represents the actions an end user would wish to take.

The list of methods for each class is far from complete. You might like to consider which additional functions we might need to print or preview the document. They exist in the *Document* class, but where else might they be needed? We will look at these operations shortly. Before that, however, we need to look at the relationships between the classes in the diagram in Fig. 10.3.

10.1.4.2 Inheritance. Object-oriented approaches to system implementation exploit a concept known as *inheritance*. The idea is that one or more classes may share common functions and attributes. If this is the case, then the shared functions need to be written only once. One of the classes, sharing the function, can be defined as being based on the other. One class is said to *inherit* data and methods from the other.

Figure 10.3 contains an example of inheritance. Class *Heading* is shown as related to class *Paragraph* by an *Is* relation. This is the relation which indicates inheritance. The *Heading* class is considered to be a refinement of the *Paragraph* class. It has all the attributes and methods of *Paragraph* plus some of its own. Any of the data and methods which it shares with *Paragraph* can be used by *Heading*. The converse is not true. No *Paragraph* object is able to use any of the methods or data specific to the *Heading* class.

Inheritance is the cornerstone of reuse in object-oriented design and implementation. The class being inherited is called the *Parent* class or *Superclass*. The class doing the inheriting is known as the *Child* class or *Subclass*. Inheritance allows a single implementation of common methods across a series of objects. The advantages are similar to those obtained by a well-structured set of subroutine calls in a function-oriented design. The advantage claimed for inheritance over subroutines is that, because a class defines all aspects, not just function, the realization of reuse is, in practice, easier. There seems to be some justification for this view. However, the design issues for inheritance are just the same as those for common subroutines, although they are expressed differently.

In our example, we have chosen to make *Heading* a subclass of *Paragraph*. For this to be a sensible choice of relationship, the classes involved must really share common semantics. The view in this design is that a heading is a specialized form of paragraph. It has the same properties and behavior as a standard paragraph, with some additions. These additions include a number indicating the heading's position in the hierarchy of the document. Whether this is a good choice or not depends on the design. It is the designer's task to look at the class hierarchy and to determine which classes should inherit from one another.

10.1.4.3 Using other classes.

There is a second kind of relationship in Fig. 10.3. This one is marked *Has*. For example, the *Document* class *Has* one or more *Sections*. The *Has* relation is a much looser one than inheritance. It implies simply that one class knows about the existence of one or more objects of another class and can use them. In our example, it could be that one of the data structures in a *Document* is a linked list of references to *Section* objects. A similar implementation could be used for the other *Has* relations in the diagram, but it is not the only possibility. The only requirement is that the class higher in the hierarchy be able to identify all such relationships in which it is a participant. In the spirit of modularity, all the information required to determine these relationships should be part of the class itself.

Because there are often one-to-many relationships between classes in object-oriented systems, there may be many places in a hierarchy where similar function is needed on a collection of objects. For example, functions to move back and forth among a list of paragraphs could be the same as those used to move along a list of words within a paragraph. Obviously, the functions would need to manipulate the list in a manner independent from the objects stored on it. This idea leads to a kind of class which has been termed a *container*. Its sole task is to hold references to a collection of other objects, typically of a single class. We might imagine having a *List* class or a *Binary tree* class, and so on. A *List* class would probably have the following methods:

next_element

previous_element

add_element

delete_element

In just the same way that generic services are supplied by subroutine libraries in function-oriented programming, *class libraries* provide generic services in object-oriented programming. Class libraries have one major advantage over subroutine libraries. In addition to using the function in the class library, the classes it implements can be used as the basis for inheriting further, more specialized classes.

10.1.4.4 Event-driven processing.

So far, all we have seen is a hierarchy of related classes and some of the services they implement. This is the static part of the design. How does this structure actually perform any work? In some ways, this example is very similar to the ticket machine we examined earlier. It responds to external events. Some user interface function, which we are not considering in this part of the design, evaluates user requests and invokes the appropriate methods in the class hierarchy to get them processed.

Consider what happens when a user requests that the document be printed. The *Document* class has a method which can be invoked to provide the required service. The question is, what is the implementation of this method? Since classes are self-contained, one obvious approach is to require that all the classes below *Document* should also provide *Print* methods. Naturally, this does not apply to classes like *Style* which do not contain printable material. Or does it? The ability to print information about the document style might be a useful additional function for the end user. Another decision from the designer is called for. For our purposes, we will ignore this possibility, concentrating instead on the task of printing the document itself.

We can add methods called *Print* to the classes *Section, Paragraph, Heading,* and *Word.* If we are careful, we should be able to arrange for *Heading* to inherit its *Print* method from *Paragraph.* Most of these methods will do little more than call the *Print* methods of the classes below them in the hierarchy. Figure 10.4 shows a structure diagram for the *Print* method of the *Section* class. We should not be surprised that we can design methods using representations from function-oriented design. After all, a method is just a function which happens to be tightly bound to a particular class. The notation does need to be extended a little, however. Rather than function names, the boxes contain class and method names. Whereas function names are always globally unique, method names do not have to be. Indeed we have already defined a number of *Print* methods. Method names do have to be unique within a class, however. The notation in Fig. 10.4. is as follows:

ClassName[MethodName]

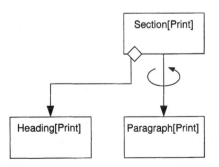

Figure 10.4   Structure diagram for *Print* section.

The *Section[Print]* method first invokes the *Heading[Print]* method, if the section has a heading. Then, for each paragraph object, it invokes the *Paragraph[Print]* method. The arrow encircling the line joining *Section[Print]* and *Paragraph[Print]* is standard structure diagram notation for a repeated operation. Just as with conditional operations, a fully annotated structure diagram would include details of the conditions of the repetition.

This delegation of the printing function can continue down the hierarchy until we reach the class which is actually rendered onto the paper. In the example, this is the *Word* class. Objects of class *Word* have all the information required to render themselves. The attributes of *Word* include information such as type style, font, and character size. A word does not, however, know where to render itself on a page. Word positioning is part of page layout. Which class is responsible for this? In the design as it stands, *Paragraph* is the place for word-positioning services. Page layout is an attribute, or rather a set of attributes, of class *Style*. *Paragraph* needs access to this information in some way. Assuming it can find information about page sizes, margins, line lengths, and so on, it is possible for *paragraph* to place words on the page, so long as it can determine the size of the word to be placed. This is the reason for the *Inquire Size* method in class *Word*.

You may decide, quite reasonably, that having *Paragraph* sensitive to attributes which really belong to a page is unreasonable. In this case, another approach might be to define a *Page* class. At first, this may seem to be an excellent approach. After all, documents consist of pages, don't they? The problem is that, in terms of the underlying structure of a document, a page is simply an artificial division of the material. Its sole purpose is to support one particular representation of the material. The difficulty can be seen quite easily by trying to fit *Page* into the hierarchy of Fig. 10.3. Does a *Section* consist of one or more *Pages* or does a *Page* consist of one or more *Sections?* The answer, of course, is both. *Page* is really independent of the logical structure of the document. While the attributes of the page used for printing are of vital importance to the word processor, they are not properties of the document itself.

There are a number of ways to solve this kind of issue. For example, the layout information could be in the *Style* class. The *Paragraph* class could inherit the information from *Style*. Alternatively, *Paragraph* could have a *Has* relationship with *Style*. This seems more natural than inheritance and is adequate since *Paragraph* will only be inquiring about the layout held in *Style*. Once again, it is only the skill and experience of the designer which can resolve this sort of issue.

This is as far as we will take the discussion of object-oriented design itself. Obviously there are many details we have not covered and there are many aspects of the example design which we have left in an unsatisfactory state.

Section 10.3 will look at using object-oriented design with MDP, but before that, we need to cover the use of function-oriented design with messaging and queuing.

10.2   Using Function-Oriented Design with Messaging and Queuing

Now that we have had a chance to look at function-oriented design and object-oriented design, it is time to try using them together with message-driven processing. We will develop a partial design for a distributed, online, company telephone directory system. In this section, we will use function-oriented design. Later, we will produce a design for the same system based on object-oriented design. As ever, we need to start with a specification, as follows:

- The system allows users at any site to inquire and modify entries in the telephone directory. The directory is divided into partitions distributed across multiple sites. Larger sites own and manage their own partitions. Small sites keep their entries in the partition owned by the nearest large site. Inquiries can be made specifying the particular partition or any group of partitions, including the entire directory.

From the specification, we can see that there are two essential activities: inquire and modify. We will assume that the term "modify" is meant in its widest context, including insertion and deletion of entries. The essential memory, in this instance, is the directory itself. It is partitioned, which means that a site may have part of the overall directory as a local resource. Readers with systems analysis experience may reasonably object to the fact that the decision about the physical partitioning of the directory has already been taken. A rigorous analysis of the whole system would start without any such preconceptions. The correct location for the data would emerge as the study continued. For the purposes of this exercise, we will assume that there are pressing reasons why the customer requires a partitioned directory. It might, for example, have to do with the cost of administration, or the network charges incurred in performing a lookup when the directory is remote.

10.2.1   Developing the data-flow diagrams

We can develop a top-level, event-partitioned, data-flow diagram from the specification as we did for the ticket machine previously. The diagram is shown in Fig. 10.5. It is split into two sections. In each section, both essential activities, lookup and update, appear. The upper section represents the system's response to a local request from an end user. The lower section shows the system's response to a request from another system. It is clear that the processing is almost identical in both cases. The difference, when handling remote requests, is the need to route replies back to the requesting location. In the local case, we do not need any information about where we should send the results and the status information. In the remote case, we clearly do. However, we can have common lookup and update functions if we allow location information to flow through them. Neither requires the information about the location making the request to perform their own processing.

For both kinds of event, the first function involved converts the request into a form which the rest of the system can process. It ana-

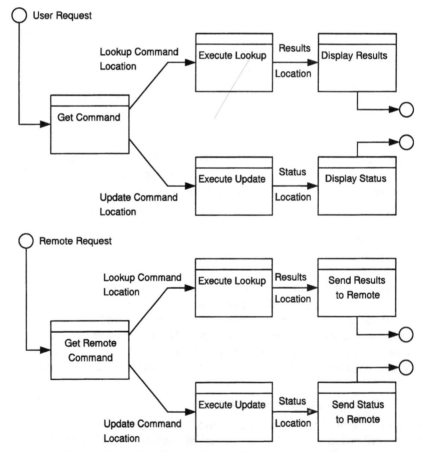

Figure 10.5 Top-level data-flow diagram for the phone directory application.

lyzes the request and routes it to either the function which performs lookup or the one which performs updates. These functions produce output: in the local case, information is displayed for the end user, and in the remote case, information is returned to the source of the request.

No essential memory is shown at this level of detail in the diagram. To see the structure of the directory, we need to go to the next level of detail. Figure 10.6 shows a more detailed view of the lookup function. The first function in this flow formulates the set of queries required to fulfil the requested operation. For example, the lookup request might contain wild-card characters, or might involve searching several different directories. An end user request might be of the form:

List all the entries for people called *sm** in Toronto, Austin, and Boulder.

Here, the names being searched for are specified as a prefix and a wild card. The term *sm** matches any surname beginning with *sm*. The job of the *Formulate Queries* box is to break down the request into one or more queries which can be handled directly within the directory. Any reasonably competent database system will be able to handle the wild-card characters in queries. However, not all enquiries will be able to run locally. *Formulate Queries* must split those which can run locally from those which must be sent to a remote site. To do this, it needs to know which site it is running at. This information is part of the directory itself. The local directory is shown as essential memory in Fig. 10.6.

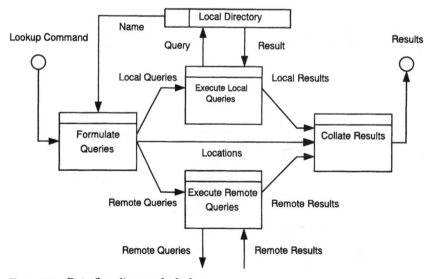

Figure 10.6   Data-flow diagram for lookup.

Local queries run against the local partition of the directory. Queries are issued and the results flow back. Remote queries are passed to the function *Execute Remote Queries,* which sends them to the appropriate remote system. Results again flow back. Local and remote results are passed to the function *Collate Results.* When this function has all the results, it outputs them as a single entity. This function might also have the responsibility of sorting the results into some specific sequence.

The design for the *Collate Results* function is shown in Fig. 10.7. This function has some interesting properties. First, it accepts multiple input events. It receives results from the local query and from all remote queries. In addition, it receives location information from *Formulate Queries.* This includes the location from which the request originated as well as all locations from which *Collate Results* should expect replies. The inputs to *Collate Results* could arrive in any order; though in practice the location information from the *Formulate Queries* would probably arrive first, it cannot be guaranteed. Also, there may be a considerable difference in the time taken to receive the results from the local and remote lookups. Since the local lookup does not involve the kind of network delay associated with the remote lookup, it is likely to complete first. *Collate Results* clearly needs to store partial results and location information while waiting for queries to complete. The essential memory for that purpose is shown in Fig. 10.7. The arrival of local results, remote results, or location information causes the appropriate data to be stored. In addition, a request to output the results is sent to the *Output Collated Results* function. This request is discarded unless all results and location information are available.

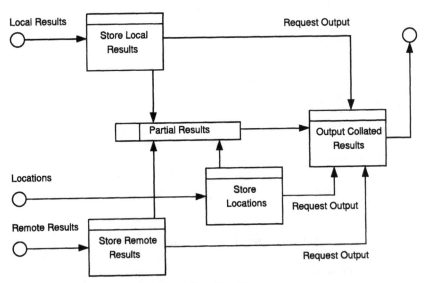

Figure 10.7   Data-flow diagram for *Collate Results.*

A couple of points need to be made about the design for *Collate Results*. First, Fig. 10.7, although expressing the operation of *Collate Results* in a convenient way, is not a true, event-partitioned, data-flow diagram. This is because it shows the behavior of the *Collate Results* function to three separate events. Second, the way essential memory is used is worthy of note. An alternative approach would be to have *Formulate Queries* write the location information into the partial results essential memory itself, rather than passing this information to *Collate Results*. Indeed, in conventional systems, this is what would happen. However, as we will see shortly, implementations based on messaging and queuing naturally handle such flows. In this particular case, the design achieves encapsulation of the partial results essential memory. Only functions inside *Collate Results* need to access it. Since the essential memory is encapsulated, it does not need to appear on the higher-level data-flow diagrams.

We now have a reasonably complete description of a design which meets the specification for the distributed telephone directory system. One interesting feature of the diagrams is that the design can handle multiple, concurrent queries. All that is necessary is that the local results, location information, and remote results flowing into *Collate Results* carry an identifier unique to the query which initiated processing. The partial results are accumulated for each query and output when complete.

The next step in the design is to partition the processes, creating units which will eventually become programs.

10.2.2 Partitioning the processes

Systems built using message queuing consist of a set of modules which execute independently and which communicate by exchanging messages. This structure maps extremely well onto traditional data-flow diagrams. Functions or groups of functions from these diagrams will constitute the system's modules. The task we face now is deciding which sets of functions in the data-flow diagram belong together and where the boundaries between modules should be placed. The larger the number of independent modules, the more flexible will be the design. However, each boundary carries with it some communications overhead, especially if the modules might execute on different machines separated by a network. Consequently, our enthusiasm for dividing the system up must be tempered by an understanding of the implications.

10.2.2.1 Modules. A good place to start investigating what constitutes a module, on a data-flow diagram, is the essential memory. Functions that share essential memory need to be more closely associated than those that do not. In particular, in a distributed application, they prob-

ably need to execute on the same machine. Essential memory is implemented as files or database entries. Functions exchange data through essential memory by reading and writing files or by manipulating the databases. Because essential memory is implemented as local disk files or databases, access is fast. Essential memory can be used freely in a design without consideration of performance implications. However, unlike disk systems, data traveling over communications connections has a relatively high latency and low bandwidth. The data-flow diagram does not consider this sort of issue, and rightly so. However, there is an implicit assumption in such diagrams that all access to essential memory is equally easy. If the system really is to be distributed, this is clearly not the case.

While distributed file systems and distributed database products are commonplace, we should not assume that they automatically provide a generally acceptable means for distributing essential memory. If we do, we may discover later that we need to send large quantities of data across the network. Instead, by clustering functions which share essential memory, at this stage, we are much more likely to end up with a system where this is not necessary. Of course, the clustering is made easier if the essential memory has been well encapsulated.

10.2.2.2 Boundaries. We have already seen that data sharing within modules is fine, whereas data sharing between modules introduces potential difficulties. Consequently, we can look for module boundaries at places in the data-flow diagram where there is little or no data sharing. Functions which do not share essential memory and which pass little data are good candidates for being the boundary between modules. We do have to be careful, though, about what we mean by passing only a small amount of data. Communications latency means that not only do we want to minimize the amount of data being passed, we also want to minimize the number of times the modules perform communication. For example, if two modules exchange 10 bytes of data but have to do it 100 times in response to a single event for the overall system, they probably should not be in different modules. Transferring 1000 bytes this way puts much more load on the communications system than a single 1000-byte transfer. In the latter case, the setup overhead is incurred only once. The actual numbers involved vary with the kind of network linking the machines. A token-ring or Ethernet-based LAN would have little trouble with our example of transferring 1000 bytes. Slow, dial-up lines, however, might. The point is that, although the communications medium may be able to handle the traffic, as designers, we should be careful with the available bandwidth. The more effectively we use network resources, the more users we will be able to support concurrently and the less we will impact other applications on the same medium.

10.2.2.3 Partitioning the example. Looking at the top-level data-flow diagram in Fig. 10.5, we can begin to see how to partition the design. The top-level diagram shows no essential memory. This means that the functions in it do not share any essential memory. Consequently, each could probably become a separate module without incurring too much performance penalty. We need to look at data traversing the links between the functions a little more closely before being certain. Data flowing between *Get Command* and the *Execute Lookup* and *Execute Update* functions is restricted to the command itself and information about the location to which results are to be returned. The lookup command is in a form close to that entered by the user. Consequently, it is compact, and these functions could occupy separate modules without impact. The same is true for remote requests. Similarly, between *Execute Update* and *Display Status,* little data flows. The situation is less clear with the boundary between *Execute Lookup* and *Display Results.* Most queries are likely to involve only a few entries from the directory flowing over the link. However, occasionally, extensive searches involving wild cards will be performed. Significant quantities of data may need to be sent in response. There is an additional consideration. Separating the module which performs the lookup, and which must have intimate knowledge of the database holding the directory, from the module controlling the user interface function, and which must have intimate knowledge of the user environment, can lead to a very flexible implementation. New user support for different terminals or client machines can be added without affecting the database handling code. Such an approach, particularly if the connection between the modules can be remote, constitutes a classic client-server arrangement appropriate for low-cost desktop machines accessing a service machine. Once again, the decision whether or not to split these modules is the designer's.

On balance then, splitting all the functions in Fig. 10.5 looks feasible. This kind of arrangement leads to maximum flexibility and the ability to move modules onto different machines at will. Looking at Fig. 10.6 and Fig. 10.7 shows that the lower-level functions in the design are naturally more tightly coupled. Essential memory is being shared between them and there are additional data flows, particularly in the case of the enquiry function. Also, with these lower-level functions, it is much less likely that there would be any reason to have additional implementations to meet user requirements. On balance, it is probably not worth splitting the functions other than at the topmost level.

Finally, it is worth noting how naturally the design becomes a message-queuing implementation. The concept of independently executing modules exchanging data via messages fits well with the organization of data-flow diagrams. Indeed, the ability of messages to carry information about where results should be routed, in addition to the data itself, makes possible the use of multiple paths through the same modules with little programming effort.

As we saw with the ticket machine example, the next step in design is to build structure diagrams for the individual functions. This work is routine and does not affect the use of message queuing in the overall implementation. All the decisions about partitioning the solution and the use of message queuing have been taken. There is more detailed work to do, and we will look at some of the tasks involved in designing messages, developing applications, and incorporating existing applications in subsequent sections. First, though, it is time to turn our attention to an object-oriented design solution to the distributed telephone directory problem.

10.3   Using Object-Oriented Design with Messaging and Queuing

In this section, we will look at an alternative design for the phone directory application. Based on the specification from Sec. 10.2, we will use an object-oriented approach to derive a design which could be implemented using messaging and queuing. As with the word processor design we looked at earlier, the first step is to define the objects from which the system will be composed. When we have done this, we will look at how the system executes and examine its responses to various events. Finally, we will show how to partition the system and see where messaging and queuing plays a role.

10.3.1   Defining the classes and objects

As we did with the word processor example, we can start searching for objects by looking for nouns in the system's specification. Based on the specification, we can derive the following list of apparently important classes.

Site

Directory

Partition

Entry

Command

Result

Port

Eagle-eyed readers will have spotted that the last three entries on the list do not correspond directly with nouns in the specification. The need for *Command* and *Result* is fairly evident from the specification, even though it does not explicitly mention them. We will justify defining them as classes, rather than as attributes, shortly. The *Port* class has been introduced to provide an abstraction for the communications which we already know will be involved in the application.

In the following sections, we will look at each class and see what it contains and how it contributes to the overall system.

10.3.1.1 Directory and partitions. Fundamental to the entire system is the directory itself. We know that it must be partitioned. We also know that one partition may be local to the application while others are remote. We can represent the directory as a hierarchy of classes. This is shown in Fig. 10.8. The *Directory* class itself provides a local abstraction of the entire directory, encapsulating both local and remote data. It provides methods for enquiry and for maintaining data within the directory. Applications using the *Directory* class are unaware of the differences between access to the local and remote data. The job of converting a request into one or more operations on different partitions is carried out within the methods of the *Directory* class. This is essentially the same function carried out by *Formulate Queries,* in the function-oriented version of the design (see Fig. 10.6).

The *Directory* class contains one or more partitions. The *Partition* class provides the same methods as does the *Directory* class, but these operate only on a particular partition. The *Partition* class is subclassed by the *Local Partition* and *Remote Partition* classes. The *Partition* class is never instantiated, all *Partition* objects being either *Local Partitions* or *Remote Partitions.* Because the interface to these two classes is identical, the *Directory* class does not need to be aware of whether a partition is local or remote. The methods within the partitions themselves deal with the appropriate way to process the requests. Because the *Directory* class is not aware of the distinction between local and remote partitions, the design supports multiple local partitions, even though the specification does not actually require it.

The *Local Partition* class encapsulates the database containing the directory itself. Interaction with the database may be complex. However, data is exchanged with the rest of the system using the *Directory Entry* class. This class encapsulates an individual entry within a partition. The need for this class is debatable. In practice, this class is little more than a data structure holding the system's view of the data in the database. However, we will persevere with it on the grounds that it is used in more than one place, and it does encapsulate the data. In general, classes which have only enquire and update methods are candidates for being simplified to attributes of other classes. However, we will see shortly that *Directory Entry* will need to provide additional methods.

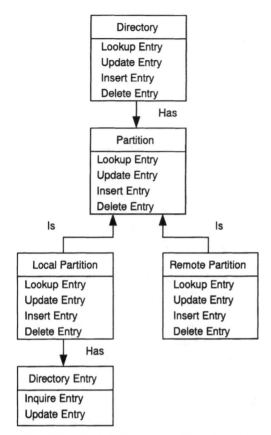

Figure 10.8 *Directory, Partition,* and *Directory Entry*
classes.

We have not yet considered what attributes the classes have. There
may be many different ones, of course, and some of these will depend on
the implementation. For example, the *Local Partition* class will need
information about the database which holds the directory entries locally.
If this is a SQL database, the information will include items such as
database and table names. For other database managers, different
attributes might be needed. However, there is one attribute which every
partition needs: the name of the location which it represents.

10.3.1.2 Commands and results. It may not be immediately apparent
why we need a class to represent the commands issued against the
directory or the results of such commands. However, in his book on
object-oriented software development, Meyer illustrates graphically
why such concepts are useful (see Meyer [1988] in the Bibliography)
Consider commands. Superficially, it seems as though a class is quite

unnecessary. A command represents some function of the underlying system and should surely map to a method rather than a class. However, this is only true if execution of the command is the only operation allowed. Some systems, for example, require an audit trail of all activity which occurs against some or all of the data. In this case, as well as being executed, a record of each command needs to be added to the audit trail. The natural place for such a function is as a method of class *Command*. The example Meyer uses is of supplying an undo facility in a text editor. Again, this requires that a number of previous operations be stored, implying additional methods for the *Command* class. The class hierarchy in Fig. 10.9 includes execute and undo methods for each command. Specific types of command are subclasses of the *Command* class. To support the undo operations, we have included a *Command List* class to hold a record of a given number of previous commands.

In Fig. 10.9, we also show a class called *Result,* which contains instances of the *Directory Entry* class. As with *Directory Entry* itself, it is not immediately obvious why we need a class rather than simply data structures which can be manipulated by methods in other classes. The answer involves being able to convert the data into an external form which can be stored or transmitted. To see how this works, we need to consider the *Port* class.

10.3.1.3 Ports and communications. One issue, which we did not cover in the word processor example but is common to many applications whether object-oriented or not, concerns the transformation between the external and internal forms of the data these applications manipulate. For example, the external representation of data manipulated by a word processor is a disk file with an appropriate format. That format may be private to the particular word processor, though, thankfully for authors, many products can now interpret each other's files. The internal representation will be quite different from that in the file. It is likely to consist of various data structures and linked lists in the main memory of the machine. Conventional programs transform between the internal and external forms of the data they manipulate with functions for the purpose. Object-oriented designs are no different, except that, in this case, the natural place for such functions is as methods of the classes which need to be transformed. Indeed, this is just the approach which the methods dealing directly with the database would take.

This same approach is equally valid when the external representation is for the purpose of communication rather than storage. We already know that it will be necessary to transmit commands and results between systems, and probably within them as well. To provide

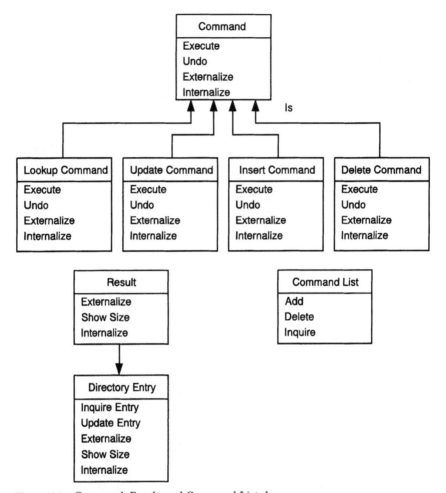

Figure 10.9 *Command, Result,* and *Command List* classes.

an abstraction of the communications, we have included a *Port* class shown in Fig. 10.10. The class offers methods for opening, closing, sending, and receiving data. We would like to be able to encapsulate the operations so that we could, for example, send a *Result* object via a *Port* object. To do this requires that *Result* and any objects it contains have methods which transform them to some external representation capable of being stored or transmitted. This is clearly a method which belongs to the class and justifies why *Result* and *Directory Entry* really are classes.

The mechanism for transmitting an object via a port involves rather low-level, implementation-specific operations. However, it is an inter-

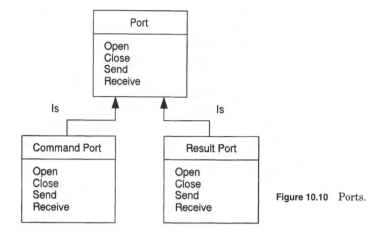

Figure 10.10   Ports.

esting example of how object-oriented approaches provide new ways of looking at old problems, so it is worth pursuing a little. The *Port* class has methods for sending and receiving. This class deals with external data representations only. It must not be able to interpret the contents of any object which it transmits or receives, for this would destroy the encapsulation provided by those other classes. However, it must be able to recognize that a particular piece of data is an object of a specific class. To this end, the data transmitted must contain this information. The *Port* class appends this class-type information on transmission and removes it on receipt. If this sounds familiar, it is because that is exactly how most low-level communications protocols work. They add data for their own use during transmission, removing it on receipt at the destination.

The *Port* class has to deal with sending buffers of data in the external representation. It must create these buffers and, consequently, needs to know how much space the external representation will require. The *Result* class has a *Show Size* method to return the space required for the entire result, including all the directory entries it contains. Naturally, it computes this value by invoking the *Show Size* methods for each of the *Directory Entry* objects which it contains, summing the results and adding the size of any data which it must itself write. The *Port* class creates a suitable buffer and invokes the *Externalize* method of the *Result* class. *Result* fills in the buffer with its own data and invokes the *Externalize* method of each *Directory Entry* object to do the same. Details such as management of the point in the buffer at which an object places its data must be handled within the *Externalize* methods. A simple approach is to let each object return the first free position in the buffer immediately following its data.

Once the *Externalize* method for the *Result* class has completed, the *Port* class has a self-defining buffer which can be transmitted to another application process or even a different machine. All the function required to transmit this buffer is contained within the methods of the *Port* class, completely insulating the rest of the application, including the objects whose data is being transmitted.

Of course, this kind of buffer, containing the external representation of the objects, can be used for more than just transmission to another process. The external representation can be used to provide external storage for the objects. An *External Storage* class could use much of the buffer-handling function of the *Port* class but provide methods for storage in a file, rather than transmission over a network. The common function could be encapsulated in an *External Representation* class, of which both *Port* and *External Storage* would be subclasses. This is yet another good example of how inheritance provides simple and powerful mechanisms for code reuse.

When receiving data, a *Port* must be able to determine the type of object represented by the data it has received, without needing to look inside the object data, which would destroy the encapsulation. The *Port* instantiates the object just received and passes the external representation to the object's *Internalize* method. This method provides the inverse of the transformation performed by *Externalize,* loading the internal representation of the object from its external representation.

This discussion of object transmission has been something of an aside. We need to return to the main function of the whole system and, in particular, now that we have the classes, to a discussion of how the system operates in total.

10.3.2   System execution

In this section, we will follow through the system's response to particular events. First, we will look at what happens as the application starts on one particular machine. Then we will go on to see how the various classes behave during a lookup operation.

10.3.2.1   Instantiating the directory.   When the system starts to run on a specific machine, there will be a number of initialization tasks required before it can respond to external events. Most important for the present discussion is the instantiation of the directory object. The internal representation of the directory and its associated objects carries a considerable amount of information. We have already seen that the names of the underlying database components and the names by which remote systems can be contacted form part of this information.

This kind of information is often referred to as the *configuration* of the system. It represents data which tailors this particular instance of the system and which must persist even while the system is not running. The natural way for an object-oriented application to store such information is as an external representation of the objects involved. We discussed this technique when we looked at external representations for communications purposes in Sec. 10.3.1.3.

As the system starts, it creates the directory object and invokes its *Load* method which results in the data being loaded and the appropriate objects being created. At the end of this operation, the entire object hierarchy, including all local and remote partitions, will have been created. Remember that the *Partition* objects do not contain the directory data itself. That is in the database. Instead, they contain all the information necessary for the data to be accessed.

Similar initialization may be necessary for the *Port* objects used by the system. The *Command List* object does not contain any information initially, but does need to be created. Individual *Command* and *Result* objects will be created and used as necessary during the processing of the system.

10.3.2.2 Executing commands. Pseudo code for the top-level function of the system is given in Fig. 10.11. Once the objects have been created, the system can start to respond to events. The *Receive* method of the *Command Port* object retrieves the next command. The *Execute* method of the command is called passing in the *Directory* object. This method uses the appropriate methods of the *Directory* object to execute the command. It returns a *Result* object which is subsequently passed to the *Send* method of the *Result Port*. The loop continues, processing requests until it is explicitly terminated. This presumably requires an additional command.

```
main(Directory Name)
Instantiate Directory(Directory Name)
Instantiate ResultPort
Instantiate CommandPort
CommandPort[open]
ResultPort[open]
Until Terminated
    CommandPort[receive](Command)
    Command[execute](Directory, Result)
    ResultPort[send](Result)
    UndoList[add](Command)
End Until
```

Figure 10.11   Top-level system execution.

This description says little about the source of requests or the target of replies. The design has concentrated on provision of a directory service. This is roughly equivalent to the combination of the *Execute Lookup* and *Execute Update* functions in the function-oriented design from Fig. 10.5. One simplifying assumption in this design is that there is only one kind of reply to a request. The *Result* object contains status information and may, in addition, contain one or more directory entries.

All of the complexity of communicating with other systems has been hidden from the majority of the code. The interaction with remote systems is within the methods of the *Remote Partition* class. If object-oriented languages explicitly supported remote operations, the hierarchical link between *Partition* and *Remote Partition* would represent the communications. Since current, popular, object-oriented languages do not directly support such concepts, the remote links need to be hidden within the methods of the local classes. The *Lookup Entry* method of the *Remote Partition* class must perform the appropriate network operations to invoke requests on the remote system in question. In the next section, we will see how this is achieved.

10.3.3 Partitioning the system

Our design has so far concentrated on the classes which constitute the directory itself. A particular class hierarchy must, in general, always belong in a single application module. Current object-oriented languages do not support any concepts akin to distributed inheritance. The hierarchy shown in Fig. 10.8 belongs in a single module. Indeed, we have already seen the code for the top-level function. Classes used with the directory also need to be instantiated in the same module. However, the main directory module, for example, is not the only one requiring the *Port* class. The design currently shows a module which accepts *Commands* and sends *Results*. One source of *Commands* is the module which implements the user interface. Another is any remote system needing access to the data in the local partitions of the directory. The user interface module will need the *Command* and *Command Port* classes as well. The idea of several different modules using the same classes is fundamentally no different from the idea of programs using libraries of subroutines. The user interface module will also need the *Result* and the corresponding *Result Port*.

The final part of the picture involves how *Remote Partitions* are able to process requests on remote directories. The *Remote Partition* class uses a *Command Port*, and probably a *Command*, to transmit requests to the particular remote system involved. We have already seen that the configuration information for each system includes names used to contact other parts of the directory, and it is here that they are used. Of course, the *Remote Partition* will also need to use *Result* and *Result Port* classes to deal with the replies from the other parts of the directory.

Figure 10.12 shows the major modules involved in the overall system. This is really not very different from the top-level data-flow diagram for the function-oriented design, shown in Fig. 10.5. The function-oriented design shows a little greater partitioning of function, though the differences are hardly fundamental.

The same issues regarding data transfers between modules affect the object-oriented design as affected the function-oriented design. There is no need to reiterate them here. We covered them fully in Sec. 10.2.2. The difference is in the notation used to describe the data flows. In an object-oriented approach, it is natural to think of sending and receiving objects. It also becomes apparent very quickly that the data flowing across the communications medium must be self-defining. In addition, the discussion about how *Ports* send and receive data showed us that the data describing each object in the data stream must be external to the object.

The module boundaries in the object-oriented design are, in fact, just about identical to those from the function-oriented design. Certainly, the data flowing across them is the same. The similarity stems from the fact that the design constraints on the transmission of data between processes is, of course, independent of the design approach used. It also probably reflects the fact that the same designer developed both designs.

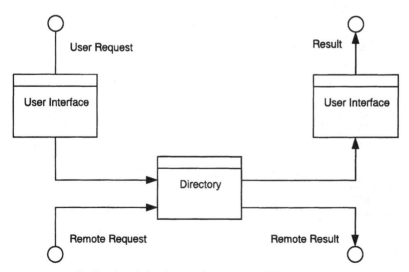

Figure 10.12 Top-level modules for the object-oriented design.

10.4   Defining the Queues and Designing the Messages

The final step in the design of a system based on message queuing is the definition of the queues which will be used and the specification of the message layouts. Once this information is available, together with the other design materials we have already developed, application programmers have all the data they need to implement the programs.

We will look briefly at the queues and messages for each of the designs we have developed.

10.4.1   Messages and queues for the function-oriented design

Figure 10.13 shows the processes and queues resulting from the function-oriented design of the distributed telephone directory system. As we saw earlier, the processes are actually the top-level functions from the data-flow diagram. The diagram structure differs from that in Fig. 10.5 because we have now allowed data to flow through a single instance of *Execute Lookup* and *Execute Update* regardless of its source. Messages must carry enough information to allow them to be routed through the application modules. In the data-flow diagrams we showed this as *Location* information.

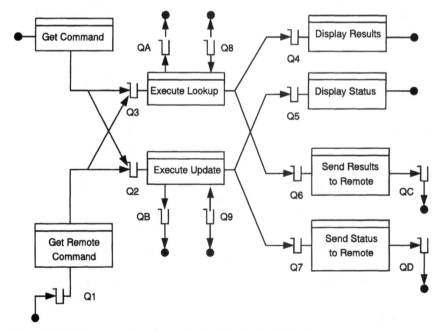

Figure 10.13   Queues and processes for the object-oriented design.

The function-oriented design is quite highly partitioned. Overall, nine queues are involved. Commands arrive locally via some user interface application, shown as *Get Command* in Fig. 10.13. Commands from other systems arrive on Q1. Incoming messages are routed, based on the type of command, to Q2 or Q3. Results of lookup operations are routed to Q4 if the request was local and to Q6 if it was remote. Likewise, status information for update requests originating locally is sent via Q5, and for those from other systems it is sent via Q7. Queues Q8 and Q9 receive replies from remote systems in response to requests which must be processed remotely. We saw the data-flow diagram which includes these paths in Fig. 10.5, for lookup operations. Updates are similar.

The queues marked as QA, QB, QC, and QD in Fig. 10.13 are not part of this system at all. They represent queues on remote systems which this instance of the system uses. Queues QA and QB are the equivalent of Q1 on a remote system. Likewise, QC is the equivalent of Q8, and QD the equivalent of Q9.

10.4.1.1 Naming the queues.

The queue names shown in Fig. 10.13 are for illustration only. When choosing queue names, it is very important to consider the scope which the name must have and the degree to which expansion is likely. In Fig. 10.13, queues Q2 to Q7 have essentially local scope. They connect components within the same application. A name including the application name as a component would probably suffice. For example, we might choose to call Q3 *PhoneDirectory.Lookup.Queue*. This would be sufficient, as long as we only ever run one copy of the application on a given system. In the case of the function-oriented version of the design, this is probably reasonable. The application already has a significant degree of internal parallelism and can handle multiple, concurrent requests when remote lookup is required. It does serialize on local requests, however. If multiple copies of the application were to be required, the queue names would need to reflect this. In addition, application components requiring services would need to know which copy of the application to contact. In a message-queuing environment, it is often better to think of a having a single point of contact for a particular application service and to use parallelism within the application rather than making it apparent to clients. This is the same philosophy used in other client-server systems (for example, DCE), where servers are often multithreaded to enable concurrent use.

In contrast, queues visible outside the phone directory application do need to include information about the directory instance in their names. The original specification explicitly required a partitioned solution, which immediately implied multiple copies of the application.

Using the partition name as part of the queue name is the most obvious approach. For example, for the queue (Q1) on which commands arrive from remote systems for execution on the San Francisco partition, we might use the name *SanFrancisco.PhoneDirectory.Command.Queue*. Having the partition name as part of the queue name allows multiple phone directory partitions to run on the same system.

10.4.1.2 Designing the messages. In the function-oriented design, the queues carry only one kind of message each. Q1, Q2, and Q3 carry commands, for example. This is a property of the application, not a restriction of MQSeries. As we shall see shortly, applications frequently need to send several kinds of message via a single queue.

The design shows essentially three kinds of message. The application uses command messages, reply messages, and status messages. Command and status messages are straightforward. They are of fixed size and are relatively small. Standard data structures suffice for these messages. Reply messages, on the other hand, require a little more thought. In principle, a query could return any amount of data up to the entire contents of all partitions. In practice, for cost and time reasons, we would probably want to restrict the maximum amount of information which a user could retrieve on a single query. This is just as well, because messaging systems usually have a maximum size associated with a single message. In more modern implementations, the limit is usually substantial, typically several megabytes. Older implementations may, however, have limits sufficiently low to prevent transmission of data in a single message. This does present the designer with a real dilemma. The simple mapping between data-flow diagrams and the structure of the message-based implementation is dependent on a single message being able to carry the data in its entirety. Obviously, such limitations can be overcome, but they add considerably to application complexity. If data is transmitted as several messages, additional error modes are introduced. There is now a possibility that interruption to the network will be made apparent to the application, since only part of the data may be received. This is highly undesirable from an application development viewpoint. It is much better if the application data can be transmitted as a single message. Then the complexity of segmenting the message for its journey over the network can be left to the messaging system to perform.

10.4.2   Messages and queues for the object-oriented design

Figure 10.14 shows the processes and queues resulting from the object-oriented design. As we remarked earlier, there is much less parallelism in this design and, consequently, there are fewer processes. Indeed, the whole directory is a single process. It accepts command messages on Q1 and sends local results and replies on Q2. QA once again represents a queue on a remote partition to which replies and commands are sent. It is equivalent to Q1 on the local system. Consequently, unlike the function-oriented design, this design receives commands and replies on its input queue. This leads to some issues. What happens, for example, if the directory is waiting for a remote site to reply when another user command arrives? Clearly, the messages must be distinguishable in some way. Newer MQSeries products allow applications to mark messages as being of a particular type. Message queues can be browsed to locate particular message types. Such facilities allow the object-oriented design to be implemented directly. Each remote request can be processed in turn, with the application waiting for the reply before starting the next query.

10.4.2.1 Naming the queues. The issues involved in choosing queue names for this design are identical to those already discussed for the function-oriented design. In this case, Q2 is the only queue which has just application scope. Q1 represents the directory partition and, consequently, needs to include the partition name.

10.4.2.2 Designing the messages. Once again, there is little to add on the subject of message design not covered already in the function-oriented design. The main difference, as we have seen, is that messages of different types do need to be distinguished on the queues. In addition, we have seen that the messages contain objects, in their external representation. Such representations need to be self-defining, with object-type information being carried independently from the object data.

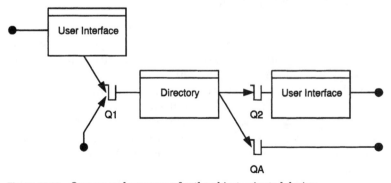

Figure 10.14   Queues and processes for the object-oriented design.

10.5   Comparing the Designs

The designs resulting from the exercises in this chapter have very different characteristics. It is time to confess that this was by no means accidental. The function-oriented design gives a highly parallel application service, capable of handling requests from many users concurrently. The object-oriented design shows a single threaded application service more suitable for a single user.

While it is true to say that function-oriented design techniques naturally tend to produce a more asynchronous design with greater parallelism, it would be unfair to accuse object-oriented design methods of being incapable of doing so. The key relationship in object-oriented design is between a particular class hierarchy and a process. Inheritance works only within a single process. Also, method invocation has the semantics of a procedure call and, consequently, is synchronous. Since we chose to represent the remote directory partition explicitly as a subclass of partition, queries against the remote partition had to be synchronous because they were invoked as methods of the partition.

The concepts of object-oriented design are particularly strong when considering single processes. To capitalize on the strengths in distributed designs is more difficult and requires explicit introduction of the notion of distribution. Function-oriented design encourages designers to think of the system as a series of independently executing units. Object-oriented design encourages an integrated view, but retains valuable modularity within a single program.

In general, it is probably easier to map function-oriented design methods to message-based solutions than it is object-oriented design methods. However, as we have shown, it is quite possible to use either approach when designing systems to be implemented using messaging and queuing.

10.6   Incorporating Existing Applications

New software products frequently need to take account of existing programs and data structures. This is as true for middleware, such as MQSeries, as it is for applications. An organization rarely, if ever, gets the opportunity to replace all its software with a new generation of products. And even if total replacement is the final goal, it invariably involves a period of time during which old and new must coexist and cooperate. Consequently, many new applications, developed using messaging and queuing, will need to interoperate with existing software. Indeed, MQSeries products can be a valuable tool in enabling such interoperation.

One approach, particularly where the existing code is maintained in-house and where the appropriate skills can be called upon, is to add support to the existing applications to use messaging and queuing. For some applications, however, this may not be a viable solution. Typical examples of such applications include large, complex, and often elderly applications which are vital to the operation of the enterprise. The risk associated with modifying such applications may be considered excessive, particularly if no staff have the detailed knowledge necessary. The documentation for such applications is rarely entirely accurate, having failed to keep up with the numerous fixes and improvements which successful systems accumulate over the years. Fortunately, alternative techniques are possible which allow these so-called *heritage applications* to be incorporated safely into new systems.

10.6.1 Heritage applications

The key concept in the safe reuse of Heritage Applications is encapsulation. Rather than modify the old application to allow it to be used in new ways, the idea is to surround it with interfaces to the new system. Figure 10.15 shows the concept. The existing application is shown in the central box. It and its resources continue to be used directly by existing users. The new, queue-based interfaces are labelled I1 to I6. Each of these represents a program which maps between data in messages on the corresponding queue and the interface provided by the heritage application. The choice of six such interfaces is entirely arbitrary. Applications using the queues to communicate with the heritage application are entirely unaware of the encapsulation. They see merely a defined set of functions provided via a particular set of message queues. This is just the same as for any conventional, message-based application. Similarly, the heritage application sees only programs which use its existing interfaces, being entirely unaware of the message queues beyond.

The job of the interface programs is essentially one of data mapping. The complexity of the mapping task depends entirely on the kind of interfaces provided by the heritage application. For example, if the existing system is based on transaction processing, the mapping might be very simple. The message formats could be chc   n to closely resemble the transaction layout. The job of the interface programs would then be particularly simple, copying the data and submitting the transactions on one side, receiving replies and copying data on the other. However, if the only interfaces available are intended to drive terminals directly, the mapping will be complex. The interface programs will have to capture screen images, emulating specific terminal types and returning data as if typed from the keyboard. Although this is a much more complex task, it is quite feasible. Indeed, products exist today for many workstation and PC-style systems to ease the task of implementing such interfaces.

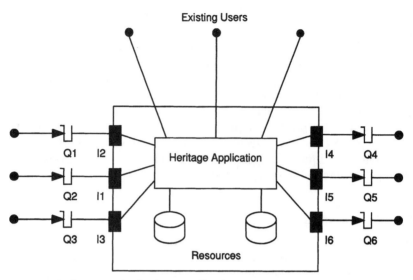

Figure 10.15   Incorporating heritage applications.

10.7   Workflow Management

By far the most common use of computer systems today is in the support of the commercial operations of businesses around the world. Although commercial applications can be extremely powerful and complex, it is still rare to find any organization whose operation is completely computerized. For example, while the sales department and the finance department may have applications to support their operations, it is usually the case that these do not interoperate in any meaningful way. In addition, it is very unusual for any single application to represent fully the operation of even a single department, let alone a sizable enterprise.

Aside from the obvious problem of disparate systems being unable to share data easily—a problem we have alluded to before—there is a more fundamental issue here. It is a problem which many organizations have recognized for some time, and a problem which is now capable of being solved by the application of technology.

Organizations can be thought of as carrying out business processes. At the grandest scale, these represent the essence of the entire organization. A construction company might build roads, for example. To achieve this overall goal requires a myriad of individual processes to be carried out within the organization. For example, the company has to purchase some equipment to be able to operate. Purchasing constitutes another business process. Actually, it probably constitutes several distinct processes. The process for purchasing five new earth-moving vehicles had better not be the same as that for ordering a box of ball-point pens if the company wishes to remain in business.

When we look at these business processes in more detail, one thing becomes apparent. They frequently involve both people and computer systems. Commonly executed processes are well understood by the people involved with them. They get plenty of practice in carrying them out. Less frequently encountered processes require support from process documentation. Usually, part of the process can be carried out with the help of a computer, but, often, manual operations are also required. For those parts of the process which are supported by a tailored application, much of the work can be automated. The appropriate business rules can be embodied in the application. For those parts of the process which are handled manually, the organization relies on the skill and experience of its employees to carry out the process correctly. While there will always be a need for the application of human skill and judgment in the operation of business processes, many organizations are looking for better ways of implementing their business processes, to assist the employees involved and to reduce the potential for confusion and error.

The ultimate aim for many organizations is to support entire business processes through their information systems, rather than supporting specific parts of some of their processes, as is the case today. The goal is to ensure the reliable and repeatable operation of their key business processes, and to reduce the amount of additional work generated by failures of their manual processes.

One approach, which has appeared in the last few years, is termed *Workflow Management*. The techniques involved map directly onto the concepts of messaging and queuing, making implementations based on products such as MQSeries an attractive proposition. Workflow Management is an emerging technique for the implementation of business processes, and its affinity with messaging and queuing makes it an interesting area for discussion of application development.

In the rest of this section, we will cover some of the basic concepts of Workflow Management and see how message-driven processing provides the necessary infrastructure to allow real systems to be implemented.

10.7.1 Work-flow concepts

The primary concept in Workflow Management is the *process model*. This model represents a particular business process. It consists of a set of interconnected *activities*. Each activity is an individual work item which must be performed. Activities might themselves be business processes which can be defined in terms of their own process model. The highest-level process model for a particular business process will probably consist of activities which are themselves all sufficiently complex to require their own process model. At the most detailed level, activities are individual operations supported directly by some application. These indivisible activities are sometimes termed *elementary activities*.

As an example of these different kinds of activity, consider the task of purchasing those five earth movers for the construction company. One activity associated with the purchase will almost certainly be tendering. Tendering is itself a business process, involving selecting potential suppliers and inviting them to quote a price for machinery meeting the required specification. By the way, development of the specification is itself another business process. One activity associated with tendering is the preparation of the letters to suppliers, inviting them to tender. This is probably an elementary activity, quite possibly based on a pro forma letter which simply needs to be updated for this particular tender.

The process model defines how individual activities, both elementary and nonelementary, are interrelated. It shows the overall flow of work through the process, and, hence, the implementation of such a model has become known as a *work flow*. Whereas elementary activities are frequently supported by computer systems in current business operations, the organization of these activities into a business process is frequently manual. The new systems support is required to automate the coordination of the existing elementary activities. Since its task is essentially to manage the flow of work through the process model, it has become known as a *work-flow manager*. The job of the work-flow manager is to ensure that the appropriate activities take place at the right time and in specific sequence, if sequence is important. In some sense, it controls execution of the process model.

The work-flow manager controls execution of activities. If the activity is already automated, this probably simply means that the work-flow manager must arrange to have the appropriate application executed. On the other hand, if the activity involves staff, the work-flow manager must arrange for the appropriate requests to be delivered to them.

10.7.2 Performing activities

Performance of activities requires three essential pieces of information. A moment's thought reveals that they are:

Who (or what) is to perform the activity

What is to be done

How is the activity to be performed

Let's take a simple example, based once more on ordering the earth-moving equipment. Suppose that the business model for tendering includes an elementary activity called *Issue Request For Tender*. The goal of this task is the physical distribution of the request for tender to those companies being invited to tender. The definition of the task might be as follows:

Who   Any secretary.

What

Activity	For each company to be invited to tender, print a copy of the tender and of the signed covering letter, and prepare an envelope. Insert the tender and covering letter and post the resulting package.
Required Data	The covering letter, the tender itself, and a list of names and addresses of the companies being invited to tender.
Required Result	One copy of the material mailed to each company being invited to tender.
Constraints	Activity completed within one working day of receipt of request.

How   Use the company word processing system to print the request for tender and the letter.

This example is deliberately simple to illustrate some specific points. Indeed, you might rightly denounce it as insulting the intelligence of any competent secretary who had been working for the company for even a short period of time. But before dismissing it, it is worth pointing out that there can be merit in defining an activity even as simple as this one. For example, it might provide much needed support for a temporary secretary brought in at short notice to cover for sudden absence.

So, recognizing that this is a trivial example, what does it show about work flow? This is an elementary activity, so the contribution of work flow is not within the activity, but rather in initiating it. The process model can contain sets of conditions which must be met before an activity can be initiated. In the case of our example, the request for tender, the covering letter, and the list of recipients must all be available before the activity can be carried out. The task of the work-flow manager, in this case, is to recognize when the conditions for initiation of the activity have been met and to issue the request to the user concerned. The appropriate data—in this case, the documents and list of names and addresses—must also be made available at the same time.

Another important aspect of the task of the work-flow manager can be in selecting the user to which a request is sent. This might be very specific. For example, further along the process of purchasing the earth movers, the business process might require that a specific company executive approve the expenditure, which might be many tens of thousands of dollars. In this case, the work-flow manager will route the request to that user. In the case of sending out the request for tender, however, the definition of the activity specified that any secretary could carry out the task. Where multiple users could perform a given activity, the possibility of routing the request to the least busy of them emerges. Monitoring the list of requests queued to each user could provide a measure of how busy each user is, as long as some account is taken of the relative lengths of time required to complete different tasks.

When the activity is complete, the user involved performs some action to tell the work-flow manager. This allows the state of the process model to be updated to reflect that of the real world. As a result, the conditions for some other activity might now be met, allowing the work-flow manager to initiate the activity.

10.7.3   Work flow and message-driven processing

The characteristics of message-driven processing are an excellent match for the implementation of work-flow manager products. Both have strong affinity with data-flow models of application systems, which we saw in Sec. 10.1.3. Indeed, business processes can be represented by diagrams very similar to the data-flow diagrams we used when looking at function-oriented design. Data flowing between activities, requests for activity initiation, and the results of activities can all be carried very naturally by queued messages. The reliability and persistent properties of messages carried by products like MQSeries are very attractive to developers of work-flow managers. Also important is the ability to interconnect many different computer systems and applications. Business processes frequently involve multiple systems, tying together existing elementary activities carried out in different parts of the organization. The ability to distribute data and requests across a heterogeneous collection of machines is a crucial requirement for practical implementations of work flow.

Vendors of work-flow products have not been slow to recognize the synergy. At the time of writing, at least two work-flow manager products were being developed to run on top of MQSeries messaging and queuing.

10.8   Summary

In this chapter, we have looked primarily at the design of applications which are to be implemented using message-driven processing. We have seen that function-oriented design is a very natural technique to use for such applications. We have also seen that object-oriented techniques can be successfully used in designing message-driven applications.

We looked briefly at some encapsulation techniques which can allow existing applications to be integrated into new client-server environments using message queuing, and, finally, we discussed the emerging technology of Workflow Management and saw that message queuing provided an excellent technology base for the development of Workflow Management products.

4

The Development

This part of the book describes how to develop messaging applications. Messaging and queuing constitutes an efficient, effective, and stable software "machine" that can be used to power new application design styles, integrate existing applications, and integrate new and existing (legacy) applications. The challenge is to recognize the uses to which this software machine can be put.

As with lots of machinery, possible uses are sometimes not so obvious and a few hints and suggestions are helpful. It is usually helpful to know a little bit about how any particular machine works and then a lot about how to operate it. The same is true with the messaging and queuing machine, and this part of the book tells you a little about the software machine (the innards of the Message Queue Manager) and a lot about how to operate it (access it with the Message Queue Interface).

The MQI provides the programmer with access to messaging and queuing services. The programmer uses the MQI while the messaging and queuing service supports (provides) the MQI (interprets and executes the CALL requests), as shown earlier in Fig. 3.9.

This part of the book describes:

- *The Message Queue Manager (MQM)*
- *The Message Queue Interface (MQI)*

11

MQSeries Messaging
and Queuing

In this chapter, we will look at the characteristics of the MQSeries Message Queue Managers. We will see how this family of products provides messaging and queuing services. We will also review most of the resources used within MQSeries and cover the terminology used. We begin by looking at the queue manager itself.

11.1 The Local Queue Manager

In MQSeries, the queue manager is the basic software system. Every queue in an MQSeries system belongs to one of the queue managers. The queue manager provides access to the queues on behalf of application programs and other MQSeries components, such as the administration tools.

Applications request services from the queue manager. They do this using an application programming interface known as the *Message Queue Interface,* or MQI. These services include the basic operations of putting a message onto a queue and getting a message from a queue. The queue manager software provides for reliable storage of queued messages. It also manages security authorization and concurrent access to the data, as well as providing specific queuing function, such as triggering, which we will look at later in this chapter.

Any application wishing to use MQSeries facilities must first connect to a queue manager. The queue manager to which an application is directly connected is known as the *local queue manager.* The local queue manager is the one which provides direct support for MQI calls made by the application. Applications can manipulate the resources owned by the local queue manager, as well as being able to put messages destined for other queue managers.

A queue manager is the basic, indivisible unit of execution in an MQSeries product. Frequently, each machine in an MQSeries network runs a single queue manager. In this case, the queue manager effectively represents the machine.

11.1.1   The default queue manager

To reduce the dependency of application code on the environment in which it is executing, MQSeries applications can connect to a queue manager without knowing its name. To do this requires a *default queue manager*. The default queue manager is the one to which a connection is made if the queue manager name is left blank on the MQI call by which applications connect to MQSeries.

Some implementations restrict a given computer system to running only a single queue manager. In this case, that queue manager is the default. Other implementations, however, allow multiple queue managers to execute concurrently on a single system. Such configurations might be used to allow production and test systems to run alongside one another. It might also be used to exploit additional performance available from multiprocessor machines. Whatever the motivation, if multiple queue managers are executing on a single system, one of them is chosen to be the default.

11.2   Queues

The basic resources managed by each queue manager are its queues. There are several different types of queue, as we shall see shortly. Every queue, regardless of its type, has attributes associated with it. The name of a queue, for example, is considered to be one of its attributes. Other attributes include the maximum number of messages allowed to be waiting on the queue, the maximum size of a message, and whether or not triggering is enabled for the queue.

The sequence in which messages are retrieved from a queue can also be controlled by an attribute on many MQSeries queue managers. Messages are usually delivered in the sequence in which they arrived on the queue. However, messages can also carry priority information, and this can be used instead to determine the sequence of delivery. Messages of the same priority are still delivered in arrival sequence. Priority delivery could be used, for example, to ensure that critical messages arrive at a server immediately, rather than having to wait behind ordinary messages.

There are a number of different kinds of queue. We will look at each of them later in this section. Before doing that, we need to examine some features which influence the behavior of queues.

11.2.1   Persistence

Messages placed on queues may be persistent or nonpersistent. Once a persistent message has been written to a queue, it survives, even if the queue manager is stopped or the system fails. Persistent messages are hardened. This means that there will be a copy of the data on disk storage. The precise mechanism used for hardening messages depends on the particular implementation of MQSeries. However, the important point is that, once the queue manager has restarted, processing of persistent messages can continue.

By contrast, nonpersistent messages do not survive across queue manager restarts. Such messages are not hardened. MQSeries guarantees that nonpersistent messages will not be retained after a restart. In some circumstances, MQSeries actually has to discard such messages. At first sight, this may seem somewhat bizarre. However, the intermediate steps in an application might exploit nonpersistent messages because of this behavior. Application designers know that the messages are discarded in the event of a failure.

Nonpersistent messages tend to be processed more quickly by queue managers, because they do not need to be hardened.

11.2.2   Syncpoints

Messages can be placed on queues or retrieved from them using syncpoint processing. Syncpoints identify the scope of logical units of work. GET and PUT operations specifying that syncpoints are to be used cause all changes to the queues to be remembered. Only when the application indicates successful termination of the unit of work, by issuing a commit call, are the changes made permanent. If the application aborts the unit of work, or if it fails, affected queues are returned to the state they were in before the logical unit of work began. Messages written to output queues are discarded and messages read from input queues reappear on them.

11.2.3   Local queues

A *local queue* is a queue which actually resides on the queue manager to which the application is directly connected. A local queue is held in the main memory and disk storage of the local system. The local queue manager controls access to its local queues.

Applications can put messages onto local queues and retrieve messages from them. In addition, they can inquire the values of the attributes of such queues and update some of them. Manipulation of queue attributes usually requires the appropriate security authorizations to have been granted.

11.2.4 Remote queues

A remote queue is one which belongs to a queue manager other than the one to which the application is directly connected. Access to this queue involves the local queue manager and the remote queue manager in some form of communication. This communication involves the Message Mover programs, which we will look at shortly.

Applications can perform operations against remote queues, but there are some restrictions. For example, messages can be placed on remote queues, but cannot be read from them. The general MQSeries model is such that queues are local to the applications which read from them.

Applications can reference remote queues in two distinct ways. The explicit method involves the application supplying both the queue manager and queue names on the MQI call used to open the queue. This requires the application to know or to be able to discover the name of the queue manager holding the queue. Because this involves the application knowing more than absolutely necessary about the configuration of the system, there is a second method. In this, a definition of the remote queue is held in the local queue manager. By opening this queue, the application is connected to the remote queue. The local definition of the remote queue holds sufficient information to allow the local queue manager to identify the appropriate remote queue manager.

The target queue on the remote queue manager does not have to be a local queue. It may itself be a local definition of yet another remote queue. Figure 11.1 shows an example. The application puts a message onto Q1. The local queue manager QM1 has a definition of Q1 as a queue on remote queue manager QM2. Together with a mover program, it transmits the message to Q2 at QM2. Queue manager QM2 has a definition of Q2 as a queue on remote queue manager QM3. Again, using a mover program, it arranges to send the message to Q3 at queue manager QM3. This is a local queue, and the message will remain on the queue until an application reads it.

There are a number of reasons why this multihop forwarding might be useful. On a network such as TCP/IP, all connected machines are directly addressable. The network itself handles the routing necessary to reach any node. However, to cross from one network type to another (for example, SNA) requires routing function in the middleware layer, as we saw in Sec. 9.1.3.5. The ability to chain remote queue definitions is the basis by which MQSeries provides such routing. In the example in Fig. 11.1, QM2 is running in a machine which bridges a TCP/IP network and an SNA network. The definition of Q2 at QM2 is all that is required to allow messages to transfer from the TCP/IP network to the SNA network.

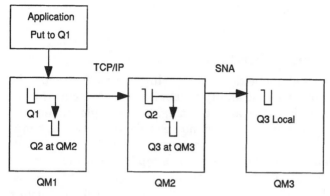

Figure 11.1 Remote queues.

In addition to providing cross-network routing, chaining remote queue definitions can be used to reduce the amount of administration required when a major reorganization of queues and queue managers is needed.

Because the use of remote queues always involves communications between queue managers, some additional resources do need to be defined. These additional resources include *channels*, which we will look at in Sec. 11.7, and *transmission queues*, which we will cover next.

11.2.5 Transmission queues

Transmission queues are the mechanism by which queue managers arrange to forward messages to remote queue managers. We saw in the previous section that the queue manager works with the mover programs when sending data to remote queues. When a queue manager receives a request to put a message on a remote queue, it actually puts it on the transmission queue associated with the queue manager which is the target for the request. The name of this queue manager may have been supplied by the application or it may have been deduced from the local definition of the remote queue.

A transmission queue forms one end of the link between a pair of queue managers. All messages whose immediate destination is a particular queue manager can be placed in the same transmission queue, regardless of the queue for which they are destined. Only one transmission queue is needed for moving messages from one queue manager to another. It is, however, possible to have multiple links between two queue managers, perhaps offering different classes of service. Each link requires a separate transmission queue.

Transmission queues are processed by the mover programs, as we will see shortly. These programs have the responsibility for transmitting messages reliably between queue managers. Movers can be thought of as special-purpose application programs which process messages on transmission queues.

11.2.6 Dynamic queues and model queues

In addition to the fixed queue definitions, which we have already discussed, MQSeries provides the ability for applications to create queues dynamically for use while they are executing. For example, an application which is a client of some service may choose to create a dynamic queue for the server to use to reply to requests it sends. The alternative would be a permanent queue definition. To simplify the set of parameters needed when creating a dynamic queue, these queues are based on a template called a *model queue*. The model queue must exist before dynamic queues based on it can be used. The model queue defines the attributes which the dynamic queue will have when it is created. A dynamic queue is created when an application issues a request to open it and references the appropriate model queue as the object being opened.

There are two kinds of dynamic queue. They differ in their lifetime and recovery characteristics. Temporary dynamic queues are deleted and any messages on them are lost when they are closed by the application that created them. Also, such queues do not support persistent messages and are not recovered if the queue manager fails. Permanent dynamic queues, on the other hand, are not deleted until an application closes them, specifying that they should be removed. This operation can be carried out by an application other than the one which created the queue. Permanent dynamic queues can be recovered after a queue manager failure and can support having persistent messages placed in them.

Distribution lists exist to allow applications to send messages to multiple destinations with only a single MQI call. A distribution list is a list containing the names of the queues to which the message is to be sent. Such lists can contain the names of local queues or of local definitions of remote queues. Using such lists, an application can send the same message to multiple destinations using a single call.

11.2.7 Dead letter queues

Occasionally, situations may arise in which the queue manager is unable to deliver a message to its intended destination. In this case it must resort to storing the message on a *dead letter queue*. The appearance of messages on dead letter queues usually indicates some problem with the system or its configuration. For example, a message could arrive at the queue manager destined for a queue which does not exist. Similarly, a message might arrive for a queue that is known to the queue manager but is in a state in which it cannot accept messages. It might be full, for example, or it might have been set not to allow messages to be added to it. Not all failing PUT operations necessarily result in messages being placed in a dead letter queue. For example, it is unnecessary to put the message in a dead letter queue when the error is related to a queue on the local queue manager of the application making the request. In this case, the error can be reported directly to the application in the return code of the call to MQSeries.

There are situations in which there is no other way to report the error. For example, the message may already have traversed the network to reach the machine on which a remote queue resides, only to find that the queue is full. The application which originally put the message executed on a machine different from the one which discovered the error. In addition, it may well not be executing any longer. When there is no other way to report the error, the queue manager makes use of dead letter queues to hold the messages, rather than simply discarding them. This way, not only is the problem visible, but there is a chance to correct the fault and redirect the messages in appropriate fashion.

As we shall see shortly, dead letter queues are the last resort for dealing with errors encountered on remote systems. Applications can make explicit use of other MQSeries facilities for monitoring the successful transmission and receipt of important messages. We will look at these facilities when we discuss the different types of report messages available, in Sec. 11.3.1.

11.2.8 Alias queues

Alias queues provide alternative naming for local queues, local definitions of remote queues, or distribution lists. They are simply a name-to-name mapping to allow queues to be referred to in alternative ways. They provide an extra level of indirection between an application and the queues it uses. MQSeries already provides significant insulation of an application from the details of the underlying system and, in particular, the configuration of network resources. Alias queues allow even the name of the underlying queue to be changed without the applications being affected.

11.3 Messages

From the point of view of the queue manager, the life of a message begins when it is first placed on a queue and ends when it is finally read from the queue. Of course, messages may pass through many queue managers as they traverse the network. Also, as we have seen, notions of when a message is considered to be actually on a queue are modified by considerations associated with syncpoint processing.

Messages carry attribute information with them as they pass through the system. This information includes the name of the queue on which replies should be sent, as well as identifiers which can allow an application to relate this message to others. Also included is the message context, and information about the origin of the message and about the user associated with the request. Almost all of the information carried with the message can be specified when the message is put onto a queue. Changing some of it, such as message context, may require that the user or application is operating with the appropriate level of security authorization.

11.3.1 Types of messages

MQSeries defines four basic types of messages. Applications are free to define additional types. The four basic types are:

Request messages

Reply messages

Datagram messages

Report messages

Requests are messages which require replies. Messages sent to a server to retrieve or update information would probably be request messages. Request messages need to carry information which will allow the reply to be routed appropriately.

Reply messages are responses to requests. Information from the associated request message determines the destination of the reply message. The applications involved in processing requests and replies control the relationship between the messages. The queue manager itself does not enforce any rules, but the information transmitted with the messages does enable applications to relate and process requests and replies.

Datagrams are messages to which no reply is anticipated or required.

Reports are messages generated in response to some error occurring in the system. They can be generated by applications, if desired. Some report messages are generated by the queue manager. This can occur, for example, if a message cannot be delivered. This situation might occur if the destination queue is on a remote machine and is full or absent. The error cannot be detected by the local queue manager of the application which put the message originally. Only later is the problem discovered, resulting in the generation of the report message. The queue manager always sends the report message to the queue specified in the original message as being for replies.

The queue manager uses report messages for other purposes, as well as for indicating errors. Messages can have expiration times associated with them. This allows the queue manager to purge them if they have been in the system for some period of time and yet have not been processed. Applications choose whether messages are to be allowed to expire and, if so, how long the expiration period should be. If a message expires, the queue manager deletes it and a report message is generated and sent to the queue for replies, specified in the expired message.

Another use of report messages is for confirmation of the arrival of a message. It is possible for applications to request that they be notified when the message actually arrives at its destination queue. This is known as *confirmation of arrival*. Similarly, an application can request to be notified when another application actually reads the message from the queue. This is called *confirmation of delivery*. In either case, the queue manager generates a report message to the queue for replies specified in the original message.

Application-defined message types can also be used. The queue manager itself takes no action based on the message type. Consequently, arbitrary values can be specified. A range of values has been set aside for application use. These values can be used, for example, to distinguish between different types of application message on the same input queue.

11.3.2 Trigger messages

Trigger messages are a particular kind of message generated by the queue manager itself. They form the core of the mechanism, provided by the queue manager, by which applications can be started automatically when work arrives for them to do. It is often preferable to start an application only when there is work for it to process, particularly if the work load is sporadic. Naturally, initial responses take longer, because the application has to be started before it can process the first message. However, unless the work load for an application is fairly constant, there will be periods of time during which it is idle but nonetheless using some system resources.

The mechanism by which applications can be started on demand is known as *triggering*. The idea is that the queue manager recognizes the arrival of messages on queues for which triggering has been enabled and informs a special-purpose application, known as a *trigger monitor*. This application is responsible for starting the application which will process the newly arrived message. The overall scheme is shown in Fig. 11.2.

In step 1, a message arrives on an application queue for which triggering is enabled. In step 2, the queue manager examines the circumstances surrounding the message arrival and determines that it

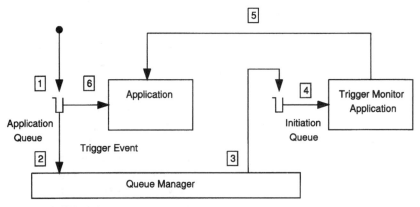

Figure 11.2   Triggering.

constitutes a trigger event. We will look at the sorts of rules which can be applied shortly. In step 3, the queue manager generates a trigger message and places it on another queue. This is the queue processed by the trigger monitor application and is known as an *initiation queue*. Its name is part of the information associated with the application queue and held because triggering is enabled. The trigger monitor application reads the message from the initiation queue at step 4. In step 5, it issues an operating-system-dependent command to start the application associated with the queue which was triggered. The association between the queue and the application to be started is held in a *process definition*. The name of this definition is one of the attributes of the queue itself. The process definition contains details which the trigger monitor needs to start the application. The information includes the name of the application, data to be passed to the application when it is started, and environment information. This information is made available to the trigger monitor in the trigger message it receives from the queue manager. Finally, in step 6, the application reads the message which had arrived on its queue.

Because a single trigger monitor can service many different queues rather than having many idle applications waiting for work, only the trigger monitor needs to be running continually. The use of process definitions means that the trigger monitor does not need to know anything about the applications it is starting. It gets all the information it needs from the trigger message. Trigger monitors can be completely generic. Conversely, if there are specialized requirements, different trigger monitors can replace the standard one supplied with MQSeries. A trigger monitor is just another MQSeries application. For triggering to be used, there must be at least one trigger monitor running. There is, however, no limit to the number of monitors which can be executing simultaneously.

11.3.2.1 Triggering rules. Because triggering might be used in a variety of ways, MQSeries tries to provide a comprehensive set of options to cover all requirements. Three basic types of trigger are defined:

- Trigger on first message
- Trigger on every message
- Trigger when specified number of messages waiting

The first two kinds of trigger leave triggering enabled. The third type disables triggering once a trigger has been generated. The first kind of triggering generates a trigger whenever a message arrives on an empty queue. This trigger is intended to be used when the application being triggered continues processing until the queue is empty and then terminates. It is triggered again when work arrives on the queue. This kind of triggering is appropriate where work arrives fairly sporadically but where the messages tend to come in batches. This might be true, for example, where remote sites have collected data over some period of time and are transmitting it to a central site to bring central records into line. Some retail operations work this way, transmitting records of the day's sales to central operations during the night.

The second kind of triggering generates a trigger whenever a message arrives on the queue. This kind of triggering is intended for applications which process a single message and terminate. The application is triggered every time a message arrives. Consequently, there is more overhead associated with processing each message, in this case, than when compared with triggering on first message. Clearly, this kind of trigger is appropriate for rare messages which must be processed immediately. If two such messages arrive in quick succession, a copy of the application will be started for each one, allowing them to be processed together. This kind of triggering is appropriate for rare messages which are, nonetheless, very important to the operation and which require immediate processing.

The final kind of triggering causes a trigger whenever there are more than a specified number of messages on the queue. In addition to generating the trigger message, this kind of triggering also disables further triggering once a trigger has been generated. This kind of triggering is intended for starting applications which will tend to run for a reasonable period of time. The arrival of messages on the queue indicates that there will probably be work to do for a reasonable period of time. The triggered application processes the messages queue to it. Once the queue is empty, it waits for more work to arrive. If no work arrives within a predetermined time, the application rearms the trigger for the queue and exits. It will be triggered again once more work arrives. We will see an example of an application which uses this kind of triggering in Sec. 12.3.

Although we have given examples here of how these triggering rules can be used, the trigger mechanism itself makes no demands on the behavior of the applications. Application designers can choose to use the trigger mechanisms in whatever way they wish.

11.3.3 Message context

In client-server systems, server applications usually perform work on behalf of client applications or end users. Servers normally execute in a context quite different from that of the applications and users for whom they are performing work. In particular, a server's security profile will often be quite different from that of the user or client on whose behalf it is performing work. Servers frequently have higher levels of authorization. To prevent unauthorized access to resources, it is obviously necessary for the server to be able to test access rights based on the capabilities of the entity submitting the work rather than its own authorizations. To enable the server to do this, *context* information can be passed in MQSeries messages. Context information describes the submitter of the request.

Context information in messages is in two parts. *Identity context* contains the information we just discussed. There is also space for accounting information, which applications can use to apportion work load to users or groups. Identity context is normally passed on unchanged if a message moves from queue to queue under application control. This allows all work to be performed in the context of the original user. Similarly, if one application processes a message and sends others in order to get work done, it can set the context in the new messages to be the same as the original. Again, all work performed will be in the context of the user who submitted the original message.

In contrast, *origin context* holds information about the application which actually put the message on the queue. Origin context is normally supplied automatically by the queue manager. Applications can override it if required.

Applications which modify or set context normally need to be given additional authorization to do so for the appropriate queues.

11.4 Processing Application Requests

MQSeries provides an application programing interface (API) as a set of functions which can be called from the C and COBOL programming languages. The API consists of a relatively small number of individual calls, known as *verbs*. Each verb has a number of basic parameters. Flexibility of operation stems from the nature of some of the parameters. These are data structures containing fields which define the options which can be set to modify the behavior of some of the calls. In every case, these parameters can be given sensible default values very simply. Consequently, basic applications can be written very easily and the subtleties of the options need be addressed only when specific behavior is required.

The kinds of function available through the MQSeries API include:

- Establishing a connection to the local queue manager

- Opening a queue to work with

- Getting a message from a queue

- Putting a message onto a queue

- Querying or changing the attributes associated with a queue

The interface uses *handles* to represent the objects being manipulated. Handles are values which are meaningless to the caller but which are used internally by MQSeries to distinguish instances of the objects being referenced. The interface is described in more detail in Chap. 12, which also includes examples of its use.

When processing API calls, MQSeries takes the same basic approach, regardless of the particular call involved. First, parameters are verified for correctness. In some cases, the values associated with one parameter may influence the values which must be supplied for another. If the parameters are correct, processing can occur. In the newer implementations of MQSeries, processing takes place in such a way that important MQSeries data, such as the contents of the queues, is isolated from the application code. This prevents application code from inadvertently or maliciously accessing and modifying the underlying data structures used internally by MQSeries.

The mechanism chosen for isolation is implementation dependent. Some systems—for example, IBM's MVS and OS/400—provide the ability to change access control authority when executing certain pieces of code. The MQSeries code executes with higher authority than the application code and, consequently, has access to data structures denied to the application code. In other systems, notably the newer UNIX-based implementations, processing of the MQSeries data structures takes place in a process separate from that in which the application is executing. Again, this isolates and protects the internal MQSeries data.

11.5 Managing the Queues

The task of maintaining the message queues involves two main kinds of operation. First, the semantics of the MQSeries API have to be performed. For example, when a message is put onto a queue for which triggering is enabled, the appropriate trigger message may need to be generated. Second, the data associated with the structure and content of the messages and the queues must be managed. This set of operations on the queue data itself is what gives each MQSeries product many of its performance and reliability characteristics. These include the ability to recover from system failures without loss of data. The techniques employed when dealing with queue-related data are often similar to those used by database or transaction managers. The aims are very similar in that data loss must be avoided without the performance overhead associated with synchronous I/O operations to disk devices. The underlying storage medium chosen to hold hardened data depends on the particular implementation. However, it is usually the file system of the underlying platform. Since file systems are rarely high-reliability components, MQSeries products take special actions when hardening data to ensure that its data is highly reliable and recoverable.

11.6 Systems Management

The management of an MQSeries system is mainly concerned with the creation, modification, and removal of queue managers and queues. In addition, the connections between queue managers also need to be maintained. These connections are known as channels. We'll cover them in more detail shortly.

The creation of a Message Queue Manager is a fundamental operation and one which typically is not carried out very often. Usually, initial queue manager creation is carried out immediately after the product itself is installed. Once a queue manager has been created and is running, its own queuing mechanisms become a component of its management. A standard systems management queue is created. Messages of appropriate format placed on this queue can be used to perform systems management. Since these messages can originate from remote machines, this mechanism provides for distributed management of MQSeries queue managers.

There are two main ways to manage the newer MQSeries queue managers. The first is the systems administration application, an interactive program providing a user interface through which management tasks can be carried out. On systems where a graphical user interface is available, the systems administration application makes use of it. Otherwise, a full-screen, panel-style interface is used. The systems administration application can be used to perform all MQSeries systems management tasks for the local queue manager and the majority of such tasks for suitable remote queue managers. This is true even where the remote queue manager is running on a machine of a different architecture.

The second mechanism for administration is based on files containing commands, rather than an interactive application. These MQSeries Script Commands are portable. The same syntax is used by any MQSeries product which supports them, regardless of the underlying machine and operating system. Consequently, a script written on one system can be executed on a different one, allowing another mechanism by which remote administration can be performed. Scripts are moved between machines by normal file transfer techniques. Scripts are particularly useful when bulk changes need to be made, especially if several queue managers are involved. In addition, they can be used to store particular sets of operations which might need to be repeated.

There is a third way to perform systems administration on some MQSeries products. Some of the operations are made available as commands, using the command interface of the underlying operating system. As with MQSeries script commands, normal commands made available this way can be combined together in files. For example, on UNIX-based implementations, these are shell scripts and on OS/2, they are command files. However, these kinds of files are less portable than MQSeries script commands. A UNIX shell script cannot be executed on OS/2, for example, whereas an MQSeries script command file would run equally well on both. Script portability was a major factor in the decision to provide MQSeries script commands as the prime mechanism for noninteractive systems administration.

11.7   Moving Messages Between Queue Managers

Messages move between queue managers via programs known as *message channel agents,* or more simply, *movers.* Movers are essentially normal, message-queuing applications. Movers read the messages to be transmitted from a queue. This queue is a transmission queue. We discussed these in Sec. 11.2.5. Messages read from the transmission queue are sent over a particular communications connection, known as a channel. At the other end of the channel is another mover program which receives the data. This program assembles the data it receives over the channel into messages and puts them onto the appropriate queues on the target queue manager. Information about the target queues flows over the channel with the messages.

11.7.1   Channels

Channels are the communications links over which data flows between queue managers. Channels are defined as point-to-point links, even though the intervening network may be very complex. For example, the route between two machines on a TCP/IP network might involve several machines and include bridges and routers. The channel definition, however, references only the machines running the queue managers being linked. A channel implies a particular network type. A single channel cannot, for example, link one machine on an SNA network with another on TCP/IP. To achieve this kind of connection, channels need to be defined from each machine to a machine which links both networks. These sets of channel definitions provide the cross network routing function within MQSeries.

Channels provide a logical view of the underlying network topology. They map MQSeries communications concepts onto the underlying network implementation. Channel definitions and the channels themselves are managed via normal MQSeries systems administration. It is possible to stop and start channels without affecting operation of the queue manager. Naturally, if a channel is stopped, work cannot be transferred over it. Messages queue up until the channel is made available once more.

11.7.2   Recovery from channel failures

The communications component of any distributed system is usually its least reliable element. Network failures occur for a variety of reasons. Components of the network, especially the medium used for wide area transmission, have to operate in the least well controlled electrical environment. Consequently, channels and their associated software are the parts of an MQSeries system most likely to be affected by an external failure. Indeed, one of the benefits of messaging and queuing is that these kinds of failures can be isolated to a component which is not intimately connected with the applications. Failures on the network affect only the channels and the message channel agents.

Detection of and recovery from channel failures is the task of the message channel agents. Typically, a network failure is detected when an operation to send or receive data fails. In addition to transmitting data between themselves, message channel agents record and transmit status information. When the channel once again becomes available, the message channel agents converse with each other to agree on the overall state of the transmission. Essentially, they need to determine the last packet of data successfully transmitted so that they can restart operations at the correct point. MQSeries ability, to guarantee that messages are not lost or transmitted twice, is heavily dependent on the process by which channel agents agree where to restart transmissions.

The only effect of a channel failure, or of the deliberate stopping of a channel, is that messages to be transmitted are delayed until the channel is available once more.

11.8 Summary

In this chapter, we have taken a very brief tour of the facilities of typical MQSeries queue managers. We looked at the queue manager itself and covered the characteristics of the various kinds of queues which are supported. We discussed various types of messages, including the reports which can be generated in response to specific events or remote errors. Finally, we looked briefly at the tasks the queue manager performs and covered the work performed by message channel agents.

In the next chapter, we will look in more detail at the application programming interface provided by MQSeries and we will see some examples of its use in practice.

12

The Message Queue Interface

The interface by which application programs access message-queuing function in MQSeries products is known as the *Message Queue Interface* or simply the MQI. In this chapter, we will look at the facilities offered by the MQI and see how to use them in some simple examples. The information in this chapter is, of necessity, incomplete. The aim is to introduce the MQI and to show how it can be used. It is not feasible to reproduce all the technical information associated with MQSeries here. Full details of the interfaces and the data structures associated with them can be found in the reference documentation associated with the MQSeries products themselves.

12.1 A Quick Tour of the Message Queue Interface

There are relatively few operations which applications need to perform when interacting with queues. Clearly, putting messages onto queues and retrieving messages from queues are the most common. However, in any comprehensive programming interface, there needs to be some mechanism by which applications can exercise control over the details of the processing. As a simple example, consider the interface to a simple, two-dimensional drawing package. The interface to the operation to draw a straight line can be extremely simple. The only information required is some definition of the coordinates of the points joined by the line. However, the line itself may have many attributes associated with it. For example, it may have a specific color, a particular style, and a width. The challenge, when designing a programming interface, is to allow the application the full range of control over the processing, without obscuring the basic simplicity of the operations.

The MQI tries to solve the interface problem by offering a relatively small number of operations, but allowing the detailed behavior of each to be modified by options and descriptors. Default values for these are provided, allowing normal operations to be carried out without the necessity for programmers to understand the subtleties of the interface until they need to be used.

We will look at the basic operations provided by the MQI first, and then we will consider some of the more important options available.

12.1.1   The basic verbs

The major functions provided by the MQI are termed *verbs*. There are 11 of these in the newer MQSeries implementations. The following sections describe each of these functions.

12.1.1.1 MQCONN—**connect to the queue manager.**   All MQSeries operations require that the application be connected to a queue manager. This connection is established via the MQCONN call. A particular queue manager can be specified by name on this call. However, applications frequently do not know and do not care which queue manager is providing their services. In this case, they can connect to the default queue manager by not specifying a name.

The result of a successful MQCONN operation is a connection handle. This is used on subsequent MQI calls to specify the particular queue manager to which the application is connected.

12.1.1.2 MQDISC—**disconnect from the queue manager.**   When an application has finished work and no longer needs a connection to the queue manager, it issues the MQDISC call. This allows the queue manager to release any resources which were being used to support the application. Any MQSeries objects which the application had been using, but which it had not explicitly released, are implicitly released by this call.

12.1.1.3 MQOPEN—**open an MQSeries object.**   In order to work with MQSeries objects such as queues, an application must first open them. This is analogous to the notion of opening a file before reading or writing to it. Queues are the most important class of MQSeries object for most applications. There are others, though. For example, there is a queue manager object, which can be opened. The class of an MQSeries object determines the kinds of operation which can be performed. For example, messages can be retrieved from MQSeries queue objects but not from queue manager objects. Both kinds of object do support inquiry of attributes, however. Applications specify the particular object to be opened by filling in an object descriptor. We will look at object descriptors in more detail in Sec. 12.1.3.

When opening an object, an application must specify what it intends to do with it. Once again, this is analogous to the situation when opening files. The intent, specified by the application, is used to determine the kind of locking and serialization which the queue manager must perform in order to protect the integrity of the data in the object. The intent is passed to the MQI in the options parameter. We will look at these options in Sec. 12.1.2.

The result of a successful MQOPEN call is an *object handle*. This is used in subsequent MQI calls to specify the object on which the requested operation is to be performed.

12.1.1.4 MQCLOSE—close an MQSeries object. When an application no longer needs to use an MQSeries object, it can issue an MQCLOSE call against it. This allows the queue manager to release resources associated with the object. Objects still open when an application disconnects from the queue manager are implicitly closed. Once again, this behavior is analogous to that of typical file systems.

12.1.1.5 MQPUT—put a message on an MQSeries queue. Applications put messages onto MQSeries queues with the MQPUT call. In addition to the data itself, applications specify a descriptor for the message. This controls a number of aspects of the processing of the message within MQSeries. The descriptor travels with the message, and the information it contains may be used to control processing anywhere on its journey through the network. For example, in the event that the message eventually proves undeliverable, the system's behavior can be specified. In addition, the message descriptor holds values associated with the message, such as its priority, type, and persistence properties. A persistent message is one which survives queue manager restarts.

Applications can also control the details of the PUT operation itself, by using the options parameter. This is used to specify, for example, whether or not the operation is part of a logical unit of work. Also, applications can use the options to specify that the context associated with the message should be taken from a source other than the application itself. This is very useful for server applications, which may need to send messages to other servers and have them processed in the context of the original request rather than that of the server itself. As we saw in Sec. 11.3.3, the context of a message includes information such as the identifier of the user associated with it.

12.1.1.6 MQGET—get a message from an MQSeries queue. Applications retrieve messages from MQSeries queues with the MQGET call. The parameters to this call are very similar to those for MQPUT. Along with the data, the message descriptor associated with the message is returned.

Once again, the application can control the details of processing using an options parameter. For example, the options control whether MQSeries should return immediately if there is no message available or wait for a message to arrive. They also control a number of other aspects, including whether or not the operation is part of a logical unit of work. In addition, many MQSeries products support conversion of the data in the message between the format used on the machine from which it originated and that used on the machine receiving it. If so, that conversion can be requested via the options parameter.

12.1.1.7 MQPUT1. This call can be thought of as a combination of the MQOPEN, MQPUT, and MQCLOSE calls. It puts a single message on a named queue, which does not have to have been opened. It is a convenience, reducing the amount of code necessary for applications which write only a single message to a particular queue. Since the queue is opened and closed automatically every time this call is made, applications which put many messages to the same queue should use the separate MQOPEN, MQPUT, and MQCLOSE calls instead.

The MQPUT1 call takes a set of parameters which is a combination of those for MQOPEN and MQPUT, allowing the application to have full control over the operation just as it would for the separate calls. Since each MQPUT1 call involves an implicit MQOPEN call and an implicit MQCLOSE call, applications which output multiple messages to a queue should open it explicitly and use the MQPUT call.

12.1.1.8 MQINQ—inquire about the attributes of an MQSeries object. The MQINQ call allows the current values of attributes of MQSeries objects to be queried by an application. MQINQ can be used to inquire on the attributes of queues, name lists, process definitions, and the queue manager itself. For example, this call can be used to determine how many messages are currently waiting on a particular queue. The call allows multiple attributes for a given object to be retrieved together. Also, some attributes have numeric values while others are represented as character strings. Consequently, the interface is a little unusual, involving two arrays and a character buffer, together with information on their size. The first array contains *selectors*. Each of these is an integer representing the particular attribute whose value is to be returned. The second array is space in which integer attributes will be returned. The buffer provides space for the values of character attributes to be returned. Character attributes are returned as blank padded, fixed-length character arrays concatenated together in the buffer. A single call to MQINQ might, for example, request five integer attributes and three character attributes. The integer attributes will be returned in the array provided by the application, and the character attributes will be placed in the character buffer, concatenated together.

The interface to MQINQ and to the related routine MQSET is designed to maximize the flexibility of manipulation of attribute values. For example, the syntax of these calls will not need to change simply because more attributes or more kinds of objects may be defined in future versions of MQSeries.

12.1.1.9 MQSET—set the attributes of an MQSeries object. The MQSET call allows the current values of attributes of MQSeries objects to be altered by an application. For example, it can be used to specify the number of messages which must be waiting on a queue before triggering occurs. The call allows several updates to be processed together. As with MQINQ, since some attributes have numeric values and others are represented as character strings, the interface to MQSET also involves two arrays and a character buffer, together with information on their size. Once again, the first array contains selectors. Each of these is an integer representing the particular attribute whose value is to be updated. The second array contains the values which will be used to update the integer attributes. The buffer contains the values which will be used to update the character attributes. These are concatenated together. A single call to MQSET might, for example, update two integer attributes and a character attribute.

12.1.1.10 MQCMIT—commit operations performed under syncpoint. MQSeries allows operations to be grouped into local logical units of work. The results of such groups of operations are not made permanent until the logical unit of work is committed. MQCMIT is the call used to commit such a unit of work. Effectively, operations carried out under syncpoint are pended until the commit occurs. Messages retrieved from queues are not deleted, but are made unavailable to other applications until the commit decision has been made. Likewise, messages put onto queues are stored, but are cannot be retrieved until the commit has been processed.

MQSeries products can also support operations being carried out as part of a distributed unit of work, under the control of a transaction manager. We will look at how how this is achieved in Sec. 12.5.

12.1.1.11 MQBACK—backout operations performed under syncpoint. The MQBACK call partners the MQCMIT call. It is used to back out a series of operations performed within a local logical unit of work. Applications call this function when a logical unit of work is being backed out. In response, the queue manager restores the state of any affected objects to the state that existed before the unit of work started. The result is as if the changes made as part of the logical unit of work had never occurred.

12.1.2 Extending the function—MQI options parameters

In this section, we look in a little more detail at the way in which the MQI operations can be tailored by the options that can be specified.

12.1.2.1 Options on MQOPEN. When opening an MQSeries object, particularly if it is a queue, the *options* parameter allows applications to specify the kind of operations they wish to perform. The following list describes some of the possibilities:

Open for input and allow the queue to be shared

Open for input exclusively for this application

Open for browsing

Open for output

Open to allow inquiry of attributes

Open to allow attributes to be updated

Open using the default options associated with the object

Some of these options are naturally exclusive. Clearly, it is not valid to specify that a queue be opened for shared read access and exclusive read access at the same time. However, it is perfectly permissible for a queue to be opened for both input and output simultaneously as well as for inquiry.

The options parameter also allows applications to request the ability to manipulate the context associated with messages which they will place on queue objects. This allows applications to send messages which will process using the context of messages they themselves received. This is useful to allow, for example, the user identifier associated with the original request to be propagated throughout all the programs involved in processing it. Applications need authority to be able to pass context, since otherwise it would be trivial to bypass security checks by claiming to be a user with the appropriate authority. An application's ability to set context is verified when it opens the queue in question. Its request to be granted the ability to manipulate context is specified by options such as:

Allow context to be specified by the application

Allow context to be set from the most recently retrieved message on another queue

An application can also request that the MQOPEN call be checked for valid authority using a user identifier other than the one under which the application is running. This allows a server to open an object with the access rights of the user submitting the request, rather than with those of the server itself.

The ability to change context and to use alternate user identifiers is dependent on the application itself having been granted authority to do so. If the application does not have sufficient authority, the MQOPEN call fails.

Finally, an application can request to be notified if the queue manager is in the process of normal shutdown when it opens an object. During a normal shutdown, the queue manager prevents new connections but allows all connected applications to complete normally before closing itself down. Most applications will probably not want to continue if the queue manager is about to stop. However, some critical operations might need to be completed before the queue manager terminates. Consequently, applications can choose whether or not the MQOPEN call should fail if the queue manager is in the process of quiescing.

12.1.2.2 Options on MQCLOSE. The options on MQCLOSE are associated with dynamic queues. Dynamic queues are queues created during the execution of an application. Such queues may be temporary, in which case they are deleted when the application terminates, or they may be permanent, in which case some method of subsequently removing them is required. The method provided by the MQI relies on options used when the queue is closed. One option will delete the queue as long as there are no messages on it and it is not in use. A second option purges any remaining messages and deletes the queue.

12.1.2.3 Options on MQGET. So far, the options we have looked at have been simple flag values specifying the detailed behavior of the MQI. The options on MQGET are a little more complex. They consist of a data structure containing both flags and value fields, some of which may be modified by the queue manager. The basic options include the following:

Wait for a message to arrive if none is available

Return immediately if there is no message available

The operation is part of a logical unit of work

Browse the first message on the queue

Browse the next message on the queue

Retrieve the message at the current browse position

Lock message

Unlock message

Accept truncated data

Convert the message data

Fail if queue manager is quiescing

The browse options allow an application to scan the messages on the queue without removing them. This way, an application can search the queue for a message meeting a particular set of criteria. The lock option, used with browse, allows an application to reserve each message that it browses. This way, it can be sure that, even if several applications are reading the same queue, the message will be available for it to read from the queue should it decide to do so. The option to accept truncated data allows an application to read the message from the queue even if the buffer it supplied to contain the message data is too short. The final option in the list allows the MQGET operation to fail if the queue manager is in the process of shutting down. This option has the same behavior as when used on MQOPEN.

In addition to these options, the get message options structure includes a field in which the length of time to wait for a message to arrive can be specified. This wait interval is specified as an integer number of milliseconds, allowing quite short delays to be requested.

Finally, the get message options structure contains a field in which the queue manager returns the resolved name of the queue from which the message was actually read. This will be different from the name specified in the MQOPEN call if the queue which was opened was an alias queue, for example.

12.1.2.4 Options on MQPUT. As with the options used when getting messages, the options on MQPUT consist of a structure containing flags and values. Once again, some of these values are returned by the queue manager. The basic options include the following:

The operation is part of a logical unit of work

There is no context associated with the message

Context should be taken from the specified queue

Context should be taken from the message descriptor

Fail if queue manager is quiescing

When context is to be taken from a queue, a valid object handle for the queue must also be specified in the appropriate field of the put message options. If this is done, the context associated with the message most recently read from that queue will be used on the message being put. If the application is setting the context itself, it fills in the appropriate fields of the message descriptor, which we will examine shortly.

Once again, the application can request that the operation fail if the queue manager is in the process of shutting down.

Finally, the put message options structure contains fields in which the queue manager returns the resolved name of the queue on which the message was put and the resolved name of the queue manager holding that queue.

12.1.3 Descriptors

In this section, we look at the two descriptors used in the MQSeries application programming interface.

12.1.3.1 Message descriptor. The *message descriptor* contains information which travels with the message across the network to its destination. The first major element of the descriptor is a set of options flags describing the kinds of report messages which are to be associated with the message. Report messages can be generated in response to various events which may occur during the life of the message within the network. The following list shows some of the events:

Deletion of the message because it was undelivered when its expiry time passed

Arrival of the message on its destination queue

Delivery of the message to an application

Occurrence of an exception

An exception might be caused, for example, if after traversing the network, the destination queue for the message is already full when it arrives.

In addition to the generation of report messages, the message descriptor also allows applications to control some of the data which those reports will contain. For example, it is possible to request that the message identifier of the message generating the report be used as the correlation identifier in the report message. This allows an application receiving the report message to identify the particular message which gave rise to the report. Similarly, the message identifier associated with the report can be taken from the message causing the report, if desired.

The message descriptor holds the message type associated with the message. This may contain a system value, indicating that the message is a request message, a datagram, and so on, or it may contain an application-defined value. Application-defined values can be used to distinguish messages of different types on a single queue. By browsing the queue, an application could, for example, retrieve only messages of a particular type.

The priority associated with the message is also held in the message descriptor. It can be given an explicit value, or it can be specified as taking the default value for the queue.

The expiry time for the message is held in the message descriptor. It can be set to unlimited, meaning that the message will never expire and will remain in the system indefinitely, even if undelivered. Alternatively, it can be used to limit the lifetime of the message. If the message has not been read from its destination queue when its lifetime expires, it becomes eligible to be discarded. If it is discarded, and the appropriate report option is set, a report message is generated. This message is sent to the destination specified in the original message as being that to which replies should be sent. Often, this will be the original source of the message.

The message descriptor also includes information used when converting the message data. Conversion is necessary if the machine architectures of the sending and receiving systems differ. It is also necessary if the national language of the systems differ. The information held in the message descriptor includes details of the encoding applied by the machine on which the message originated. It also includes details of the character set used to encode character data. Since knowledge of the structure of the data in the message is necessary to allow conversion to be carried out, a format name is also sent in the message descriptor. This allows the conversion service in the queue manager to apply the appropriate conversions. Some format names are reserved. They are used to define messages which queue managers themselves use to communicate with one another.

The message descriptor contains a field indicating the persistence properties of the message. Persistent messages survive queue manager restarts, whereas nonpersistent messages do not. As with priority, each queue has a default value for persistence and messages can be specified to use this value.

The message identifier field of the message descriptor allows a specific value to be associated with the message. This is normally used to identify reply messages and to allow them to be associated with the original message which gave rise to them. Applications can specify this field or allow the queue manager to generate a unique value for it. In addition to the message identifier, the correlation identifier is also in the message descriptor. This identifier is usually used to associate messages with one another. Both the message identifier and the correlation identifier can be used independently or together to select which messages are retrieved during MQGET operations.

To enable correct routing of the messages which result from processing of a particular message, the message descriptor also contains fields in which the queue name and queue manager name for replies to the message can be specified. If the queue manager name is not given explicitly, the queue manager from which the reply will be sent must be able to determine the target queue manager name from its own tables in the usual way.

Context information travels with the message in the message descriptor. The message context includes information about the user and the application creating the message, and the date and time when the MQPUT operation occurred.

12.1.3.2 Object descriptor. The object descriptor is used when opening an object with the MQOPEN call or when putting a single message on a queue with MQPUT1. This descriptor defines the name and type of the object to be used. The object type could, for example, be queue, queue manager, or process definition. If the object does not belong to the local queue manager, the name of the queue manager owning the object is also specified in the object descriptor.

To allow applications to open objects using authority other than that associated with the application itself, the object descriptor contains a field which can be used for an alternate user identifier. This will be used in authority checking if the appropriate flags were set in the open options and if the application has appropriate authority to request use of an alternate identifier.

To support the use of dynamic queues, the object descriptor contains a field to hold a dynamic queue name. To open a dynamic queue, the appropriate model queue is specified in the object name field and the name for the new dynamic queue is given in the dynamic queue name field.

12.2 Using Basic MQI Functions

In this section, we look at the use of the basic MQI functions, including connecting to the queue manager, opening the appropriate queue, putting messages onto queues, and getting messages from queues. In a later section, we will look at more complex operations, including triggering and the use of logical units of work.

To provide a framework for the discussion, we will use a simple application program. This program transfers small files. Although not particularly representative of typical MQSeries applications, it does illustrate the basic operations in a very simple way. The example is written in C. However, the code is simple and the explanatory text is written with the needs of those who do not program in C in mind.

12.2.1   The file transfer example application

The file transfer application has a single function. It transfers a file from one place to another. To do this, it makes use of two MQSeries application programs. One, called mqftp, reads the file and sends its data as an MQSeries message. The other, mqftpr, reads these MQSeries messages and writes them to the destination files.

To perform the transfer, several pieces of information are required. First, the name of the file to be sent is needed. Second, the name to be used to store the file in its new location is required. Third, the destination queue manager needs to be known. Finally, the name of the queue to be used must be known.

To get this information, the application takes the following approach. The name of the file to be transferred, its name after the transfer, and the target queue manager are all taken as input parameters to the sending program, mqftp. The queue to be used is fixed. The receiver program, mqftpr, always reads a particular queue. File data is always sent on that queue.

We said that this application is not representative of typical MQSeries applications. It is worth examining one of the reasons here, before we look at the source code in detail. One of the main reasons that the example application is untypical is that it uses a fixed queue name. Also, the user must specify the queue manager to which the file will be transferred. While this kind of information is required to transfer files to remote computer systems and, indeed, is very similar to that needed by the standard TCP/IP-based ftp application, it is unusual for MQSeries. Many MQSeries applications do not care which queues they service. In addition, when replying to requests, all the information necessary to determine where to send replies is contained in the input messages they receive. MQSeries provides the means by which application code can be completely independent of the particular queues being used. Triggering, for example, removes the need even to know which input queue is to be serviced. Exploiting these features of MQSeries makes for more flexible systems and obviates the need to change program code in order to alter the flow of messages through it. However, for simplicity in the example application code, we have ignored these advantages of MQSeries to allow us to give concrete examples of the use of the basic MQI functions.

It is time to look at the first component of the application in detail. We will start with the sender program.

12.2.2   The sender program mqftp

The sender program is shown in Fig. 12.1. At the top of the listing is a block of comments identifying the program. The real code starts immediately after the comment labeled (A). The lines which begin #include cause the compiler to read the files mentioned to access standard definitions frequently used in C programs. These definitions allow the use of standard C facilities, for example, reading and writing files. The #include statement after point (B) in the code causes the compiler to read the cmqc.h file, which contains all the definitions required for an application to use the MQI.

At point (C) in the code, the #define statements associate literals with specific values. For example, the compiler will associate MAX_FILE_SIZE with the actual value 4096. This is the value we use to limit the size of the files which are transferred by the program. Consequently, this application will transfer whole files up to 4096 bytes in length, or the first 4096 bytes of any longer files.

The next section of code, at point (D), defines the correct usage for the function named TerminateProg. We will see the implementation of the routine later. This function gathers together the actions needed to end the program cleanly, whether normally or in error. The C language allows the return value and the parameters to a function to be defined separately from the function itself. That is being done here. The definition is known as a *function prototype*. It allows the compiler to verify that, elsewhere in the code, the function is being invoked correctly. It is especially useful when the implementation and the call are in separate compilation units. In this case, the definition shows that the function takes four parameters. The first is a character string called endmsg. The specification char* indicates that endmsg is a pointer to a character. This is the normal way to reference strings in C. A string is a sequence of characters. The end of the string is marked by a character with a binary value of zero. The second parameter to TerminateProg is of type PMQHOBJ. This is a type defined by the MQI header file, cmqc.h, and is used to hold a pointer to an object handle. This is the kind of handle used to identify an object, such as a queue, to a queue manager. The third parameter to TerminateProg is of type PMQHCONN. This is also a type defined by the MQI header file, and is used to hold a pointer to a connection handle. This is the kind of handle used to identify a connection to a queue manager. The final parameter is of type MQLONG. Once again, this is a type defined by the MQI header file. It represents the completion code of the program.

The main entry point to the program is at point (E) in the code. This is where the code starts to execute. The definition of the main entry point is fixed for C programs. The parameters are argc, a count of the number of parameters passed to the program, and argv, an array of character strings, one for each parameter. As we shall see shortly, the program will use these strings to identify the file to be copied and its new name, as well as the queue manager to which it will be copied.

```
/*****************************************************************/
/* Module Name: mqftp.c                                         */
/*                                                              */
/* Function:   This program is the sender half of the file transfer */
/*             example. It reads the specified file and sends its   */
/*             contents as a series of messages to the              */
/*             File.Transfer.Queue on the specified queue manager   */
/*                                                              */
/* Usage:      Invoked from command line as                     */
/*                                                              */
/*  mqftp <queue_manager> <from_file> <to_file>                 */
/*                                                              */
/* where                                                        */
/*  <queue_manager>  is the queue manager to send the file to   */
/*  <from_file>      is the file to send                        */
/*  <to_file>        is the name to use for the new copy of the file */
/*                                                              */
/* NOTES:                                                       */
/* 1) This code is only an example. For simplicity it performs only */
/*    very basic error checking and has a crude user interface. */
/*****************************************************************/

/*****************************************************************/
/* (A) Include definitions for standard C language functions    */
/*****************************************************************/
#include        <stdio.h>
#include        <stdlib.h>
#include        <string.h>

/*****************************************************************/
/* (B) Include definitions for MQSeries                         */
/*****************************************************************/
#include        <cmqc.h>

/*****************************************************************/
/* (C) Define constants used in this program                    */
/*****************************************************************/
#define MAX_FILE_SIZE   4096

/*****************************************************************/
/* (D) Prototypes for local functions                           */
/*****************************************************************/
void TerminateProg(char     *endmsg,
                   PMQHOBJ  pQueue_Handle,
                   PMQHCONN pConnection_Handle,
                   MQLONG   End_Reason) ;

/*************************/
/* (E) Main Entry Point  */
/*************************/
int main(int argc, char**argv)
{
    /*****************************************************************/
    /* (F) Declare Structures used in the MQ API calls. Also, initialize */
    /* the descriptors and options.                                 */
    /*****************************************************************/
    MQOD            Queue_Descriptor    = {MQOD_DEFAULT} ;
    MQMD            Message_Descriptor  = {MQMD_DEFAULT} ;
    MQPMO           Put_Message_Options = {MQPMO_DEFAULT} ;
    MQHCONN         Connection_Handle   = MQHC_UNUSABLE_HCONN ;
```

Figure 12.1   The sender program for the file transfer example.

```
MQHOBJ        Queue_Handle        = MQHO_UNUSABLE_HOBJ ;
MQLONG        Open_Options ;
MQLONG        CompCode ;
MQLONG        Reason ;

/***********************************************************************/
/* (G) Declare other variables used in the program                   */
/***********************************************************************/
FILE          *fp ;
int           nbytes ;
char          *qmgr_name ;
char          *from_file ;
char          *to_file ;

struct
{
  char          to_file[128] ;
  unsigned int  Data_Length ;
  MQBYTE        Buffer[MAX_FILE_SIZE] ;
} F_Transfer_Msg ;

/***********************************************************************/
/* (H) Retrieve command line arguments.                              */
/*     argv[1] is the name of the target queue manager               */
/*     argv[2] is the name of the file being sent                    */
/*     argv[3] is the name of the to_file                            */
/***********************************************************************/
if (argc != 4)
{
  printf("Usage:\n") ;
  printf(" mqftp queue_manager_name from_file to_file\n") ;
  exit(16) ;
}
qmgr_name = argv[1] ;
from_file = argv[2] ;
to_file   = argv[3] ;

/***********************************************************************/
/* (I) Open the file to be transferred                               */
/***********************************************************************/
fp = fopen(from_file, "r") ;
if (fp == NULL)
{
  printf("Could not open from_file %s\n", from_file) ;
  exit(16) ;
}

/***********************************************************************/
/* (J) Connect to the local queue manager (i.e. default)             */
/***********************************************************************/
MQCONN ("",
        &Connection_Handle,
        &CompCode,
        &Reason) ;
if (CompCode == MQCC_FAILED)
{
  TerminateProg("Could not connect to default queue manager",
                &Queue_Handle,
                &Connection_Handle,
                Reason) ;
```

Figure 12.1 *(Continued)*

```
  exit(16) ;
}

/**********************************************************************/
/* (K) Construct the descriptor for the file transfer queue      */
/**********************************************************************/
strncpy(Queue_Descriptor.ObjectName,
        "File.Transfer.Queue",
        MQ_Q_NAME_LENGTH) ;

strncpy(Queue_Descriptor.ObjectQMgrName,
        qmgr_name,
        MQ_Q_MGR_NAME_LENGTH) ;

/**********************************************************************/
/* (L) Open the queue on which we will send the data            */
/**********************************************************************/
Open_Options = MQOO_OUTPUT ;
MQOPEN (Connection_Handle,
        &Queue_Descriptor,
        Open_Options,
        &Queue_Handle,
        &CompCode,
        &Reason) ;

if (CompCode == MQCC_FAILED)
{
  TerminateProg("Could not open File.Transfer.Queue",
                &Queue_Handle,
                &Connection_Handle,
                Reason) ;
}

/**********************************************************************/
/* (M) Send the content of the file                            */
/**********************************************************************/
strncpy(F_Transfer_Msg.to_file,
        to_file,
        sizeof(F_Transfer_Msg.to_file)) ;


/**********************************************************************/
/* (N) Read the data from the file                             */
/**********************************************************************/
nbytes = fread(F_Transfer_Msg.Buffer, 1, MAX_FILE_SIZE, fp) ;
if (nbytes == MAX_FILE_SIZE)
{
  printf("WARNING: Copy of file may have been truncated\n") ;
}

if (nbytes != 0)
{
  /**********************************************************************/
  /* (O) Send the file data via the queue                        */
  /**********************************************************************/
  F_Transfer_Msg.Data_Length = nbytes ;

  MQPUT (Connection_Handle,
         Queue_Handle,
         &Message_Descriptor,
```

Figure 12.1 *(Continued)*

```
              &Put_Message_Options,
              (sizeof(F_Transfer_Msg) - MAX_FILE_SIZE + nbytes),
              &F_Transfer_Msg,
              &CompCode,
              &Reason) ;

    if (CompCode == MQCC_FAILED)
    {
      TerminateProg("Could not send file data",
                    &Queue_Handle,
                    &Connection_Handle,
                    Reason) ;
    }
  }

  /****************************************************************/
  /* (P) Close the file                                         */
  /****************************************************************/
  fclose(fp) ;

  /****************************************************************/
  /* (Q) Terminate with all work finished                       */
  /****************************************************************/
  TerminateProg("Processing Completed Normally",
                &Queue_Handle,
                &Connection_Handle,
                MQRC_NONE) ;

  return(MQRC_NONE) ;
}
/****************************************************************/
/* (R) Function to terminate the program                      */
/****************************************************************/
void TerminateProg(char     *endmsg,
                   PMQHOBJ  pQueue_Handle,
                   PMQHCONN pConnection_Handle,
                   MQLONG   End_Reason)
{
  MQLONG CompCode ;
  MQLONG Reason ;
  MQLONG Close_Options ;

  /****************************************************************/
  /* (S) Issue the message associated with termination          */
  /****************************************************************/
  printf("Sender: %s. Reason was %ld\n",
         endmsg,
         End_Reason) ;

  /****************************************************************/
  /* (T) Close the queue                                        */
  /****************************************************************/
  if (*pQueue_Handle != MQHO_UNUSABLE_HOBJ)
  {
    Close_Options = MQCO_NONE ;
    MQCLOSE (*pConnection_Handle,
             pQueue_Handle,
             Close_Options,
             &CompCode,
```

Figure 12.1 *(Continued)*

```
            &Reason) ;

    printf("Sender: Queue Close Reason Code was %ld\n",
          Reason) ;
}
/*********************************************************************/
/* (U) Disconnect from the local queue manager                    */
/*********************************************************************/
if (*pConnection_Handle != MQHC_UNUSABLE_HCONN)
{
    MQDISC (pConnection_Handle,
           &CompCode,
           &Reason) ;

    printf("Sender: Disconnect Reason Code was %ld\n",
          Reason) ;
}
  exit(End_Reason) ;
}.
```

Figure 12.1 *(Continued)*

At point (F) are declarations of data structures used within the main program itself. For example, the first line declares an object descriptor. Its type, MQOD, is defined in the MQI header cmqc.h. The structure contains fields which we discussed in Sec. 12.1.3.2. Because we will use it when manipulating the queue on which we will send the messages, we call the variable itself Queue_Descriptor. In this declaration and the two which follow it, the structures are initialized with default values. These values are defined in the MQI header file, cmqc.h. In the case of the object descriptor, the default values are represented by the literal MQOD_DEFAULT, defined in cmqc.h. The braces which surround the literal are a required part of the C language syntax for initialization of a variable which is a structure. This method of assigning a default value to an MQI structure can be used only where the variable is being declared. Later in the program, if we should need to reset individual fields within the structures, we will need to assign values explicitly.

We initialize the Connection_Handle variable to the value MQHC_UNUS-ABLE_HCONN, defined by the MQI header file, showing that we do not yet have a usable connection. Likewise, we initialize the Queue_Handle. During program termination, we can test for these values to see if we managed to open the queue or if we managed to connect to the queue manager.

The next section of the code, at point (G), declares other variables used in the program. The variable F_Transfer_Msg is a type we have not met before. It is known as a *structure,* which is why it begins with the keyword struct. A structure is a collection of individual variables which are treated as a single entity. F_Transfer_Msg is the structure which holds the information which we will send as an MQSeries message when transferring a file. Within F_Transfer_Msg there is a field called to_file in which the name of the target file will be placed. This is an area 128 bytes in length. The field Data_Length will hold the amount of file data being transferred. Finally, the Buffer field, an area MAX_FILE_SIZE in length, will hold the data from the file itself.

Program execution proper starts at (H), where the command line arguments are retrieved. The test on the count of arguments, argc, establishes whether the correct number has been supplied. In C, the term != is used to mean "not equal to." The count is always one more than the actual number of parameters. The first argument is always the name of the program itself. If the count is wrong, the program prints a simple message about how it should be used. The printf function is used to do this. In this form, it prints two simple character strings. The notation \n, used within the strings, causes a new line to be started after the values have been printed.

The lines which follow the if statement cause the variables qmgr_name, from_file, and to_file to be set from the corresponding input parameters. The notation, involving square brackets, used in these assignment statements selects the appropriate element of the argv array to assign to each variable. At (I), the program attempts to open the file to be copied. The name of the file is in variable from_file. The fopen function will open the file and return a file pointer if successful. The value of this pointer will be stored in variable fp. The second parameter to fopen indicates the mode in which the file is being opened. The value "r" indicates that the file is being opened for reading. If fopen fails, the value of fp is set to the special value NULL. The program detects this with the if statement at point (I). The term == is used in C to express equality in tests, whereas = is used to specify assignment. If the file could not be opened, the program prints an error message, using printf, and terminates, passing a return code of 16 to the operating system. The printf for the error message illustrates the way in which values as well as literals can be output. The term %s within the string indicates a substitution point. In this case, it is for the string variable from_file. Consequently, the error message will contain the name of the file which could not be opened.

At (J), we finally reach code which calls the MQI. The first call has to be to connect to the local queue manager. The first parameter to MQCONN is the name of the queue manager. As this is an empty string, we are requesting connection to the default local queue manager. The second parameter is the connection handle we declared earlier. This is filled in by the MQCONN call and will allow us to identify the connection on subsequent MQI calls. The leading ampersand on the reference to this parameter, and indeed to CompCode and Reason as well, indicates that the address of the variable is being passed into MQCONN. This is necessary because of the mechanism which C uses to update parameters passed in by callers. The final two parameters appear on all MQI calls. CompCode is the completion code from the result and Reason is an associated code giving more detail about any error that may have occurred. For example, if CompCode contains the value MQCC_WARNING, Reason might contain MQRC_ALREADY_CONNECTED, indicating that a request to connect has been ignored because the application is already connected to the queue manager. This is only a warning. Execution can continue, and the connection handle returned will be valid. On the other hand, if CompCode contains the value MQCC_FAILED, it indicates that a real error has occurred. Reason, in this case, might contain MQRC_Q_MGR_STOPPING, indicating that the queue manager is refusing connections because it is about to stop. If the connection call fails, MQSeries sets the connection handle to the value MQHC_UNUSABLE_HCONN to mark it as incapable of being used.

If the connection operation does fail, for any reason, the program ends by calling the TerminateProg function. We will see how this function works later. In essence, it issues the message passed as its first parameter, ends the use of any MQSeries resources in use, and finally exits from the program.

The next block of code, at point (K), sets up two of the important values for the object descriptor to be used to open the queue. The name of the queue and the name of the queue manager are filled in. Although the program is connected to the default queue manager, the queue to be opened can be on any queue manager which is known about. Remember that the queue manager to which we are connected has tables which allow it to route messages to the correct destination. As long as there are appropriate definitions which specify where the messages are to be sent, the open call will succeed. The mechanism to set the names involves the standard function strncpy. This copies character strings, but limits the maximum length which can be transferred. In this case, the queue name is a literal. The queue is called File. Transfer.Queue. The maximum length of queue names is given by MQ_Q_NAME_LENGTH. Once again, this is defined in cmqc.h. The queue manager name, on the other hand, is one of the input parameters. We saved it earlier in the variable qmgr_name. The maximum length of a queue manager name is MQ_Q_MGR_NAME_LENGTH.

Now that we have established the object descriptor values, we can open the queue. The MQOPEN call is at point (L). The open options are set to MQOO_OUTPUT, requesting that we can put messages onto the queue. The first parameter to the MQOPEN call itself is the connection handle we received when connecting to the queue manager. The second parameter is the object descriptor which specifies the queue we want to open. Next come the open options and then the address of the variable which the MQOPEN call will update with a handle for the object, if the call is successful. Once again, the final pair of parameters are the completion code and reason. If the open call fails, we invoke the `TerminateProg` routine, as before, to end the program.

At point (M), we start to set up variables for the transfer of the file. The `strncpy` function is used to copy the variable `to_file`, containing the name of the file to be copied to, into the `to_file` field of the `F_Transfer_Msg` data structure, which represents the message. The notation `F_Transfer_Msg.to_file` specifies this field of the structure.

At point (N), we read the data from the file into the field `Buffer` of the message. The `fread` function does this. Variable `fp` is the file pointer we received when we opened the input file back at point (I). The term `MAX_FILE_SIZE` specifies the maximum number of bytes we want `fread` to transfer from the file. The literal 1, which precedes it in the parameter list to `fread`, indicates that specification of how much data can be transferred is in bytes. Variable `nbytes` is set by the call to the number of bytes actually transferred from the file. If this count is less than `MAX_FILE_SIZE`, it means we have read the entire file. However, if the count equals `MAX_FILE_SIZE`, there may be more data in the file than we can transfer. In this case, we issue a warning about possible file truncation by using `printf`. The text of the warning is slightly vague, since, if the file is exactly `MAX_FILE_SIZE` in length, we will issue the warning without truncating the file. There are ways to distinguish these cases, but, as this is an example of MQSeries programming rather than of file handling, we will not concern ourselves with them here.

If data has been read from the file, we issue the `MQPUT` call at point (O). We use the connection handle `Connection_Handle` and the object handle `Queue_Handle` given to us by MQSeries in response to previous successful calls. We use the default message descriptor and put message options which we initialized back at point (F). The expression `(sizeof(F_Transfer_Msg) - MAX_FILE_SIZE + nbytes)` deserves some study. We need to tell MQSeries the total size of the message being sent by the `MQPUT` call. We could simply specify the total size of the `F_Transfer_Msg` structure. However, for files smaller than `MAX_FILE_SIZE`, this results in the transfer of unused data across the network, which is wasteful. The expression evaluates the amount of data which must be transferred in order to send the contents of the file. It works like this. The term `sizeof(F_Transfer_Msg)` causes the compiler to calculate the number of bytes needed to represent the entire `F_Transfer_Msg` structure. This includes the `Buffer` field. If we subtract the size of `Buffer`, we are left with the size of the rest of the structure. Finally, adding `nbytes`, the size of the data we read from the file, yields the size of the part of `F_Transfer_Msg` which we do need to send. If the `MQPUT` operation fails, we end the program via `TerminateProg`. Once the file has been sent, at (P) we close the input file.

We have now finished the work the program has to do. We terminate by calling `TerminateProg` at point (Q). The value `MQRC_NONE` passed as return code is the one used by MQSeries when there has been no error.

The `TerminateProg` routine itself begins at point (R). The first statement defines the function entry point, specifying the parameters it expects. This is very closely related to the function prototype we saw back at point (D).

The first executable statement is at point (S), where `printf` is used to display the termination message passed to the routine as the `endmsg` parameter. This call to `printf` also displays the value of the `End_Reason` parameter. The specification `%ld` in the first parameter to `printf` tells it that `End_Reason` is a long integer, allowing `printf` to format and print it appropriately.

At point (T), we close the queue, if it was successfully opened. We do this by testing the value of the queue handle pointed to by the `pQueue_Handle` parameter. The value `MQHO_UNUSABLE_HOBJ` is the one we used to initialize the object handle originally. It is also the value which MQSeries uses to mark an object handle when it is not open, for example, after a failing `MQOPEN` call or after a successful `MQCLOSE`. We display the final return code from the `MQCLOSE` call using `printf`.

Although we have shown it in the example, it is not strictly necessary to close the open MQSeries resources explicitly this way, though it is recommended. Disconnection from the queue manager will implicitly close any open resources. Such implicit close operations use default values for the close options. As a consequence, there may be situations in which an application does need to issue `MQCLOSE` calls explicitly, because it needs to control the close options. Think of `MQCLOSE` in much the same way as you would think of using `fclose` or any other file closing function. It is not strictly necessary to use these functions, but it is good programming practice to do so wherever possible.

At point (U) we disconnect from the queue manager. Once again, we test the relevant handle to see if we did manage to connect. In this case, we check to see if the `pConnection_Handle` parameter has the value `MQHC_UNUSABLE_HCONN`. If not, we did manage to connect. The return code from the `MQDISC` call is displayed using `printf`.

Finally, the `exit` call stops the program and returns `End_Reason` to the operating system as the program's termination code.

12.2.3   The receiver program `mqftpr`

The listing of the receiver program is given in Fig. 12.2. The code from point (A) to point (F) is essentially identical to that in the sender program `mqftp`. At (G), where the additional variables used in the program are declared, we see the first real differences. We do not need variables to hold the name of the file being copied from or the queue manager name. There is a new variable called ended, however. This will be used to control a loop later in the program.

At point (H), the program verifies the command line with which it was started. In contrast to mqftp, this program takes no command line parameters. If it detects that any were supplied because argc is other than 1, it issues an error message via printf and terminates.

Connection to the default queue manager takes place at point (I), with any failure causing the program to terminate via TerminateProg.

At point (J), the object descriptor for the file transfer queue is constructed. In contrast to the way this is done in mqftp, only the queue name itself is specified here. The queue must be local to the mqftpr program. It is a rule within MQSeries that queues from which messages are being read are local to the application reading them. Consequently, we do not need to specify a queue manager when opening the queue.

The program opens the queue at point (K). The options for the MQOPEN call are set to MQOO_INPUT_EXCLUSIVE, requesting that only this program be allowed to read the queue. As before, any failure is handled by calling TerminateProg.

At point (L), the program starts to execute a loop. The while statement causes all the statements between the { immediately following it and the matching } just before point (Q) to be executed continuously while the expression in parentheses is true. We will see shortly the conditions under which ended becomes nonzero.

At point (M), a message is read from the input queue. The options for the MQGET call are set to cause the program to wait until a message is available. The length of time to wait is set to be 30 seconds. The value assigned to the WaitInterval field of the Get_Message_Options structure is in milliseconds. If no message becomes available within this period, the MQGET call will fail. The code immediately following the MQGET call detects the failure and will cause the loop to terminate by setting ended to 1. Of course, this does not actually take effect until the while statement is again executed. However, all of the rest of the processing in the loop is in the else clause of the if statement which tests for a failure in the MQGET call. Consequently, none of this code is executed if the MQGET fails.

The message identifier and correlation identifier in the message descriptor are also initialized before the MQGET call. This is necessary because they are input and output fields which are updated after a successful operation. Their values can affect subsequent calls if the message descriptor is reused, as it is in this program. If you are not explicitly using these identifiers, it is a good idea to get into the habit

```
/**************************************************************************/
/* Module Name: mqftpr.c                                                */
/*                                                                      */
/* Function:    This program is the receiver half of the file transfer */
/*              example. It reads messages from a queue, interpreting  */
/*              them as data to be placed into files. The program reads */
/*              File.Transfer.Queue on the local queue manager.         */
/*                                                                      */
/* Usage:       Invoked from command line as                           */
/*                                                                      */
/*   mqftpr                                                             */
/*                                                                      */
/* NOTES:                                                               */
/* 1) This code is only an example. For simplicity it performs only    */
/*    very basic error checking.                                       */
/**************************************************************************/

/**************************************************************************/
/* (A) Include definitions for standard C language functions            */
/**************************************************************************/
#include        <stdio.h>
#include        <stdlib.h>
#include        <string.h>

/**************************************************************************/
/* (B) Include definitions for MQSeries                                 */
/**************************************************************************/
#include        <cmqc.h>

/**************************************************************************/
/* (C) Define constants used in this program                            */
/**************************************************************************/
#define MAX_FILE_SIZE   4096

/**************************************************************************/
/* (D) Prototypes for local functions                                   */
/**************************************************************************/
void TerminateProg(char      *endmsg,
                   PMQHOBJ  pQueue_Handle,
                   PMQHCONN pConnection_Handle,
                   MQLONG   End_Reason) ;

/***************************/
/* (E) Main Entry Point    */
/***************************/
int main(int argc, char**argv)
{
    /**************************************************************************/
    /* (F) Declare Structures used in the MQ API calls. Also, initialize  */
    /* the descriptors and options.                                        */
    /**************************************************************************/
    MQOD          Queue_Descriptor    = {MQOD_DEFAULT} ;
    MQMD          Message_Descriptor  = {MQMD_DEFAULT} ;
    MQGMO         Get_Message_Options = {MQGMO_DEFAULT} ;
    MQHCONN       Connection_Handle   = MQHC_UNUSABLE_HCONN ;
    MQHOBJ        Queue_Handle        = MQHO_UNUSABLE_HOBJ ;
    MQLONG        Data_Length ;
    MQLONG        Open_Options ;
    MQLONG        CompCode ;
    MQLONG        Reason ;
```

Figure 12.2   The receiver program for the file transfer example.

```
/**************************************************************************/
/* (G) Declare other variables used in the program                      */
/**************************************************************************/
int             ended = 0 ;
FILE            *fp ;
int             nbytes ;

struct
{
  char          to_file[128] ;
  unsigned int  Data_Length ;
  MQBYTE        Buffer[MAX_FILE_SIZE] ;
} F_Transfer_Msg ;

/**************************************************************************/
/* (H) Verify command line                                              */
/*     There are no command line arguments                              */
/**************************************************************************/
printf ("Receiver Starting\n") ;

if (argc != 1)
{
  printf("Usage:\n") ;
  printf(" mqftpr\n") ;
  exit(16) ;
}

/**************************************************************************/
/* (I) Connect to the local queue manager (i.e. default)                */
/**************************************************************************/
MQCONN ("",
        &Connection_Handle,
        &CompCode,
        &Reason) ;
if (CompCode == MQCC_FAILED)
{
  TerminateProg("Receiver: Could not connect to default queue manager",
                &Queue_Handle,
                &Connection_Handle,
                Reason) ;
  exit(16) ;
}

/**************************************************************************/
/* (J) Construct the descriptor for the file transfer queue             */
/**************************************************************************/
strncpy(Queue_Descriptor.ObjectName,
        "File.Transfer.Queue",
        MQ_Q_NAME_LENGTH) ;

/**************************************************************************/
/* (K) Open the queue on which we receive data                          */
/**************************************************************************/
Open_Options = MQOO_INPUT_EXCLUSIVE ;
MQOPEN (Connection_Handle,
        &Queue_Descriptor,
        Open_Options,
        &Queue_Handle,
        &CompCode,
```

Figure 12.2 *(Continued)*

```
        &Reason) ;

if (CompCode == MQCC_FAILED)
{
  TerminateProg("Could not open File.Transfer.Queue",
            &Queue_Handle,
            &Connection_Handle,
            Reason) ;
}

/*********************************************************************/
/* (L) Read messages from the queue                                */
/*********************************************************************/
while (ended == 0)
{
  /*****************************************************************/
  /* (M) Retrieve a message from the queue                       */
  /*****************************************************************/
  Get_Message_Options.Options = MQGMO_WAIT ;
  Get_Message_Options.WaitInterval = 30000 ;
  memcpy(Message_Descriptor.MsgId, MQMI_NONE, sizeof(MQMI_NONE)) ;
  memcpy(Message_Descriptor.CorrelId, MQCI_NONE, sizeof(MQCI_NONE)) ;

  MQGET (Connection_Handle,
         Queue_Handle,
         &Message_Descriptor,
         &Get_Message_Options,
         sizeof(F_Transfer_Msg),
         &F_Transfer_Msg,
         &Data_Length,
         &CompCode,
         &Reason) ;

  if (CompCode == MQCC_FAILED)
  {
    ended = 1 ;
  }
  else
  {
    /*************************************************************/
    /* (N) Report what is happening                            */
    /*************************************************************/
    printf("Receiver: Transferring file to %s\n",
           F_Transfer_Msg.to_file) ;

    /*************************************************************/
    /* (O) Open the file to be transferred.                    */
    /*************************************************************/
    fp = fopen(F_Transfer_Msg.to_file, "w") ;
    if (fp == NULL)
    {
      printf("Receiver: Could not open output file %s\n",
             F_Transfer_Msg.to_file) ;
      printf("   Data will be discarded\n") ;
    }
    else
    {
      /***********************************************************/
      /* (P) Write out the data.                               */
      /***********************************************************/
```

Figure 12.2 *(Continued)*

```
        nbytes = fwrite (F_Transfer_Msg.Buffer,
                         1,
                         F_Transfer_Msg.Data_Length,
                         fp) ;
        if (nbytes != F_Transfer_Msg.Data_Length)
        {
          printf("Receiver: Problem writing to file %s\n",
                 F_Transfer_Msg.to_file) ;
        }
        fclose(fp) ;

      }
    }
  }    /* End while(ended = 0) */


/*****************************************************************/
/* (Q) Terminate the program                                  */
/*****************************************************************/
  TerminateProg("Processing Completed",
                &Queue_Handle,
                &Connection_Handle,
                Reason) ;

  return(Reason) ;
}

/*****************************************************************/
/* (R) Function to terminate the program                      */
/*****************************************************************/
void TerminateProg(char      *endmsg,
                   PMQHOBJ   pQueue_Handle,
                   PMQHCONN  pConnection_Handle,
                   MQLONG    End_Reason)
{
  MQLONG CompCode ;
  MQLONG Reason ;
  MQLONG Close_Options ;

  /*****************************************************************/
  /* (S) Issue the message associated with termination          */
  /*****************************************************************/
  printf("Receiver: %s. Reason was %ld\n",
         endmsg,
         End_Reason) ;

  /*****************************************************************/
  /* (T) Close the queue                                        */
  /*****************************************************************/
  if (*pQueue_Handle != MQHO_UNUSABLE_HOBJ)
  {
    Close_Options = MQCO_NONE ;
    MQCLOSE (*pConnection_Handle,
             pQueue_Handle,
             Close_Options,
             &CompCode,
             &Reason) ;

    printf("Receiver: Queue Close Reason Code was %ld\n",
           Reason) ;
```

Figure 12.2 *(Continued)*

```
}
/*********************************************************************/
/* (U) Disconnect from the local queue manager                     */
/*********************************************************************/
if (*pConnection_Handle != MQHC_UNUSABLE_HCONN)
{
  MQDISC (pConnection_Handle,
          &CompCode,
          &Reason) ;

  printf("Receiver: Disconnect Reason Code was %ld\n",
          Reason) ;
}
exit(End_Reason) ;
}
```

Figure 12.2 *(Continued)*

of resetting them just before making the call that uses the message descriptor. Programmers new to the MQI frequently fall foul of this aspect of the interface.

After a successful MQGET operation, at point (N) the program reports that it has received a request to transfer a file and displays the target file name, taken from the message which was received.

At point (O), the file is opened using fopen, as we saw in mqftp. The difference here is that the second parameter is specified as "w". This requests the ability to write to the file. If the file already exists, any data already in it is deleted. The name of the file to be opened is taken from the message structure. Any problems encountered in opening the file cause an error message to be issued but do not stop the program. The receiver program behaves like a server, staying active and processing many requests. Failure of a single request does not stop it executing.

At point (P), the file data is written out to the file from the Buffer field of the F_Transfer_Msg structure which now contains the message we obtained from the earlier MQGET call. The length of the file data is taken from the Data_Length field of the structure. This was filled in when the file was read, in the sender program mqftp. Any problems in writing the file data are reported via a printf call.

Just before point (Q) is the end of the while loop which began back at point (L). The program will remain in this loop until the variable ended becomes nonzero. This happens on any failure of the MQGET call at point (M), as we saw earlier.

At point (Q), TerminateProg is called to end the program. Since the reason that the loop ended is a failure in the MQGET call, there will always be a nonzero reason code, and this is passed to TerminateProg. Normal termination of this program is when no message has been received for 30 seconds. Under these conditions, the final reason code will be MQRC_NO_MSG_AVAILABLE.

At point (R) is the implementation of the TerminateProg function, which is essentially identical to that in mqftp. It prints the termination message, closes the queue if open, and disconnects from the queue manager if connected.

12.2.4 Running the example

To execute the file transfer example requires some resources to be defined on one or more queue managers. For the case where files are being transferred between machines, the receiver requires a local queue definition and the sender a remote queue definition. In addition, a channel needs to be defined between the machines. While less realistic, it is perfectly possible to run the example entirely within one queue manager by executing the sender and receiver on the same machine. This requires only that we define a single local queue. Figure 12.3 shows an MQSC script which defines a suitable queue. In this script, lines beginning with an asterisk are comments. The + characters at the end of lines indicate continuation.

The script contains a single DEFINE command creating a local queue called File.Transfer.Queue. This is the queue referenced by both sender and receiver programs. The REPLACE clause specifies that any existing definition of the queue be overwritten by this one. This is useful when experimenting with different queue attributes, though obviously it must be used carefully in a production environment. The LIKE clause requests that the queue inherit all the attributes from the existing queue called SYSTEM.DEFAULT.LOCAL.QUEUE. This default queue definition is normally created at the same time as the queue manager itself. A number of other default objects are created, including a remote queue, an alias queue, and a channel. These default objects have useful attributes set, making it easy to create other, similar objects. Indeed,

```
************************************************************************
*                                                                    *
* Script Name: mqftpsc.mqs                                           *
*                                                                    *
* Description: Define the resources needed by the file transfer      *
*              example program                                       *
*                                                                    *
************************************************************************
************************************************************************
* (A) FILE.TRANSFER.QUEUE                                            *
************************************************************************
   DEFINE QLOCAL('File.Transfer.Queue') REPLACE +
* The queue on which to base the definition
         LIKE('SYSTEM.DEFAULT.LOCAL.QUEUE') +
* Text description of the queue
         DESCR('Queue for file transfer example')
```

Figure 12.3 An MQSC script defining the resources for the file transfer example.

the only attribute we have changed is the textual description of the queue. The resulting queue will be enabled for PUT and GET operations, but only a single application will be able to open the queue for GET. This is the behavior we desire for the receiver program.

We create the queue, by running the MQSC program and specifying the commands in Fig. 12.3. Since the MQSC program is a command, the precise syntax used to run it is operating-system-specific. On UNIX or OS/2, for example, the syntax is:

```
runmqsc < mqftpsc.mqs
```

The command is runmqsc and the name of the file containing the MQSC script is mqftpsc.mqs. Users of PC systems and UNIX workstations will be familiar with the < symbol indicating that the command is to take its input from the specified file. As it executes, MQSC generates a report. By default, this appears on the terminal, though it can be redirected to a file or printer.

Once the queue has been created, we can execute the application programs. For example, to start the receiver on a UNIX system, the following command is all that is necessary:

```
mqftpr &
```

On OS/2, the equivalent is

```
start mqftpr
```

The difference here is simply in the syntax which these operating systems use to specify that the program be run in the background. This just means that while running, the receiver does not tie up the keyboard and display. This is important, because we need them in order to run the sender program.

Suppose that on a UNIX system, a user called jim wants to transfer a file to temporary disk storage. Also, suppose that the queue manager on which we defined File.Transfer.Queue is called qm1. The command issued would be:

```
mqftp qm1 /home/jim/file2go /tmp/newfile
```

On UNIX systems, the /tmp directory is usually provided for temporary work space for any user. This command copies file file2go from the user's own disk storage into the temporary space and renames it to newfile.

While the receiver program is running, any number of transfers can be initiated using the `mqftp` command. If the receiver is idle for 30 seconds, it stops.

If the receiver is not running when the sender is executed, the messages will accumulate on the queue. The sender program runs to completion normally. When the receiver is subsequently started, the transfers are completed normally. This is a very simple example of *time-independent processing*. It also illustrates that a server does not have to be available for a client to be able to execute.

Copying local files this way is, of course, not terribly useful. Operating systems have facilities for doing this sort of operation. However, the application works just as well if the receiver is on a remote machine. As we pointed out earlier, remote operation requires that a channel be available between the two queue managers involved. Although a discussion of the finer points of channel definitions is beyond the scope of this book, we will extend the file transfer example shortly when we look at triggered applications.

12.2.5 Programming style

The example was written to be easy to read. Because of this, there is considerable duplication between `mqftp` and `mqftpr`. By keeping the code as a single entity, each program can be read without the need for continual cross references to other programs. However, in real applications, common definitions, such as that for the `F_Transfer_Msg` structure and the value of `MAX_FILE_SIZE`, would be held in a single file. This file would be included by both the application programs, using a `#include` statement similar to those at point (A) in `mqftp`. Likewise, functions common to both programs would be placed in a separate file, compiled independently from the main programs and linked with them to form the final executable files. The `TerminateProg` function could easily be modified to be common for `mqftp` and `mqftpr`, and indeed also to the trigger monitor program we will look at shortly.

12.3 Triggered Applications

Having covered the basic use of the MQI, it is time to look at some more advanced topics. In this section, we will extend the file transfer application to make the receiver program be triggered into life by the arrival of work for it to do.

Triggering is the mechanism which MQSeries products provide to allow applications, and particularly servers, to be started when there is work for them to do. We looked at this mechanism in Sec. 11.3.2.

Basically, the idea is that when some condition is met on a particular queue, the queue manager writes a trigger message to an initiation queue. A special application program, called a *trigger monitor,* reads this message and uses the information it contains to start the appropriate program to process the queue. Many application queues can be triggered by a single trigger monitor application.

To convert the file transfer example to triggered operation, we need to do several things. We must change the definition of the File.Transfer.Queue to enable triggering. We must also define the initiation queue which will be used by the trigger monitor. We must create a process definition for the receiver program. This definition provides information the trigger monitor needs to start the receiver. Remember that, in the general case, the trigger monitor must be capable of starting many different applications to run against many different queues. Finally, we must write the trigger monitor application.

12.3.1   The new definitions

Figure 12.4 shows the new MQSC commands we need to run to define the resources for the triggered version of the application. At (A), the first definition in the file is for the file transfer queue. The definition now includes extra fields which establish how the triggering operates. The TRIGGER keyword enables triggering for the queue. The clauses TRIGTYPE(DEPTH) and TRIGDPTH(1). define the conditions under which triggering occurs. In this case, the appearance of the first message on the queue will cause a trigger message to be generated. However, once this message has been generated, triggering will be disabled. This prevents a flood of trigger messages causing copies of the receiver application to be started. Remember that the receiver program does not terminate as soon as it has finished transferring a file. Instead, it waits for 30 seconds to see of there is more work to do. As we shall see later, the application reenables triggering on the queue just before terminating, ensuring that the subsequent arrival of work will cause it to be started once more. There are many options related to triggering. This particular combination is very convenient for simple server-type applications with a sporadic work load.

The queue definition contains three further clauses related to triggering. The TRIGDATA clause defines data which is passed from the queue definition to the trigger monitor when the queue is triggered. It has no meaning as far as the queue manager is concerned. It could be used to pass specific information to an application which might be triggered to process one of a number of different queues. The PROCESS clause specifies that the details of the application to be started when this queue is triggered will be found in a process definition called MQFTPR2. We will look at the process definition in detail shortly. Finally,

```
*************************************************************************
*                                                                       *
* Script Name: mqftpsc2.mqs                                             *
*                                                                       *
* Description: Define the resources needed by the file transfer        *
*              example program with triggering enabled                  *
*                                                                       *
*************************************************************************
*************************************************************************
* (A) FILE.TRANSFER.QUEUE                                               *
*************************************************************************
  DEFINE QLOCAL('File.Transfer.Queue')  REPLACE                      +
* The queue on which to base the definition
         LIKE('SYSTEM.DEFAULT.LOCAL.QUEUE')                          +
* Text description of the queue
         DESCR('Queue for file transfer example with triggering')    +
* Enable triggering for this queue
         TRIGGER                                                     +
* Trigger when first message arrives, then disable triggering
         TRIGTYPE(DEPTH)                                             +
         TRIGDPTH(1)                                                 +
* Trigger Data to be passed to trigger monitor
         TRIGDATA('File Transfer Example Application')               +
* Process definition associated with trigger
         PROCESS(MQFTPR2)                                            +
* Trigger regardless of message priority
         INITQ('INITIATION.QUEUE')


*************************************************************************
* (B) INITIATION.QUEUE                                                  *
*************************************************************************
  DEFINE QLOCAL('INITIATION.QUEUE')  REPLACE                         +
* The queue on which to base the definition
         LIKE('SYSTEM.DEFAULT.LOCAL.QUEUE')                          +
* Text description of the queue
         DESCR('Initiation Queue')


*************************************************************************
* (C) Receiver Process Definition                                      *
*************************************************************************
  DEFINE PROCESS(MQFTPR2) REPLACE                                    +
* Descriptive text
         DESCR('Triggered File Transfer Receiver')                   +
* Application Type
         APPLTYPE(0)                                                 +
* Applicaton Identifier
         APPLICID('mqftpr2 &')                                       +
* User Data
         USERDATA(' ')                                               +
* Environment Data
         ENVRDATA(' ')
```

Figure 12.4 An MQSC script defining the resources for the triggered example.

the INITQ clause specifies that trigger messages should be sent to the queue called INITIATION.QUEUE. This queue is defined at (B) in Fig. 12.4. There is very little to say about this definition. Initiation queues do not have any special attributes. They are normal queues used for a special purpose. In this case, we take all default attributes from the SYSTEM.DEFAULT.LOCAL.QUEUE apart from the description.

At (C) is the process of definition which carries the information used by the trigger monitor program when starting the receiver application. The definition of File.Transfer.Queue references this process definition. The fields in the process definition have no particular meaning to the queue manager, and could be used for almost anything. However, by convention, each field is used for a specific purpose.

The DESCR clause holds descriptive text which helps in system administration. The APPLTYPE clause records the application type. This is needed for environments where different kinds of application need to be started in different ways. For example, a normal program, such as the file transfer receiver, can be started by issuing a command. However, if the application to be started runs under the control of a transaction manager, such as CICS, the mechanism of invoking it might be entirely different. User-defined values can be used here to allow a specific customer installation to have its own mechanisms for initiating applications via the trigger monitor. For the file transfer example, the application type is unused. The trigger monitor supports only one kind of application, namely those which can be started as commands.

The APPLICID clause holds the details of the application to be started by the trigger monitor. In this case, the clause actually specifies the command to be issued to start the receiver program. In Fig. 12.4 we show the UNIX form of the command.

The function of the user data field, defined in the USERDATA clause, is to hold additional information which is passed to the application when it starts. Typically, this is passed to the application via its command line. The file transfer receiver does not require any parameters, and, consequently, this field is blank.

The final clause in the process definition is the ENVRDATA. Its function is to hold any data used in establishing the environment in which the application is to run. For example, on UNIX or OS/2, this might include environment variable values to be set before execution. Once again, the file transfer example does not require any such environment. In fact, as we shall see, the trigger monitor application ignores this data.

12.3.2   The trigger monitor program trigmon

The listing of the trigger monitor program is given in Fig. 12.5. The initial code is once again very similar to that which we have seen before. One difference, at (F), is that the variable Trigger_Message is declared as being of type MQTM. This is the special data type associated with trigger messages. Another difference, at (H), is that we pass the name of the initiation queue to the trigger monitor program as a parameter. This is in the spirit of trigger monitors, the idea being that the code should be as general as possible.

```
/******************************************************************/
/* Module Name: trigmon.c                                         */
/*                                                                */
/* Function:    This program is a trigger monitor. It is capable of */
/*              starting the mqftpr file transfer receiver program  */
/*              when messages arrive on its queue.                */
/*                                                                */
/* Usage:       Invoked from command line as                      */
/*                                                                */
/*   trigmon <initiation_queue>                                   */
/*                                                                */
/* NOTES:                                                         */
/* 1) This code is only an example. For simplicity it performs only */
/*    very basic error checking.                                  */
/******************************************************************/

/******************************************************************/
/* (A) Include definitions for standard C language functions      */
/******************************************************************/
#include        <stdio.h>
#include        <stdlib.h>
#include        <string.h>

/******************************************************************/
/* (B) Include definitions for MQSeries                           */
/******************************************************************/
#include        <cmqc.h>

/******************************************************************/
/* (C) Define constants used in this program                      */
/******************************************************************/


/******************************************************************/
/* (D) Prototypes for local functions                             */
/******************************************************************/
void TerminateProg(char    *endmsg,
                   PMQHOBJ  pQueue_Handle,
                   PMQHCONN pConnection_Handle,
                   MQLONG   End_Reason) ;

void CvtCArray2Strg(char    *string_version,
                    char    array_version[],
                    MQLONG  array_length) ;

/**************************/
/* (E) Main Entry Point   */
/**************************/
int main(int argc, char**argv)
{
    /******************************************************************/
    /* (F) Declare Structures used in the MQ API calls. Also, initialize */
    /* the descriptors and options.                                 */
    /******************************************************************/
    MQOD        Queue_Descriptor    = {MQOD_DEFAULT} ;
    MQMD        Message_Descriptor  = {MQMD_DEFAULT} ;
    MQGMO       Get_Message_Options = {MQGMO_DEFAULT} ;
    MQHCONN     Connection_Handle   = MQHC_UNUSABLE_HCONN ;
    MQHOBJ      Queue_Handle        = MQHO_UNUSABLE_HOBJ ;
    MQTM        Trigger_Message ;
```

Figure 12.5 The trigger monitor program for the file transfer example.

```
MQLONG          Data_Length ;
MQLONG          Open_Options ;
MQLONG          Close_Options ;
MQLONG          CompCode ;
MQLONG          Reason ;

/**********************************************************************/
/* (G) Declare other variables used in the program                  */
/**********************************************************************/
char            *initiation_queue ;
char            QNameStrg[MQ_Q_NAME_LENGTH + 1] ;
char            ProcessNameStrg[MQ_PROCESS_NAME_LENGTH + 1] ;
char            TriggerDataStrg[MQ_TRIGGER_DATA_LENGTH + 1] ;
char            ApplIdStrg[MQ_PROCESS_APPL_ID_LENGTH + 1] ;
char            EnvDataStrg[MQ_PROCESS_ENV_DATA_LENGTH + 1] ;
char            UserDataStrg[MQ_PROCESS_USER_DATA_LENGTH + 1] ;
char            CommandStrg[MQ_PROCESS_APPL_ID_LENGTH +
                           MQ_PROCESS_USER_DATA_LENGTH + 2] ;

/**********************************************************************/
/* (H) Verify command line                                          */
/*     argv[1] is the name of the initiation queue on which         */
/*             trigger messages will arrive                         */
/**********************************************************************/
printf ("Trigger Monitor Starting\n") ;

if (argc != 2)
{
  printf("Usage:\n") ;
  printf(" trigmon <initiation_queue>\n") ;
  exit(16) ;
}
initiation_queue = argv[1] ;

/**********************************************************************/
/* (I) Connect to the local queue manager (i.e. default)            */
/**********************************************************************/
MQCONN ("",
        &Connection_Handle,
        &CompCode,
        &Reason) ;
if (CompCode == MQCC_FAILED)
{
  TerminateProg("Trig Mon:Could not connect to default queue manager",
                &Queue_Handle,
                &Connection_Handle,
                Reason) ;
  exit(16) ;
}

/**********************************************************************/
/* (J) Construct the descriptor for the initiation queue            */
/**********************************************************************/
strncpy(Queue_Descriptor.ObjectName,
        initiation_queue,
        MQ_Q_NAME_LENGTH) ;

/**********************************************************************/
/* (K) Open the initiation queue                                    */
```

Figure 12.5 *(Continued)*

```
/******************************************************************/
Open_Options = MQOO_INPUT_AS_Q_DEF   |
               MQOO_FAIL_IF_QUIESCING ;
MQOPEN (Connection_Handle,
        &Queue_Descriptor,
        Open_Options,
        &Queue_Handle,
        &CompCode,
        &Reason) ;

if (CompCode == MQCC_FAILED)
{
  TerminateProg("Could not open initiation queue",
               &Connection_Handle,
               &Queue_Handle,
               Reason) ;
}

/******************************************************************/
/* (L) Read messages from the queue                             */
/******************************************************************/

while (CompCode == MQCC_OK)
{
  /****************************************************************/
  /* (M) Retrieve a message from the queue                      */
  /****************************************************************/
  Get_Message_Options.Options = MQGMO_WAIT |
                                MQOO_FAIL_IF_QUIESCING ;
  Get_Message_Options.WaitInterval = 60000 ;
  memcpy(Message_Descriptor.MsgId, MQMI_NONE, sizeof(MQMI_NONE)) ;
  memcpy(Message_Descriptor.CorrelId,MQCI_NONE,  sizeof(MQCI_NONE)) ;

  MQGET (Connection_Handle,
         Queue_Handle,
         &Message_Descriptor,
         &Get_Message_Options,
         sizeof(Trigger_Message),
         &Trigger_Message,
         &Data_Length,
         &CompCode,
         &Reason) ;

  if (CompCode == MQCC_OK)
  {
    /**************************************************************/
    /* (N) Extract the contents of the trigger message as NULL  */
    /*     terminated strings                                   */
    /**************************************************************/
    CvtCArray2Strg(QNameStrg,
                   Trigger_Message.QName,
                   sizeof(Trigger_Message.QName)) ;

    CvtCArray2Strg(ProcessNameStrg,
                   Trigger_Message.ProcessName,
                   sizeof(Trigger_Message.ProcessName)) ;

    CvtCArray2Strg(TriggerDataStrg,
                   Trigger_Message.TriggerData,
```

Figure 12.5 *(Continued)*

```
                         sizeof(Trigger_Message.TriggerData)) ;

        CvtCArray2Strg(ApplIdStrg,
                       Trigger_Message.ApplId,
                       sizeof(Trigger_Message.ApplId)) ;

        CvtCArray2Strg(EnvDataStrg,
                       Trigger_Message.EnvData,
                       sizeof(Trigger_Message.EnvData)) ;

        CvtCArray2Strg(UserDataStrg,
                       Trigger_Message.UserData,
                       sizeof(Trigger_Message.UserData)) ;

        /****************************************************************/
        /* (O) Display the contents of the trigger message            */
        /****************************************************************/
        printf("Trigger message received:\n") ;
        printf("  Triggered queue:    %s\n", QNameStrg);
        printf("  Process Definition: %s\n", ProcessNameStrg);
        printf("  Trigger Data:       %s\n", TriggerDataStrg);
        printf("  Application ID:     %s\n", ApplIdStrg);
        printf("  Environment Data:   %s\n", EnvDataStrg);
        printf("  User Data:          %s\n", UserDataStrg);

        /****************************************************************/
        /* (P) Perform the operation defined by the Application ID and */
        /* User Data                                                  */
        /****************************************************************/
        strcpy(CommandStrg, ApplIdStrg) ;
        strcat(CommandStrg, " ") ;
        strcat(CommandStrg, UserDataStrg) ;

        system(CommandStrg) ;
     }
  } /* End while MQCC == OK */

  /****************************************************************/
  /* (Q) Disconnect with all work finished                      */
  /****************************************************************/
  TerminateProg("Processing Completed",
                &Connection_Handle,
                &Queue_Handle,
                Reason) ;

  return(MQRC_NONE) ;
}

/****************************************************************/
/* (R) Function to convert a character array used by the MQI into */
/*     a NULL terminated string                               */
/* NOTE: string_version must be 1 byte longer than array_version */
/****************************************************************/
void CvtCArray2Strg(char    *string_version,
                    char    array_version[],
                    MQLONG  array_length)
{
  int i ;

  /****************************************************************/
```

Figure 12.5 *(Continued)*

```
/* Copy the whole array into the string version                */
/*************************************************************/
memcpy(string_version, array_version, array_length) ;

/*************************************************************/
/* Find the start of any trailing blanks                       */
/*************************************************************/
for (i = (array_length-1) ; i >= 0 ; i--)
{
  switch(string_version[i])
  {
    case '\0':
      /*******************************************************/
      /* Its already NULL terminated so we need do no more */
      /*******************************************************/
      i = -1 ;
      break;

    case ' ':
      /*******************************************************/
      /* Its a blank, convert to NULL                      */
      /*******************************************************/
      string_version[i] = '\0' ;
      break;

    default:
      /*******************************************************/
      /* Its non-blank. Stamp a NULL in the following      */
      /* position and end                                  */
      /*******************************************************/
      string_version[i+1] = '\0' ;
      i = -1 ;
      break;

  }
}
return ;
}

/*************************************************************/
/* (S) Function to terminate the program                       */
/*************************************************************/
void TerminateProg(char     *endmsg,
                   PMQHOBJ  pQueue_Handle,
                   PMQHCONN pConnection_Handle,
                   MQLONG   End_Reason)
{
  MQLONG CompCode ;
  MQLONG Reason ;
  MQLONG Close_Options ;

  /*************************************************************/
  /* (T) Issue the message associated with termination       */
  /*************************************************************/
  printf("Trig Mon: %s. Reason was %ld\n",
         endmsg,
         End_Reason) ;

  /*************************************************************/
  /* (U) Close the queue                                     */
```

Figure 12.5 *(Continued)*

```
/*************************************************************************/
if (*pQueue_Handle != MQHO_UNUSABLE_HOBJ)
{
  Close_Options = MQCO_NONE ;
  MQCLOSE (*pConnection_Handle,
           pQueue_Handle,
           Close_Options,
           &CompCode,
           &Reason) ;

  printf("Trig Mon: Queue Close Reason Code was %ld\n",
         Reason) ;
}
/*************************************************************************/
/* (V) Disconnect from the local queue manager                         */
/*************************************************************************/
if (*pConnection_Handle != MQHC_UNUSABLE_HCONN)
{
  MQDISC (pConnection_Handle,
          &CompCode,
          &Reason) ;

  printf("Trig Mon: Disconnect Reason Code was %ld\n",
         Reason) ;
}
exit(End_Reason) ;
}
```

Figure 12.5 *(Continued)*

At (J), the object descriptor for the initiation queue is initialized. The name of the queue is taken from the parameter passed in when the application is started. At (K), when the queue is opened, the program specifies some options we have not seen before. The value MQOO_INPUT_ AS_Q_DEF replaces explicit specification of shared or exclusive read. It means that this program is prepared to read the queue using whatever rule was specified when the queue was defined. Since we took default values in the script for defining the queue (Fig. 12.4), we will, in fact, have exclusive access to the initiation queue. The value MQOO_FAIL_ IF_QUIESCING means that the MQOPEN call should fail if the queue manager is attempting to shut down. This is a good option to use in general. Although a queue manager will reject new connections while it is trying to quiesce, applications already connected are allowed to continue to completion. Obviously, some applications may have critical tasks to perform before the queue manager terminates. In most cases, however, applications can simply stop, safe in the knowledge that any queued messages will be held until the queue manager restarts. The vertical bar sign | between the options is the symbol which C uses for a logical OR operation. This combines the options together in a way which means that the MQOPEN call will honor them both. This pair of options is used here simply to illustrate some different possible values for the open options. The values used by the receiver program would work just as well here.

At (L), the loop for reading messages begins. At (M), the options for the MQGET call are established. As before, we specify that the call is to wait if no messages are available. The timeout period is set to 60 seconds. In practice, trigger monitor applications would probably use an indefinite wait period, specified by the MQWI_UNLIMITED value. As on the MQOPEN call, we request that the MQGET call fail if the queue manager is trying to stop. As in the receiver program, any failure of the MQGET call causes the program to end, via TerminateProg. This includes the case when the MQGET times out with no trigger message available. This is convenient for experimentation, because it means that the trigger monitor will terminate cleanly if we simply leave it alone.

If a trigger message is successfully retrieved, the code at (N) extracts the information it contains. One small difficulty with using the MQI from C language programs is that the MQI represents character data as fixed-size arrays, blank padded on the right, rather than C language strings. As we have already mentioned, C language strings have variable length, the end being marked by a character with value of decimal zero. The function CvtArray2Strg performs the required conversion, stripping the unnecessary trailing blanks and properly terminating the strings. Its action will not be described here. Readers familiar with C should have no trouble understanding its operation. Readers unfamiliar with C need not worry about the details. The result is that data from the trigger message is converted into a form in which it can be manipulated naturally by a C program.

The trigger monitor program displays the data it received from the trigger message at point (O), using the printf function we have already seen many times. At point (P), it actually starts the application which has been triggered. To do this, it creates a string which will be passed to the system function. This function causes the string passed to it to be executed as a command. The effect is as if the command were typed in from the command line. The trigger monitor creates this command in a variable called CommandStrg. First it copies in the value of the application identifier retrieved from the trigger message. It does this using the strcpy function. For the definitions in Figure 12.4, this field will have the value mqftpr &. The trigger monitor appends a blank character after the application identifier, using the strcat function and then appends the contents of the user data retrieved from the trigger message. For our example, this field is blank. As we pointed out earlier, for simplicity, this trigger monitor ignores any environment data in the trigger message.

Once the command has been issued and, assuming that it is run in the background as with our example, the trigger monitor regains control and reads the next trigger message back at point (M).

The rest of the program is very similar to the receiver program and will not be described again here.

12.3.3 The modified receiver
program mqftpr2

To work with the form of triggering we specified in Fig. 12.4, the receiver program needs to be modified slightly. A full listing is given in Fig. 12.6. Only the modifications need to be described. The first of these is at point (K) where we open the queue specifying the additional option of MQOO_SET. This option allows the receiver program to alter the attributes of the queue, as well as being able to read messages from it. We will use this ability to rearm the triggering of the queue just before the program terminates. This rearming takes place within the Termi-nateProg function at point (T). This is the other change from the previous version of the program.

In TerminateProg, the object handle for the queue is tested. If it does not have the value MQHO_UNUSABLE_HOBJ, it means that the queue is actually open and so the attribute can be set. The MQSET call allows multiple attributes to be set in one call. In this case, we need to set only one, but even so we need to set up the arrays and counts expected by the call. The variable Selectors is an array of numbers, each of which uniquely identifies an attribute to be updated. The values of the attribute numbers, used by MQSeries, are defined in cmqc.h. In this case, we need to update just one attribute. Its attribute number is MQIA_TRIGGER_CONTROL and it controls whether triggering is currently enabled for the queue. The value we wish to set it to, namely MQTC_ON, goes into the variable IntAttrs at the same array index as the corresponding attribute number goes into Selectors. Since there is only one attribute to be updated, these values go into array index 0. In C, array indices always start from 0. The first value of 1 in the parameter list to MQSET indicates that just a single attribute is to be set. The second 1 in the list is the number of integer attributes to be set. MQSET can simultaneously update integer and character attributes. The 0 and the NULL parameters following IntAttrs indicate to MQSET that we are not updating any character attributes. The final two parameters are the now familiar completion and reason codes.

This code for rearming triggering works together with the particular kind of triggering we specified in Fig. 12.4 to give the following behavior. When a message arrives on File.Transfer.Queue, the queue manager generates a trigger message and puts it onto INITIATION.QUEUE. At the same time, it disables triggering of File.Transfer.Queue. This allows us to prevent the trigger monitor program from starting multiple copies of the receiver. One copy is started and processes any messages on the queue. If the queue becomes empty, the receiver waits for up to 30 seconds and then terminates itself. Before ending, however, it rearms the trigger, via the MQSET call, so that the trigger monitor can restart it should more work become available.

```
/*********************************************************************/
/* Module Name: mqftpr2.c                                           */
/*                                                                  */
/* Function:    This program is the receiver half of the file transfer */
/*              example. It reads messages from a queue, interpreting  */
/*              them as data to be placed into files. The program reads */
/*              File.Transfer.Queue on the local queue manager.        */
/*                                                                  */
/*              This is the triggered version                        */
/*                                                                  */
/* Usage:       Invoked via the trigger monitor                     */
/*                                                                  */
/*    mqftpr                                                        */
/*                                                                  */
/* NOTES:                                                           */
/* 1) This code is only an example. For simplicity it performs only  */
/*    very basic error checking.                                     */
/*********************************************************************/

/*********************************************************************/
/* (A) Include definitions for standard C language functions        */
/*********************************************************************/
#include        <stdio.h>
#include        <stdlib.h>
#include        <string.h>

/*********************************************************************/
/* (B) Include definitions for MQSeries                             */
/*********************************************************************/
#include        <cmqc.h>

/*********************************************************************/
/* (C) Define constants used in this program                        */
/*********************************************************************/
#define MAX_FILE_SIZE  4096

/*********************************************************************/
/* (D) Prototypes for local functions                               */
/*********************************************************************/
void TerminateProg(char     *endmsg,
                   PMQHOBJ   pQueue_Handle,
                   PMQHCONN  pConnection_Handle,
                   MQLONG    End_Reason) ;

/*************************************/
/* (E) Main Entry Point   */
/*************************************/
int main(int argc, char**argv)
{
    /*********************************************************************/
    /* (F) Declare Structures used in the MQ API calls. Also, initialize */
    /* the descriptors and options.                                      */
    /*********************************************************************/
    MQOD        Queue_Descriptor    = {MQOD_DEFAULT} ;
    MQMD        Message_Descriptor  = {MQMD_DEFAULT} ;
    MQGMO       Get_Message_Options = {MQGMO_DEFAULT} ;
    MQHCONN     Connection_Handle   = MQHC_UNUSABLE_HCONN ;
    MQHOBJ      Queue_Handle        = MQHO_UNUSABLE_HOBJ ;
    MQLONG      Data_Length ;
    MQLONG      Open_Options ;
```

Figure 12.6 The modified receiver program for the triggered example.

```
MQLONG          CompCode ;
MQLONG          Reason ;
/**************************************************************************/
/* (G) Declare other variables used in the program                      */
/**************************************************************************/
int             ended = 0 ;
FILE            *fp ;
int             nbytes ;
char            to_file[128] ;

struct
{
  char          to_file[128] ;
  unsigned int  Data_Length ;
  MQBYTE        Buffer[MAX_FILE_SIZE] ;
} F_Transfer_Msg ;


/**************************************************************************/
/* (H) Verify command line                                              */
/*     There are no command line arguments                              */
/**************************************************************************/
printf ("Receiver Starting\n") ;

if (argc != 1)
{
  printf("Usage:\n") ;
  printf(" mqftpr2\n") ;
  exit(16) ;
}

/**************************************************************************/
/* (I) Connect to the local queue manager (i.e. default)               */
/**************************************************************************/
MQCONN ("",
        &Connection_Handle,
        &CompCode,
        &Reason) ;
if (CompCode == MQCC_FAILED)
{
  TerminateProg("Could not connect to default queue manager",
                &Queue_Handle,
                &Connection_Handle,
                Reason) ;
  exit(16) ;
}

/**************************************************************************/
/* (J) Construct the descriptor for the file transfer queue            */
/**************************************************************************/
strncpy(Queue_Descriptor.ObjectName,
        "File.Transfer.Queue",
        MQ_Q_NAME_LENGTH) ;

/**************************************************************************/
/* (K) Open the queue on which we receive data                         */
/**************************************************************************/
Open_Options = MQOO_INPUT_EXCLUSIVE |
               MQOO_SET ;
MQOPEN (Connection_Handle,
```

Figure 12.6 *(Continued)*

```
                &Queue_Descriptor,
                Open_Options,
                &Queue_Handle,
                &CompCode,
                &Reason) ;

if (CompCode == MQCC_FAILED)
{
  TerminateProg("Could not open File.Transfer.Queue",
                &Queue_Handle,
                &Connection_Handle,
                Reason) ;
}

/*******************************************************************/
/* (L) Read messages from the queue                              */
/*******************************************************************/
while (ended == 0)
{
   /*************************************************************/
   /* (M) Retrieve a message from the queue                   */
   /*************************************************************/
   Get_Message_Options.Options = MQGMO_WAIT ;
   Get_Message_Options.WaitInterval = 30000 ;
   memcpy(Message_Descriptor.MsgId, MQMI_NONE, sizeof(MQMI_NONE)) ;
   memcpy(Message_Descriptor.CorrelId, MQCI_NONE, sizeof(MQCI_NONE)) ;

   MQGET (Connection_Handle,
          Queue_Handle,
          &Message_Descriptor,
          &Get_Message_Options,
          sizeof(F_Transfer_Msg),
          &F_Transfer_Msg,
          &Data_Length,
          &CompCode,
          &Reason) ;

   if (CompCode == MQCC_FAILED)
   {
     ended = 1 ;
   }
   else
   {
      /*******************************************************/
      /* (N) Report what is happening                       */
      /*******************************************************/
      printf("Receiver: Transferring file to %s\n",
             F_Transfer_Msg.to_file) ;

      /*******************************************************/
      /* (O) Open the file to be transferred.               */
      /*******************************************************/
      fp = fopen(F_Transfer_Msg.to_file, "w") ;
      if (fp == NULL)
      {
        printf("Receiver: Could not open output file %s\n",
               F_Transfer_Msg.to_file) ;
        printf("   Data will be discarded\n") ;
      }
      else
```

Figure 12.6 *(Continued)*

```
   {
     /*************************************************************/
     /* (P) Write out the data.                                 */
     /*************************************************************/
     nbytes = fwrite (F_Transfer_Msg.Buffer,
                      1,
                      F_Transfer_Msg.Data_Length,
                      fp) ;
     if (nbytes != F_Transfer_Msg.Data_Length)
     {
       printf("Receiver: Problem writing to file %s\n",
              F_Transfer_Msg.to_file) ;
     }
     fclose(fp) ;

   }
 }
}      /* End while(ended = 0) */


/*****************************************************************/
/* (Q) Terminate the program                                   */
/*****************************************************************/
TerminateProg("Processing Completed",
              &Queue_Handle,
              &Connection_Handle,
              Reason) ;

return(Reason) ;
}

/*****************************************************************/
/* (R) Function to terminate the program                       */
/*****************************************************************/
void TerminateProg(char      *endmsg,
                   PMQHOBJ    pQueue_Handle,
                   PMQHCONN   pConnection_Handle,
                   MQLONG     End_Reason)
{
  MQLONG CompCode ;
  MQLONG Reason ;
  MQLONG Close_Options ;
  MQLONG Selectors[1] ;
  MQLONG IntAttrs[1] ;

  /*****************************************************************/
  /* (S) Issue the message associated with termination           */
  /*****************************************************************/
  printf("Receiver: %s. Reason was %ld\n",
         endmsg,
         End_Reason) ;

  /*****************************************************************/
  /* (T) Rearm the trigger for next time                         */
  /*****************************************************************/
  if (*pQueue_Handle != MQHO_UNUSABLE_HOBJ)
  {
    Selectors[0] = MQIA_TRIGGER_CONTROL ;
    IntAttrs[0]  = MQTC_ON ;
    MQSET (*pConnection_Handle,
```

Figure 12.6 *(Continued)*

```
              *pQueue_Handle,
              1,
              Selectors,
              1,
              IntAttrs,
              0,
              NULL,
              &CompCode,
              &Reason) ;
   }

   /*******************************************************************/
   /* (U) Close the queue                                           */
   /*******************************************************************/
   if (*pQueue_Handle != MQHO_UNUSABLE_HOBJ)
   {
      Close_Options = MQCO_NONE ;
      MQCLOSE (*pConnection_Handle,
              pQueue_Handle,
              Close_Options,
              &CompCode,
              &Reason) ;

      printf("Receiver: Queue Close Reason Code was %ld\n",
             Reason) ;
   }

   /*******************************************************************/
   /* (V) Disconnect from the local queue manager                   */
   /*******************************************************************/
   if (*pConnection_Handle != MQHC_UNUSABLE_HCONN)
   {
      MQDISC (pConnection_Handle,
             &CompCode,
             &Reason) ;

      printf("Receiver: Disconnect Reason Code was %ld\n",
             Reason) ;
   }
   exit(End_Reason) ;
}
```

Figure 12.6 *(Continued)*

12.3.4 Arming the trigger

Although we have said that the examples are not complete robust
applications, they do illustrate some of the issues facing developers of
MQI applications. One of these came to light when testing the programs. We cover it here because it is a generic problem facing anyone
using the kind of triggering we specified for the file transfer example.
Recall that once the queue has been triggered, the queue manager disables triggering, a feature we use to prevent multiple copies of the
receiver from executing. We rely on the receiver program itself rearming the queue when it terminates. If the receiver program ever terminates without rearming the queue—for example, because of some
serious programming or system error—the queue will remain with

triggering disabled. It will never be reenabled, since the code which does that is in the receiver program which runs in response to a trigger. One solution to the problem is to have an independent method of rearming the trigger. An MQSC script to do this is shown in Fig. 12.7. The ALTER command can be used to change any queue attribute. Specifying the TRIGGER keyword rearms the trigger for the queue.

12.4 Local Logical Units of Work

Most MQSeries queue managers provide the ability to perform sets of operations on queues as local logical units of work. To use this feature, applications specify MQGMO_SYNCPOINT as one of the options on each MQGET which is to be part of the unit of work. Similarly, they specify MQPMO_SYNCPOINT as one of the options on each MQPUT call. The first MQGET or MQPUT call specifying that the operation is part of a local unit of work causes the unit of work to be started. While the unit of work is in progress, the application can see the effects of its changes, but other applications cannot.

The application completes the unit of work by issuing either an MQCMIT or MQBACK call. The MQCMIT call commits all the changes, which occurred during the unit of work, causing them to be made permanent. The MQBACK call discards all the changes, returning the system to the state it was in before the unit of work started. An MQCMIT or MQBACK call ends the unit of work.

12.5 Participating in Transactions

Many MQSeries queue managers are able to participate in distributed logical units of work. As we saw in Sec. 9.4.2, distributed units of work required a coordinator to control the two-phase commit process. MQSeries queue managers do not provide unit-of-work coordination.

```
***********************************************************************
*                                                                    *
* Script Name: mqftpsc3.mqs                                          *
*                                                                    *
* Description: Arm the triggering of the file transfer queue         *
*                                                                    *
***********************************************************************
***********************************************************************
* (A) FILE.TRANSFER.QUEUE                                            *
*                                                                    *
***********************************************************************
   ALTER QLOCAL('File.Transfer.Queue')                           +
* Enable triggering for this queue
       TRIGGER
```

Figure 12.7 MQSC script to enable triggering for the file transfer example.

Instead, they act as resource managers in the same way as most relational databases do. Support for distributed logical units of work requires that resource managers interact directly with the logical-unit-of-work coordinator. Applications commit or roll back the unit of work by calls to the unit-of-work coordinator rather than to the resource managers themselves. The coordinator then works directly with the resource managers to complete the unit of work.

When working under a unit-of-work coordinator, applications do not issue MQCMIT or MQBACK calls. They do, however, need to specify MQGMO_SYNCPOINT for MQGET operations and MQPMO_SYNCPOINT for MQPUT operations. When the application requests a commit or rollback operation via the unit-of-work coordinator, the appropriate action will be taken by the queue manager in response to direct calls from the unit-of-work coordinator.

Although the applications have very little to do to support distributed units of work, there is some administration involved in defining the relationship between the queue manager and the unit-of-work coordinator. The exact details of this vary from system to system, but the work involved is, in general, similar to that for establishing the relationship between a database manager and a unit-of-work coordinator.

12.6 Summary

In this chapter, we looked at the MQI, the application programming interface to MQSeries products. We discussed the basic verbs provided by the interface and described some of the options available when using them. We looked at an example application which illustrated the use of the basic MQI calls, and saw how it could be extended to be triggered. Finally, we examined some of the other, more advanced MQI calls which support transaction-oriented applications.

Appendixes

This part of the book provides additional materials related to
messaging and queuing. Each of these appendixes provides
additional detailed information that you may find useful. The
appendixes are as follows:

A. *Messaging Examples (some representative uses of
   messaging)*

B. *A Summary of the Message Queue Interface (a "cheatsheet"
   of sorts)*

C. *MQSeries Products and platforms (the IBM messaging and
   queuing products available as of this writing)*

D. *Mail and Online Messaging (a contrast and comparison)*

Messaging Examples

In Chap. 5, we looked at a number of kinds of information flow patterns. In this appendix, we will discuss some specific examples of distributed applications and see the flow patterns that they represent. We will also show how these flow patterns are created using message queuing.

For the purposes of this appendix, we will tend to discuss the arrangements in terms of their use in client-server environments. The terminology is becoming widely understood and conveniently describes roles which can be ascribed to programs in the examples. It is worth pointing out, however, that the client-server approach is, in fact, a special case of a more generalized model of the way in which programs can interact with each other. Message queuing provides excellent support for client-server-based applications, as we shall see shortly. However, it is equally applicable to other interaction models, such as peer-to-peer. Although our examples tend to concentrate on client-server processing, other styles, such as peer-to-peer, may be important in specific applications and are handled just as easily by message queuing.

The examples in this section vary in complexity. We start with a very simple case involving merely a single transfer and work our way via trees and lattices to a final discussion of how messaging and queuing can be used to overcome one of the perennial data processing problems—namely, the limitations of the so-called *batch window*.

A.1 Data Entry

The data entry task is one faced by many organizations, despite the increasing use, in recent years, of techniques such as image processing and character recognition. Many organizations receive data in forms incompatible with the computer systems they use. Most commonly, the information is on paper, frequently handwritten. A good example is the application forms received by insurance companies for new business or for claims on existing policies. Other examples include payments of credit card bills, payments received by public utilities, and orders received by mail-order houses.

Whatever the data, the data entry task involves keying information extracted from the paper copy into the relevant computer system. The task is carried out by data entry clerks. Naturally, this process is error-prone. The usual approach is to have the data keyed multiple times by different clerks. If the same data is keyed each time, it is considered to be correct.

Figure A.1 shows a client-server approach to supporting the data entry task. The client program, running on machine A, accepts the clerk's input. This machine is typically a low-powered PC. It can make basic verification checks, such as that the content of numeric fields is indeed numeric. Once the data has passed these basic editing tests, it is queued for the server machine B. This machine has responsibility for storing the data and for performing the verification that multiple copies of the same data are indeed identical. Although the diagram shows only one client, typically there would be a large number of client machines attached to the server.

The data flow in this example is a simple transfer. There is no flow from the server back to the client. Errors detected in the data, indicated by differences between multiple copies, are handled by having the data keyed in again later, not by having the clerk correct the entry.

The MQSeries calls involved in such a flow are naturally very simple. The client program A puts each completed entry as a message on queue Q1 to the server. The server simply waits on the queue, taking each message as it arrives, storing the data, and deciding whether the data is to be considered as correct or not by checking it against other copies of the same entry.

Although only a single client is shown in Fig. A.1, it is natural to have many client machines serviced by a single server.

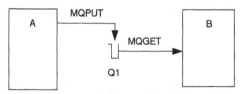

Figure A.1   Data entry using basic transfer.

A.2   Client-Server Processing with Message Pairs

The message pair is the basis for the support of traditional, synchronous client-server processing using message queuing. In this example, program A requests some service from program B. Program A requires a reply from program B in order to complete its processing. For example, if the request from program A were for the retrieval of some data, the reply would probably include the data itself. Figure A.2 illustrates the example. Program A sends requests to program B via queue Q1, while the replies are returned to it from program B via Q2. Because program A requires the reply before it can complete its operation, this example is essentially synchronous. Although program A may perform processing after it has issued the message to program B, there will come a point at which it needs the reply and will have to wait for it if the reply has not already been sent.

Naturally, message pairs can be used to connect multiple client applications to a single server. The arrangement for two clients is shown in Fig. A.3. Though both clients send requests to the single server queue QS, the server correctly distributes replies back to their source. This is achieved in the server using the reply-to-queue information, which flows with the messages. The clients specify their own input queues, QC1 and QC2 in Fig. A.3, as their reply-to-queues. The server sends replies to the specified queues. Naturally, because MQSeries completely hides the underlying network from the server, the clients might be communicating with the server by different and incompatible forms of networking. For example, one might be using TCP/IP, while the other uses SNA.

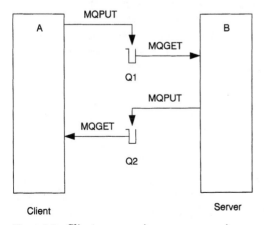

Figure A.2   Client-server using a message pair.

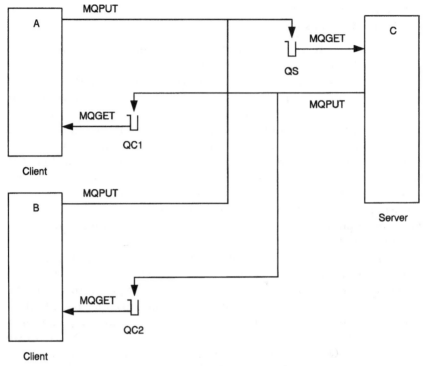

Figure A.3 Client-server with independent message pairs.

Because of its synchronous characteristics, the message pair is very similar to the traditional remote procedure call. Both client and server have to be available simultaneously, as does the network which connects them. For client-server arrangements which must be synchronous, the message pair is the natural choice. However, it does not exploit any of the features of MQSeries which lead to true time independence. We will shortly see arrangements which do.

A.3   Client-Server Processing with Open Chains

A simple open chain or relay is shown in Fig. A.4. Although superficially very similar to the message pair, the open chain arrangement provides true time independence. In the figure, program A puts a message onto queue Q1. This message is read and processed by program B. However, instead of returning the reply to the original source of the request, it is passed to a third program C via queue Q2. Since program A does not have to wait for the reply from program B, there is no time synchronization between them. The overall flow is time-independent. If program B can process the request immediately and send the reply, program C will be able to process it very soon after the request was sent. However, if some system or network problem delays processing of the request, program C may not receive the reply for some significant period of time. However, in the open chain arrangement, unlike the message pair, there is no application waiting for a specific reply. Instead, program C may not even be triggered to run until the reply is available.

We can contrast the open chain with the message pair arrangement shown in Fig. A.2. In Fig. A.4, we can consider program A to be a client making a request of the server program B, but not waiting for the reply to the request. Program C forms a second part of the client program whose job is to process replies from the server. By splitting the client into these two parts, we achieve time independence, removing the restrictions of synchronous processing associated with the message pair and related techniques, such as remote procedure call.

The key difference between the message pair and the open chain, in this example, can be summarized as follows. In the message pair, the communications and processing associated with the messages happen while the client application is executing. Consequently, the client is aware of delays. In the open chain, the work happens while the client applications are idle. Consequently, they are unaware of delays.

A.4   Client-Server Processing with Trees

In previous examples, there has been only one server application. Once multiple servers are servicing a client's requests, we start to see more

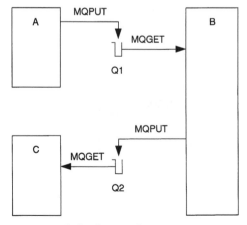

Figure A.4   A simple open chain.

complex topologies emerge. The tree is one such structure. In Fig. A.5, program A is a client and there are three server programs: B, C, and D. The client program A requests service from the server B by sending a message via queue Q1. Similarly, the request to server C is a message on queue Q2, and the request to server D is a message on queue Q3. All messages specify the same reply-to-queue, namely Q4. Consequently, the client program A can read the replies from the servers from this single queue.

The tree arrangement introduces additional parallelism into the processing of the application. Not only may the client perform additional processing while requests are being handled by the servers, but the servers themselves are operating in parallel. It would not be uncommon for the servers to be running on separate machines.

This tree arrangement has the same kind of synchronous processing characteristic we saw earlier in the example based on a message pair. Consequently, it is eminently suitable for those situations in which synchronous operation is required. The client issues the requests and

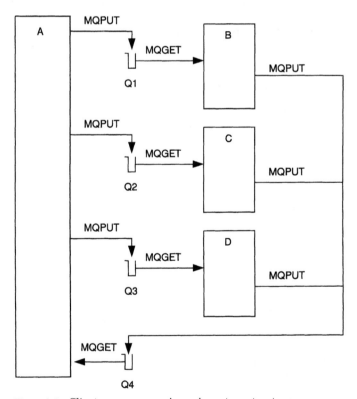

Figure A.5   Client-server processing using a tree structure.

then waits for all the replies to arrive. As with the message pair, network delays or failures will be apparent to the client. Since the servers may be running on different remote machines, the time taken for their replies to arrive may be very different.

The tree is a classic structure for accessing multiple servers from a single client. However, it is a synchronous processing structure. We will see a time-independent version of this kind of structure in the next section.

A.5   Client-Server Processing Using a Split-Join Structure

Figure A.6 shows a split-join structure which is roughly equivalent to the tree structure of Fig. A.5. As in the tree structure, the client program A issues requests to the server programs B, C, and D using their respective queues Q1, Q2, and Q3. The servers process the requests in parallel, and each sends its reply on to the same reply-to-queue, namely Q4. The difference now is that Q4 is no longer read by client program A. Indeed, program A has probably already terminated, having sent its requests to the servers. The reply-to-queue Q4 is processed by program E. The overall processing of the requests from client program A is now time-independent.

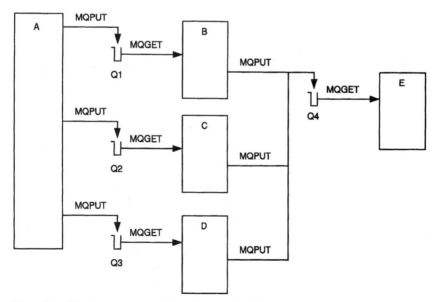

Figure A.6   Client-server processing using a split-join structure.

Program E plays very much the same role as did the function *Collate Results,* which we saw in the design for the distributed telephone directory in Fig. 10.6. It sweeps up replies from the servers, collates them, and when all replies for a specific operation have been received, completes the processing.

A.6   Reducing the Problem of the Batch Window

Most large, commercial data processing centers have sets of operations which must be carried out when there is no other activity on the systems involved. These operations are known as *batch processing.* At the end of each day's processing, application programs may need to run against the enterprise's data. Update operations may have to be run to merge the changes accumulated during the day with data from previous days. Frequently, the operations required consist of a set of application programs linked together by disk files. The upper part of Fig. A.7 shows a typical arrangement. Batch application A takes data from file F1 and creates files F2 and F4. File F2 is processed by application B, producing file F3. Together with file F4 this file is processed by application C to produce the final file F5.

One major challenge for batch processing like that shown in Fig. A.7 is that the length of time during a day in which it can be carried out is shrinking. Increasingly, users are demanding almost continuous availability of online applications. The pressures are particularly intense

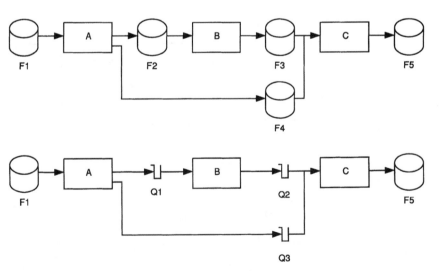

Figure A.7   Reducing the problem of the batch window.

when applications are international and may be accessed by users in different time zones. The time during which batch operations can be carried out on a computer system is often known as the *batch window*. Message queuing can offer some help in reducing the length of this window.

A major problem with batch operations, such as those illustrated in Fig. A.7, is that they are serial and single-threaded. The intermediate files are produced in their entirety before the next step is started. This is done at least partly for protection. Should one step in the overall job fail, all the information required for it to be rerun is available, reducing the time taken to complete the overall task. It is not necessary to restart the job from the beginning. An obvious consequence of the serial nature of the task is that the total length of time taken to execute the batch operations is the sum of the times needed for the individual steps. In the message-queuing version of the task, shown in the lower half of Fig. A.7, the intermediate files have been replaced by MQSeries queues. All the application programs now execute concurrently, reading data from queues and writing data to queues. The initial input and the final output can, of course, still be files. The overlap of processing made possible by the concurrent execution of the multiple steps provides the possibility for reducing the overall execution time of the batch operation.

In some cases, the batch window length is rigidly fixed (for example, 4:00 p.m. to 5:00 p.m. for some stock market closing operations) and the windows problem is that there is just too much specialized processing to be accomplished within the fixed time period. MQSeries queues can transform some of the serialized processing into parallel processing so that everything can be accomplished within the fixed-time batch window.

A Summary of the Message Queue Interface

This appendix contains a brief summary of the Message Queue Interface (MQI), the interface by which application programs access the messaging and queuing functions in MQSeries products. A more complete description is given in Chap. 12, which includes examples of its use. In this appendix, we only describe the structure of the interface and show the syntax of the calls.

There are just 11 basic operations, or verbs, in the MQI. This makes the MQI a very simple interface in comparison with typical operating system interfaces, such as those for OS/2 or Windows, which offer many hundreds of separate verbs. Despite its simplicity, it is a very powerful interface, offering application programmers considerable control over the processing performed. This control is exercised using options and descriptors within the calls. Default values for these mean that programmers need only be concerned with the particular options they actually need to use, disregarding others.

B.1 The Structure of the Message Queue Interface

The eleven MQI verbs can be grouped according to the kinds of operations they perform. We will look at each of the categories in the sections that follow. Some verbs deal with messages, some with queue managers, some with queues, and others with units of work.

The following sections describe the purpose of each of the verbs.

B.1.1   Verbs for manipulating messages

The verbs for manipulating messages are the most important part of the MQI. The two basic verbs for message manipulation (dealing with messages) are MQPUT and MQGET, as shown in Fig. B.1. The queue descriptor X stores information about queue X, including the location of the space where messages are stored. Likewise, for queue descriptor Y. The queue manager uses the queue descriptors.

Most of the work of the MQI is performed through just these two verbs, MQPUT and MQGET. They are used to fill message queues (MQPUT messages into queue space X) and empty message queues (MQGET messages from queue space Y).

The two most vital parameters for these two verbs identify the message and the queue. That is, these parameters tell the messaging and queuing program where the output message is and which output queue it should be placed in, or from which input queue an input message should be taken and where the message should be placed. These are simple operations.

B.1.1.1   MQPUT. Applications put messages into queues with the MQPUT call. As well as the data itself, applications specify additional information, which travels across the network with the message data. This additional information describes the message in a number of different ways, allowing the applications that will process it to understand

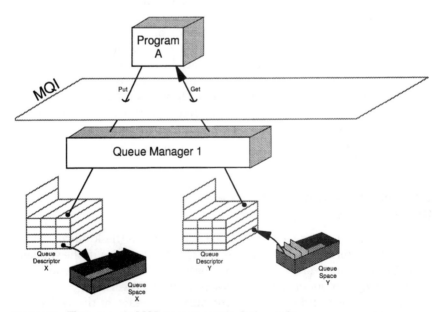

Figure B.1   The two main MQI message manipulation verbs.

its characteristics and to act accordingly. Default values are available which make constructing this additional information very simple.

MQPUT places a message, transparently to the program, into either a local or a remote queue. The messages are automatically moved to where the queue is located. One or more programs can be filling the same queue.

B.1.1.2 MQGET. Applications get messages from queues with the MQGET call. In addition to returning the message data, this call also retrieves the additional information about the message which travels with it.

MQGET gets a message from a local queue.* The messages can be taken from the queue either sequentially or selectively out of sequence. One or more programs can be emptying the same queue.

B.1.1.3 MQPUT1. There is one special-case, message-oriented verb. The MQPUT1 call puts a single message in a queue. It combines a number of individual operations into a single call. It is a convenience function intended to be used by applications to put messages into queues that they use only very infrequently. MQPUT1 is also used for an intentional one-way-only message transfer. It can be used for that situation whereby a message is to be placed into a queue and no response is expected (that is, a one-way distribution).

B.1.2 Additional MQI verbs

There are eight additional MQI verbs, as shown in Fig. B.2. Two of these verbs deal with the queue manager, four deal with queues, and two deal with units of work.

In Fig. B.2, you can notice that the arrows reach progressively deeper beneath the MQI to touch just the queue manager, to touch just the queue descriptor, or to touch the message queue space.

The upward arrow on the extreme right side of Fig. B.2 reflects an MQGET operation. The downward arrow to the left of the MQGET upward arrow reflects an MQPUT (or MQPUT1) operation.

B.1.2.1 Queue-manager-related verbs. There are two queue-manager-related verbs. These verbs are concerned directly with queue managers. They deal with the relationship that an MQI messaging application program has with a particular queue manager.

---

* Some degree of asymmetry exists across the entire computer industry with regard to getting and putting messages. Most messaging systems provide putting to local or remote queues but provide getting only from local queues since getting from remote queues is quite complex!

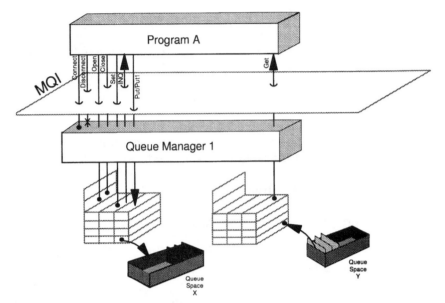

Figure B.2   The eleven MQI verbs.

The queue manager supports the MQI, as shown in Fig. B.3. The queue manager, represented by the large, transparent box in Fig. B.3, has several disjoint responsibilities:

- Interpreting and executing the MQI verbs from the application program.
  (See the MQI plane at the top of Fig. B.3.)
- Managing local queues (including their attributes, space allocations, and such).
  (See the right side of Fig. B.3, local queuing.)
- Moving messages to and from remote queues quickly.
  (See the bottom of Fig. B.3, messages out and messages in.)

Following the dotted lines drawn from the PUT arrow at the top left of Fig. B.3 shows that messages are either handled by local queuing or they are sent out by message moving. Messages in are handled by local queuing.

MQCONN.   The MQCONN call connects an application to a queue manager. All MQI operations require that the application be connected to a queue manager. This connection is established via the MQCONN call. A particular queue manager can be specified by name on this call. However, applications frequently do not know and do not care which

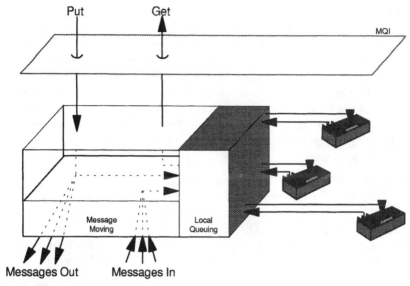

Put   Get

Figure B.3   The queue manager.

queue manager is providing their services. In this case, they can connect to the default queue manager by not specifying a name.

MQCONN associates a program to a particular queue manager. Thereafter, the queue manager can honor the MQPUT and MQGET verbs.

There can be, in the most general case, multiple queue managers within a single local environment. One example of the convenience for such a choice is the dual existence of pilot application programs and production application programs within the same application environment. One queue manager could deliver only to pilot programs, while the other delivers only to production programs.

MQDISC. The MQDISC call disconnects an application from the queue manager. When an application has finished work and no longer needs a connection to the queue manager, it issues the MQDISC call to tell the queue manager that it no longer needs the resources which it was using.

MQDISC disassociates a program from a particular queue manager. Thereafter, the queue manager no longer honors the MQPUT and MQGET verbs to the queue previously opened.

B.1.2.2 Queue-related verbs.   There are four queue-related verbs. These four verbs either deal with the relationship that an MQI messaging application program has with a particular queue or they deal with the attributes of a particular queue.

MQOPEN. MQOPEN gives a program access to a particular queue. The MQOPEN call opens MQSeries objects. In order to work with MQSeries objects such as queues, an application must first open them. Queues are the most important class of MQSeries objects for most applications. There are others, though. For example, there is a queue manager object, which can be opened. The class of an MQSeries object determines the kinds of operations which can be performed. For example, messages can be retrieved from MQSeries queue objects but not from queue manager objects. Both kinds of objects do support inquiry of attributes, however.

MQCLOSE. MQCLOSE takes away a program's access to a particular queue. The MQCLOSE call closes MQSeries objects. When an application no longer needs to use an MQSeries object, it can issue an MQCLOSE call against it. This allows the queue manager to release resources associated with the object. Objects still open when an application disconnects from the queue manager are implicitly closed.

MQINQ. MQINQ retrieves the attributes of a particular queue. The MQINQ call allows the current values of attributes of MQSeries objects to be queried by an application. MQINQ can be used to inquire on the attributes of any objects, including queues, process definitions, and even the queue manager itself. For example, this call can be used to determine how many messages are currently waiting on a particular queue. The interface to MQINQ and to the related MQSET call is designed to maximize the flexibility of manipulation of attribute values. For example, the syntax of these calls will not need to change simply because more attributes or more kinds of objects may be defined in future versions of MQSeries products.

MQSET. MQSET sets the attributes of a particular queue. The MQSET call allows the current values of attributes of MQSeries objects to be altered by an application. As with the MQINQ call, the attributes of any MQSeries object can be set via this call. For example, it can be used to specify the number of messages which must be waiting on a queue before triggering occurs.

B.1.2.3 Verbs relating to logical units of work

MQCMIT. MQSeries allows operations to be grouped into logical units of work. The results of such groups of operations are not made permanent until the logical unit of work is committed. MQCMIT is the call used to commit such a unit of work. Effectively, operations carried out within a unit of work are pended (that is, made transparent) until the commit occurs. Messages retrieved from queues are not deleted, but are made unavailable to other applications until the commit decision has been made. Likewise, messages that are put to queues are stored, but are unavailable to be gotten until the commit has been processed.

MQBACK. The MQBACK call is a partner to the MQCMIT call. It is used to back out a series of operations performed within a logical unit of work. Applications call this function when a logical unit of work is being backed out. In response, the queue manager restores the state of any affected objects to that which existed before the unit of work started. The result is as if the changes made as part of the logical unit of work had never occurred. MQBACK is used only if, for any reason whatsoever, MQCMIT does not result in the commit of a unit of work.

B.2 The Syntax of the Message Queue Interface

In this section, we list the MQSeries calls and show their syntax. Further details of the use of the calls and the meaning of the parameters are contained in Chap. 12.

All MQI verbs are related to an association of the MQI messaging application program (pgm) to an entity of some sort. First, a program must become associated with a queue manager (qm) and then one or more queues (q). MQCONN and MQDISC make and break associations with queue managers; MQOPEN and MQCLOSE make and break associations with individual message queues. MQINQ and MQSET monkey with the attributes of a message queue. Then, a program can do real work: MQPUT, MQPUT1, and MQGET deal with messages (msg); MQCMIT and MQBACK deal with units of work (uow).

The following sections identify the syntax of the 11 MQI verbs in somewhat the same sequences as they are used in a program.

B.2.1 The queue manager-related verbs

The two MQI verbs that deal with queue managers are MQCONN (connect to a queue manager) and MQDISC (disconnect from a queue manager).

B.2.1.1 MQCONN syntax

```
MQCONN (Name, Hconn, CompCode, Reason)
```

Name   name of the queue manager to connect to

Hconn   returned connection handle

CompCode   returned completion code

Reason   returned reason code

B.2.1.2 MQDISC syntax

```
MQDISC (Hconn, CompCode, Reason)
```

Hconn	connection handle
CompCode	returned completion code
Reason	returned reason code

B.2.2 The queue-related verbs

The four MQI verbs that deal with queues are MQOPEN (open a message queue), MQCLOSE (close a message queue), MQINQ (inquire about the attributes of a message queue), and MQSET (set the attributes of a message queue).

B.2.2.1 MQOPEN syntax

```
MQOPEN (Hconn, ObjDesc, Options, Hobj, CompCode, Reason)
```

Hconn	connection handle
ObjDesc	object descriptor
Options	open options
Hobj	returned object handle
CompCode	returned completion code
Reason	returned reason code

B.2.2.2 MQCLOSE syntax

```
MQCLOSE (Hconn, Hobj, Options, CompCode, Reason)
```

Hconn	connection handle
Hobj	object handle
Options	close options
CompCode	returned completion code
Reason	returned reason code

B.2.2.3 MQINQ syntax

```
MQINQ (Hconn, Hobj, SelectorCount, Selectors, IntAttrCount,
       IntAttrs, CharAttrLength, CharAttrs, CompCode, Reason)
```

Hconn	connection handle
Hobj	object handle
SelectorCount	number of attributes to be retrieved

Selectors	array of ordinals defining which attributes are to be retrieved
IntAttrCount	number of integer attributes to be retrieved
IntAttrs	array to receive integer attribute values
CharAttrLength	length of buffer for character attributes
CharAttrs	buffer to receive character attribute values
CompCode	returned completion code
Reason	returned reason code

B.2.2.4 MQSET syntax

```
MQSET (Hconn, Hobj, SelectorCount, Selectors, IntAttrCount,
       IntAttrs, CharAttrLength, CharAttrs, CompCode, Reason)
```

Hconn	connection handle
Hobj	object handle
SelectorCount	number of attributes to be set
Selectors	array of ordinals defining which attributes are to be set
IntAttrCount	number of integer attributes to be set
IntAttrs	array of integer attribute values
CharAttrLength	length of buffer containing character attributes
CharAttrs	buffer containing character attribute values
CompCode	returned completion code
Reason	returned reason code

B.2.3 The main PUT and GET verbs

The two MQI verbs that do most of the work are MQPUT (put a message) and MQGET (get a message).

B.2.3.1 MQPUT syntax

```
MQPUT (Hconn, Hobj, MsgDesc, PutMsgOpts, BufferLength, Buffer,
       CompCode, Reason)
```

Hconn	connection handle
Hobj	object handle
MsgDesc	message descriptor
PutMsgOpts	put message options
BufferLength	length of buffer containing message
Buffer	buffer containing message
CompCode	returned completion code
Reason	returned reason code

B.2.3.2 MQGET syntax

```
MQGET (Hconn, Hobj, MsgDesc, GetMsgOpts, BufferLength, Buffer,
       DataLength, CompCode, Reason)
```

Hconn	connection handle
Hobj	object handle
MsgDesc	message descriptor
GetMsgOpts	get message options
BufferLength	length of buffer for message
Buffer	buffer for message
DataLength	actual length of retrieved message
CompCode	returned completion code
Reason	returned reason code

B.2.4 A special-case PUT verb

The special-case verb that puts a one-way message is MQPUT1 (put a single message).

B.2.4.1 MQPUT1 syntax

```
MQPUT1 (Hconn, ObjDesc, MsgDesc, PutMsgOpts, BufferLength,
        Buffer, CompCode, Reason)
```

Hconn	connection handle
ObjDesc	object descriptor
MsgDesc	message descriptor
PutMsgOpts	put message options
BufferLength	length of buffer containing message
Buffer	buffer containing message
CompCode	returned completion code
Reason	returned reason code

B.2.5 The logical unit of work verbs

The two MQI verbs that deal with logical units of work (UOWs) are MQCMIT (commit) and MQBACK (backout).

B.2.5.1 MQCMIT syntax

```
MQCMIT (Hconn, CompCode, Reason)
```

Hconn	connection handle
CompCode	returned completion code
Reason	returned reason code

B.2.5.2 MQBACK syntax

```
MQBACK (Hconn, CompCode, Reason)
```

Hconn   connection handle
CompCode   returned completion code
Reason   returned reason code

B.3 General Facilities of the Message Queue Interface

Attributes for both messages and queues must be maintained.

B.3.1 Message attributes

The attributes found in the message header help control the delivery of the message to its destination queue (see Fig. B.4). The message can be outbound as a request, outbound as a response, outbound as a status message, or outbound as a one-way distribution message.

Online messaging accomplishes asynchronous (time-independent) communication. Messaging programs run as fast as they can, independently from all other programs, so that messaging programs naturally accomplish parallel operations. Messaging (MQI) programs never have connections; only the queue managers have connections. Messaging programs use queues in lieu of connections, thus operating independently of networking protocols. The networking burden is borne by the queue managers, not the messaging programs. Figure B.5 illustrates these messaging characteristics.

A simple example of using the MQI is shown in Fig. B.6. In this example, program A uses MQPUT to place an invoice in the queue Invoice-Q.

Message = Header + User Data

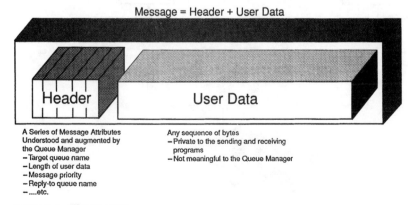

Header   User Data

A Series of Message Attributes
Understood and augmented by
the Queue Manager
 – Target queue name
 – Length of user data
 – Message priority
 – Reply-to queue name
 – ....etc.

Any sequence of bytes
 – Private to the sending and receiving
   programs
 – Not meaningful to the Queue Manager

Figure B.4   The message.

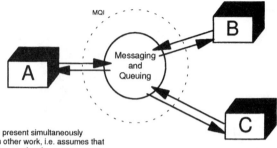

- **Asynchronous**
    - – Both programs need not be present simultaneously
    - - Sending program continues with other work, i.e. assumes that receiving program is busy, or not operational
    - - Receiving program is 'event-driven' to perform task based on the content of the received message
        - – Synchronous-like operation can be achieved by appropriate application design

- **Enables Parallelism**
    - – Independently operating programs
    - – Multiple (non-serial) sub-tasks

- **Connectionless**
    - – Messaging programs never use private connections
    - – Connections are used between Queue Managers

- **Easy to Use Programming Interface**
    - – Simple model, few verbs
    - – Network protocol independent

Figure B.5   Messaging characteristics.

Program B, another local program, uses MQGET to get an invoice from the queue Invoice-Q. In the general case, the queue Invoice-Q is filled with at least one message most of the time so that program B may not instantly get the message that program A just placed in the queue, it might take a few milliseconds or more.

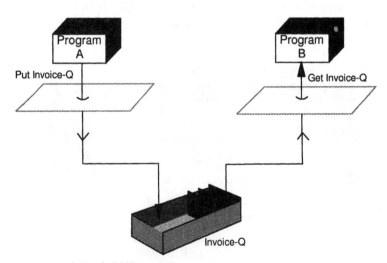

Figure B.6   A simple MQI example.

Local and remote queues are transparent to the application program, as shown in Fig. B.7. The messaging and queuing queue manager automatically moves the message to either a local queue or to a remote queue without additional information through the MQI from the application program.

B.3.2 Queue attributes

Queues have particular attributes. Queues have both a descriptor and some allocated space, as shown in Fig. B.8. The descriptor controls operations against the queue; the allocated space is used to store the messages. The queue descriptor stores attributes (information) about the queue, as shown in Fig. B.9. "Queue" usually implies both the descriptor and its associated storage space, as shown by the dotted oval in Fig. B.8. Neither is usable without the other.

The queue descriptor can have an alias (substitute) descriptor, as shown in Fig. B.10. The alias queue descriptor merely points to the real queue descriptor and gives it an alternate label. The real queue descriptor contains the information about the queue; the alias queue descriptor usually contains just a pointer or alternate versions of what is in the real queue descriptor.

Multiple queue descriptors can exist for the same queue, as shown in Fig. B.11. Multiple queue descriptors give different programs a choice of how to access a particular queue. Different programs can access the

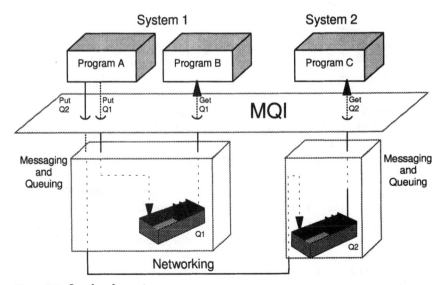

Figure B.7 Local and remote queues.

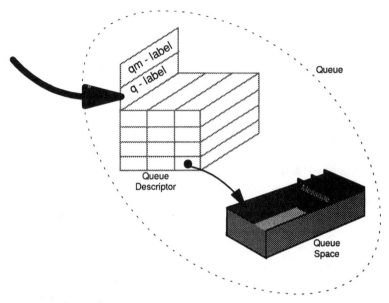

Figure B.8 Queue descriptor, queue space.

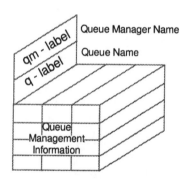

Figure B.9 Queue descriptor.

same queue differently. In Fig. B.11, choice 1 uses the real queue descriptor, choice 2 uses one of the alias queue descriptors, and choice 3 uses the other alias queue descriptors.

Local and remote queue descriptors are different, as shown in Fig. B.12.

Local queue descriptors (see the bottom left dotted oval of Fig. B.12) describe local queues. For every remote queue descriptor (see the top right dotted oval of Fig. B.12) that describes a remote queue, there is (at another location) a corresponding local queue descriptor for that same queue.

Queue managers manage and use queue descriptors, as shown in Fig. B.13. The queue descriptors, which are just control blocks, have to

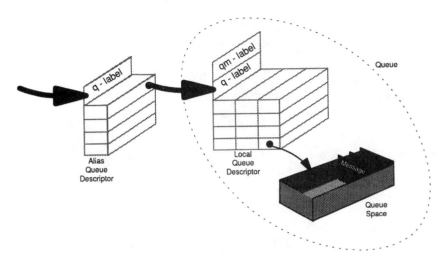

Figure B.10   Alias queue descriptor.

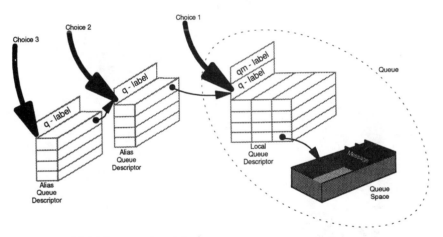

Figure B.11   Multiple queue descriptors.

belong to some queue manager. Queue descriptors are vital to a queue manager in the management of queues to know which programs are associated with which queues, where the storage space is located, which access constraints are in effect (for example, exclusive use), what size thresholds are established, and all such control information.

With messaging, messages are sent to queues and not to programs! The queue manager helps associate queues to programs for both MQPUT and MQGET. Application programs can have exclusive use of a particular queue (if such is defined for that queue), as shown in

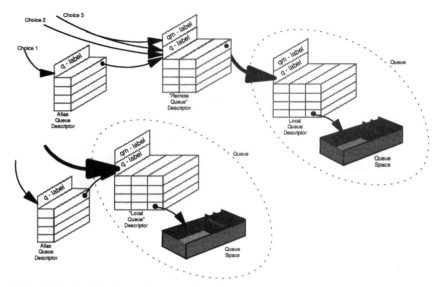

Figure B.12   Local and remote queue descriptors.

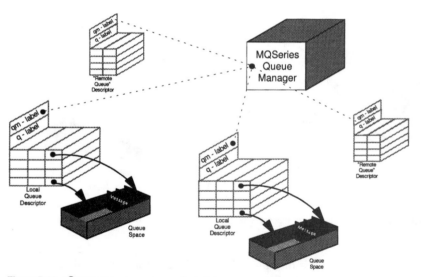

Figure B.13   Queue manager, queue descriptors.

Fig. B.14. For each program there may be just one queue (see left side of Fig. B.14) or many queues (see right side of Fig. B.14).

Application programs can also share particular queues, as shown in Fig. B.15. Two programs can share one queue where one program is filling the queue while the other is emptying it (as shown at the extreme left side of Fig. B.15). This accomplishes a one-way message flow. Two

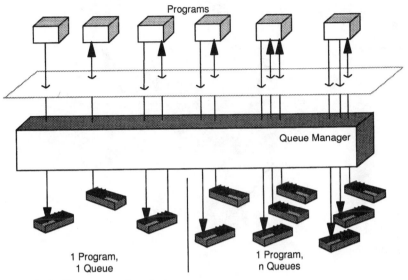

Figure B.14   Programs and exclusive-use queues.

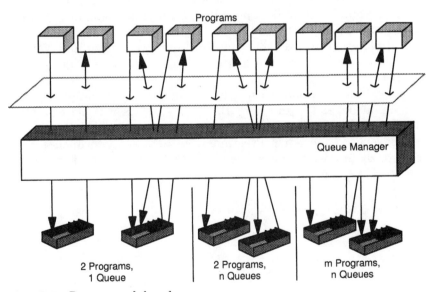

Figure B.15   Programs and shared-use queues.

programs can also share one queue where they are both filling and emptying the same queue, but selectively getting messages, as shown at the left side of Fig. B.15, "2 Programs, 1 Queue."

Two programs can exchange data by using two queues, where one program fills one queue and empties another while the other program

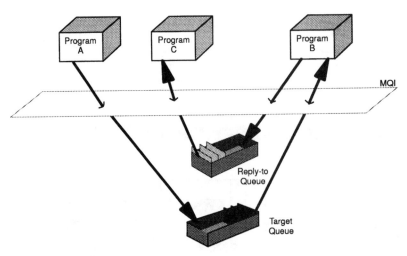

Figure B.16   Reply-to queue.

does the opposite operations on the same queue, as shown in the middle of Fig. B.15, "2 Programs, n Queues." For example, Program A fills Queue 1 while Program B empties it, and Program A empties Queue 2 while Program B fills it.

Multiple programs can share multiple queues in an infinite variety of exchange and distribution patterns, as shown at the right side of Fig. B.15.

Application programs can specify that responses be sent to a special reply-to queue, as shown in Fig. B.16. Here, Program A places a message in the target queue but the reply message is placed by Program B in the reply-to-queue for Program C to get. Reply-to queues help accomplish unidirectional message flows (that is, the open chains described in Chap. 6).

C

MQSeries Products

The messaging and queuing functions that have been described in this book are currently available in a family of products known as the MQSeries product family. This MQSeries family covers a wide range of operating systems and networking protocols. This is necessary because for middleware, such as MQSeries, to be effective it must be available on all the platforms in an enterprise. Few large enterprises have an IT (information technology) strategy that is single-sourced.

While the MQSeries products run on different operating systems, they provide the same set of functions across all the platforms. The products all supply the same API, that is, the Message Queue Interface (MQI). This is a flexible, high-level application programming interface that provides the functions described in this book. The languages that are supported vary with the operating system. The most common commercial languages used in each operating system are currently supported. However, the C language is generally available to enable a degree of portability.

Not only are there common functions and a common API, but the different platforms all interconnect and interoperate with each other. This is achieved by the use of a Message Channel Agent (MCA) that provides the routing and the safe movement of messages from one platform to the next. In this process, it uses its own internal queue, the transmission queue, to hold the messages while it transfers messages over the network. This network can use a variety of protocols, SNA LU6.2, TCP/IP, and others. It is also possible to combine various protocols within the same MQSeries application network. The MCA handles the protocols and the movement of messages, which is transparent to the application program. All the application does is PUT to a queue; the messaging middleware, MQSeries, handles the movement.

The messages are moved using the Message Channel Protocol (MCP). This is a sync-level-2 protocol that gives the assured delivery capability of the product. The sync-level-2 protocol means that the messages are held in the *sender's* Transmission queue until the *receiver's* MCA has acknowledged that the complete message has been received and stored in the target queue.

Also provided is a set of administration functions that allow the queues and subsystems to be defined, tailored, and maintained. These administrative functions are generally available in either panel or command-driven form.

As MQSeries provides an application-to-application interface, each product supplies a set of sample programs that show the use of the MQI. While these programs are not a complete solution, they can be used to understand the techniques used in the style of programming used by messaging and queuing communication middleware.

The MQSeries products were originally developed with an alliance partnership between IBM and Systems Strategies Inc. (SSI), which later became Apertus Technologies, Inc. Under the agreement, the SSI product, ezBRIDGE Transact Version 2.4, was developed to support the MQI, the MCA, and the other functions of the MQSeries products, and was packaged as Version 3. This means that the MQSeries products were then made up of products from two suppliers, IBM and SSI. The products provided the same API, a similar level of function, and the same level of service. By taking this approach of modifying an existing product range and, in parallel, developing a new set of products, the operating system coverage was greatly extended and the time to market was reduced. The ezBRIDGE Transact products also supported an earlier API and communications protocol. A migration aid was included in the Version 3 products that allowed simple migration to the MQSeries product. This partnership enabled a wide platform coverage very early in the product's life and helped to establish it. These products have been consolidated and are all owned and developed now by IBM.

Please note that these products are still under development and that this summary is accurate as of this writing, but please refer to the product literature for a description of the currently available functions and operating system coverage.

C.1   Product Summary

Products operate with the same MQI for each of the operating systems identified in this section.

C.1.1 MVS/ESA

IBM Message Queue Manager MVS/ESA Version 1 Release 1 runs on any IBM System/370 or System/390 under MVS/ESA. It provides messaging and queuing support for CICS/ESA, IMS/ESA, TSO, and batch applications. This function is provided by a new subsystem that is capable of being accessed from the MVS application environments by a set of subsystem adapters. Some of the highlights of the subsystem are:

- The adapters allow the messaging and queuing subsystem to participate with the syncpoint management supplied by CICS/ESA and IMS/ESA. The communication within the operating system is achieved by cross-memory communication facilities. This gives very good performance when used for internal cross-subsystem communication.

- The messaging and queuing paradigm is ideal for parallel processing. It provides the ability to overlap processes in different environments and bring the results together at the completion of the operation.

- The use of the subsystem offers a high degree of system integrity, as the application code is kept separate from the MQM MVS/ESA system code.

- The subsystem supports applications written in C, COBOL, PL/1, and Assembler.

- The subsystem is now capable of sending and receiving messages from IBM and non-IBM platforms via both SNA and TCP/IP.

- The subsystem has comprehensive recovery and restart facilities to ensure that messages are delivered only once and are not lost including:

  Sophisticated logging mechanism

  Single startup and recovery procedure for use at normal and abnormal shutdown

  Automatic recovery from transaction, system, and (provided suitable backups are available) storage media failures

  Use of System Management Facility (SMF) to record system statistics

- The system can be used in conjunction with any security product that complies with the System Authorization Facility. It allows for several choices of security levels (by subsystem, connection, resource, or context).

- Extensive system management facilities help the system operator or system administrator set up, examine, and manipulate the queues that are controlled by MQM MVS/ESA. These facilities are provided through panels and a command interface.

C.1.2 OS/400

The IBM Message Queue Manager/400 runs on any IBM AS/400 system under OS/400 Version 2 Release 3. It provides the full functions of the Message Queue Interface and provides for time-independent communications between programs both within the operating system and between connected heterogeneous systems.

The highlights of the product are:

- It is capable of sending and receiving messages from IBM or non-IBM platforms via both TCP/IP and SNA.

- If CICS/400 is installed, MQM/400 can interoperate with CICS/400 transactions and link them to non-CICS environments.

- Syncpoint management and unit-of-work facilities are provided by OS/400 and supported by MQM/400.

- There is a comprehensive set of restart and recovery facilities used:

  Message logging

  Single startup procedure for use after both normal and abnormal shutdowns

  Automatic recovery from transaction and system failures

  Recovery of persistent messages on queues damaged by media failures

- There are systems management facilities that help the system operator or system administrator set up, examine, and manipulate the queues that are controlled by MQM/400. A single point of control may be used to administer a network of AS/400s using MQM/400. The facilities are provided through both panels and the command line interface.

- Applications written in RPG, C, or COBOL can use the MQI in OS/400.

C.1.3 OS/2

MQSeries for OS/2 Version 2.0 runs on an IBM Personal System/2 supported by IBM Operating System/2 in the server configuration. In the client configuration it can run in OS/2, IBM's DOS, AIX, or Microsoft Windows. Each client supplies the MQI functions.

The highlights are:

- Provides full MQI functions including memory-based messages, message priority, message expiration, and confirmation of message arrival and delivery

- Can set message queues to trigger application programs to start remotely, giving improved automation

- Supports TCP/IP, LU 6.2, NetBIOS, and IPX (via NetBIOS emulation) communication

- Provides a single point of control for administering MQSeries application networks on OS/2 AIX, MVS, and OS/400 systems, improving systems management

- Transports data from one MQSeries platform, converting it, if requested, so that it can be used on another

- Can operate with CICS for OS/2, providing coordination between MQSeries and CICS resources by allowing CICS for OS/2 transactions to issue MQSeries calls

C.1.4 AIX/6000

MQSeries for AIX Version 2.0 runs on an IBM RISC System/6000 supported by AIX/6000 Version 3 in the server configuration. In the client configuration it can run in OS/2, IBM's DOS, AIX, or Microsoft Windows. Each client supplies the MQI functions.
The highlights are:

- Provides full MQI functions including memory-based messages, message priority, message expiration, and confirmation of message arrival and delivery

- Can set message queues to trigger application programs to start remotely, giving improved automation

- Gives greater freedom to reconfigure your setup by being able to exploit DCE directories for queue names

- Provides a single point of control for administering MQSeries application networks on OS/2 AIX, MVS, and OS/400 systems, improving systems management

- Transports data from one MQSeries platform, converting it, if requested, so that it can be used on another

- Can operate with CICS for AIX, providing coordination between MQSeries and CICS resources by allowing CICS for AIX transactions to issue MQSeries calls

C.1.5 VSE/ESA

ezBRIDGE Transact on VSE/ESA Version 3 runs on IBM System/370 or System/390 under VSE/ESA and executes in the CICS VSE/ESA environment. As such, it benefits from all the advantages provided by the CICS environment in the areas of availability, recovery, and security.

The highlight of the product is:

- Administration panels are provided to assist with the management of the queues.

C.1.6   Tandem

ezBRIDGE Transact on Tandem Version 3 runs on any Tandem non-stop system that supports Guardian 90 operating system. As such, it is able to add to this nonstop environment the capability of the generalized messaging and queuing system.

The highlight of the product is:

- Messaging value is added to nonstop systems.

C.1.7   System 88

ezBRIDGE Transact on System 88 Version 3 runs on any IBM System 88 or Stratus system that supports the VOS operating system.

The highlight of the product is:

- Messaging value is added to S/88.

C.1.8   DEC

ezBRIDGE Transact on DEC runs on any Digital Equipment Corporation (DEC) VAX system capable of running the VMS operating system.

The highlights of the product are:

- Queuing is supported over DECNet.
- The general-purpose queuing mechanism can be used within the VMS operating system to enable program-to-program communications between processes.

C.2   Interoperation

MQSeries products interoperate effectively.

C.2.1   Interoperation between MVS/ESA and other platforms

The flow of messages between MQM MVS/ESA and other non-ESA platforms (including non-IBM platforms) is very similar to that described in the previous section. Figure C.1 illustrates communication between MVS/ESA and the following platforms:

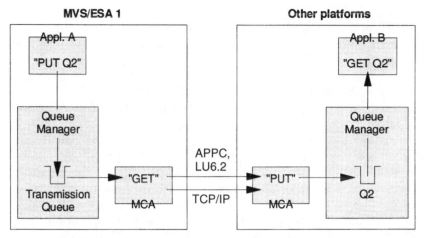

Figure C.1   MQSeries interoperation between MVS/ESA and other platforms.

OS/2, DOS, Windows 3.1

OS/400

VSE/ESA (though CICS ISC is used on the VSE/ESA side as well)

System/88

Stratus

Tandem

The RISC System/6000 with AIX, Message Queue Manager/400, and MQM MVS/ESA provide a TCP/IP connection, as well as an SNA LU6.2 connection. The DEC/VAX uses DECnet between VAX systems, but supports LU6.2 connections to other systems.

If MVS/ESA-to-MVS/ESA communications is required and CICS/ESA is available on both systems, then CICS ISC can be used to support the MCA. CICS/ESA is not a requirement but is an option that may ease the operations and improve performance.

Each MQSeries family member will provide an MCA application, supporting the MCP protocol previously described. These will support the communication sessions and provide the safe-movement facilities and the assured once-and-once-only delivery previously described. One difference to note is that, in ezBRIDGE Transact implementations of the MQSeries family, all messages are currently classified as persistent, meaning that they will be held as disk-resident.

Note that with MQM MVS/ESA and MQM/400, a message can be classified as nonpersistent, saving some overhead if delivery of the message is not critical. Nonpersistent messages may be memory-resident and are not logged. A nonpersistent message will be removed from the outgoing transmission queue if the system goes down. However, if only the communication line fails, the message will be retransmitted if a return acknowledgment has not been received. Another difference is that the message batch size (see discussion in the previous section) is set to 1 on all MQSeries products other than MQM MVS/ESA and MQM/400.

As an aid to data conversion between EBCDIC and ASCII, and between other system architectures, MQSeries will provide conversion of message header information. The receiving application may then use this header information to convert the body of the message itself. An exit is provided in the MCA program application so that the application designer may create or use a specific conversion program.

Finally, a gateway was provided with MQM MVS/ESA to allow interconnection with SSI ezBRIDGE Transact systems that do not support the MQI. SSI ezBRIDGE Transact 2.4 products have a READ/WRITE programming interface different from the IBM MQI, though similar in concept. Also, these systems employ a different communication protocol: the ezBRIDGE Transact MLP (Message Layer Protocol). By providing this MLP gateway on MQM MVS/ESA, messages can be sent and received by MQM MVS/ESA to and from ezBRIDGE Transact 2.4 applications.

C.2.2 Interoperation between MQM/400 platforms

Interoperation between MQM/400-supported AS/400s is identical to that described in the MVS/ESA section with the difference that a TCP/IP MCA support is also available. See Fig. C.1.

The Message Channel Agent program can be configured with either an LU6.2 mover or a TCP/IP mover. Thus, existing or new TCP/IP networks can be used between AS/400 systems. For example, messages can be passed along a TCP/IP network to a gateway MQM/400 that can, in turn, forward the message to a host MVS/ESA system through SNA LU6.2 communication.

The TCP/IP support will also allow AS/400 TCP/IP connection to an RISC System/6000 and to an OS/2-based LAN.

C.2.3 Interoperation in PC LAN environments

In client-server environments, applications reside on different systems, often sharing a common file system. Applications are driven by the client, usually a workstation. The server responds to requests, carrying out additional application services, updating local databases, or acting as a gateway to host systems or other LANS.

In Fig. C.2, MQSeries for OS/2 is installed first in the server/gateway OS/2 system at the top, with client support downloaded to each client system. The file server and communication gateway functions may be on separate systems or may be combined in a single system, as illustrated here. In either case, they must be IBM PS/2s or equivalent with OS/2 2.0. The client systems, however, can be a mixture of DOS, Windows, or OS/2 systems. Note that the DOS systems are supported on DOS 3.2 and above (though DOS 5.0 is recommended), Windows on Windows 3.1, and OS/2 on OS/2 Version 2.0. Note also that a PS/2 or equivalent can operate in a non-LAN stand-alone environment, with an SNA LU6.2 connection to a host system. In this case, OS/2 2.0 is required.

MQSeries for OS/2 may use SNA LU6.2 and TCP/IP connections to other systems and LANs.

The applications are driven from the client workstations, with requests being made to the OS/2 server/gateway, and perhaps through this gateway to a host, to other LANS, or to other systems in the network. All queues in the LAN are resident in the LAN file server, not in the client node, so that both application clients and the communication server are treated as clients of the LAN file server. However, through the LAN workstation software, the files in the server are often shared by the application in the client. The client applications access these files in the same way as if they resided directly on the client workstation.

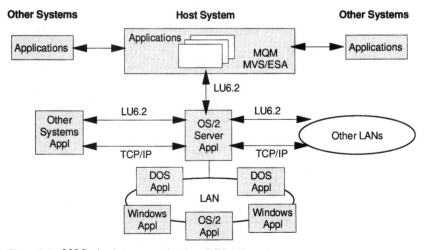

Figure C.2 MQSeries interoperation in a PC LAN environment

It is convenient to think of a LAN configuration with a file server as a single system. Within the LAN, all queues are local with respect to the applications running on each node, since each has access to the queues stored on the file server.

Client applications link to an MQI stub provided with the product. (The MQI stub then resolves the queue name requested by the application programmer to the actual target queue name.)

The client application issues an MQ PUT. The message is placed on a queue in the server by the message queue management code residing at the client node. If the message is destined elsewhere, such as to the host or another LAN, the server's message queue manager safely sends the message on to the Message Queue Manager in the receiving system. The application on the receiving system then issues an MQ GET to retrieve the message.

Note that messages can be exchanged between applications on different nodes in the same LAN, in the same manner as if they were applications residing together on a single system. A message is PUT to a (shared) queue by an application in Client 1, and then retrieved by a GET issued by the application in Client 2. Each client has its own message queue management facilities, which work with the shared files on the server where the actual queues reside. Thus, what might be considered a remote queue (for another node) by a particular application, is actually a shared file on the server, and is treated as a local queue.

A single LAN may support multiple servers and/or communication gateways. This concept is illustrated in Fig. C.3, showing multiple local queue managers residing on a single LAN. Client applications can be running on DOS, Windows 3.1, or on OS/2, and a local queue manager may be supported on any of these clients. These local queue managers communicate outside the LAN via the MCA (Message Channel Agent) program on the OS/2 server/gateway. When a user logs on to a particular node, the application client at this workstation is associated with a local queue manager and with the LAN file server.

Clients communicate with associated file servers using Novell IPX or other similar LAN support. Communication server-to-communication server linkage is accomplished through Message Channel Agents using LU6.2.

C.2.4   Interoperation between RISC System/6000 AIX platforms

Interoperation between MQSeries-supported RISC System/6000s (see Fig. C.1) is identical to that described in Sec. C.2.1, with the difference that a TCP/IP (Ethernet) Message Channel Agent is also available.

Domain: Queue Manager 1   **Domain: Queue Manager 2**

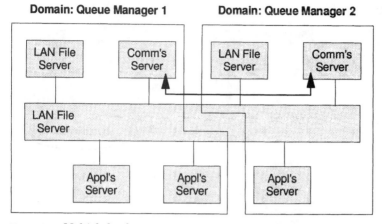

Figure C.3   Multiple local queue managers on one logical LAN.

The Message Channel Agent program can be configured with either an LU6.2 mover or a TCP/IP mover. Thus, existing or new TCP/IP networks can be used between RISC System/6000 or MQM/400 nodes (and, after fulfillment of the Statement of Direction, to OS/2-based PC LANs). For example, messages can be passed along a TCP/IP network to a gateway RISC System/6000 that can, in turn, forward the message to a host MVS/ESA system through SNA LU6.2 communication.

C.2.5   Client-server support

The RISC System/6000 with AIX can also support client-server environments. As described in the previous section on OS/2, the AIX server can provide both file server and gateway functions. Clients can be other RISC System/6000 AIX systems with MQSeries support. Display terminals can also be supported by AIX applications running in the RISC System/6000.

In Fig. C.4, MQSeries for AIX/6000 is installed first in the server/gateway AIX system at the top, with client support downloaded to each client system. The file server and communication gateway functions may be on separate AIX/6000 systems, or they may be combined in a single system, as illustrated here.

The client systems are also AIX/6000 systems. Both the server and client systems are supported by AIX 3.2 and above, with LAN connectivity hardware.

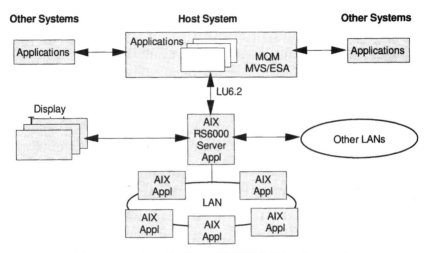

Figure C.4   MQSeries interoperation in an AIX/6000 LAN environment.

Because the server above also functions as a gateway, it requires communication hardware supporting IBM's SNA/LU6.2, with SNA Services/6000 software. (Note that AIX/6000 systems can also communicate with other AIX/6000s and AS/400s using TCP/IP.)

The applications are driven from the client workstations, with requests being made to the AIX server/gateway and perhaps, through this gateway to a host, to other LANS, or to other systems in the network. Users at display terminals can also drive applications that can use messaging and queuing functions to communicate throughout the network.

Platform	PU	Notes
MVS/ESA	5	
CICS/VSE	5	
PC LAN	2.1	
RS/6000 AIX	2.1	
AS/400 OS/400	2.1	
S/88 VOS	2.1	
Stratus VOS	2.1	
VAX/VMS	2.1	Using DEC SNA Gateway
VAX/VMS	2.0	Using SSI Peer-to-Peer
Tandem/Guardian	2 or 5	

Figure C.5   SNA native PU-type support, by platform.

	MVS/ESA	VSE/ESA	PC LAN	AIX/6000	exB/400	MQM/400	S/88	Stratus	Dec Vax	Tan Guard
MVS/ESA	TCP/IP									
VSE/ESA										
PC LAN	TCP/IP		TCP/IP	TCP/IP		TCP/IP				
RJS/6000	TCP/IP		TCP/IP	TCP/IP		TCP/IP				
exB/400										
MQM/400	TCP/IP		TCP/IP	TCP/IP		TCP/IP				
S/88										
Strat										
Dec Vax									DecNet	
Tan Guard										

Figure C.6 Software required for system interoperation.

All queues in the LAN are resident in the LAN file server, not in the client node. However, through the LAN workstation software, the files in the server are often shared by the application in the client. The client applications access these files in the same way as if they resided directly on the client workstation.

It is convenient to think of a LAN configuration with a file server as a single system. Within the LAN, all queues are local with respect to the applications running on each node, since each has access to the queues stored on the file server.

Client applications link to an MQI stub provided with the product. (The MQI stub then resolves the queue name requested by the application programmer to the actual target queue name.)

The client application issues an MQ PUT. The message is placed on a queue in the server by the message queue management code residing at the client node. If the message is destined elsewhere, such as to the host or another LAN, the server's message queue manager safely sends the message on to the message queue manager in the receiving system. The application on the receiving system then issues an MQ GET to retrieve the message.

Note that messages can be exchanged between applications on different nodes in the same LAN, in the same manner as if they were applications residing together on a single system. A message is PUT to a (shared) queue by an application in Client 1, and then retrieved by a GET issued by the application in Client 2. Each client has its own message queue management facilities, which work with the shared files on the server, where the actual queues reside. Thus, what might be considered a remote queue (for another node) by a particular application is actually a shared file on the server and is treated as a local queue.

C.2.6 Intersystem software requirements

For each MQSeries-supported system platform, the underlying communication transport will often be SNA LU6.2. There are differences, however, in PU (physical unit) type support between systems. For this reason, additional support is also provided so that "any-to-any" connectivity can be achieved. Figure C.5 shows the PU types provided in the native LU6.2 support. To resolve the inconsistencies shown in Fig. C.5, alternative transport layers are supported on the appropriate platforms, as shown in the "any-to-any" support table in Fig. C.6.

D

Mail and Online Messaging

Mail messaging and online messaging are often confused with one another since they are both referred to as "messaging." This appendix describes the distinction between mail messaging (of the X.400 or e-mail type) and online messaging (of the MQI type). This chapter explains that the transport networking services normally interconnecting mail systems could be easily replaced by messaging and queuing communication middleware in many cases. As mail and online messaging both evolve, they may in the future become blended for some message traffic in some special cases, but, as of this writing and for some time to come, the two types of messaging are and will remain quite different* in most cases.

D.1 A Spectrum of Messaging

Mail (stored mailbox) messaging and MQI (online queue) messaging are radically different. They were developed for different purposes and operate with different characteristics. Mail messaging is at one end of a spectrum; MQI messaging is at the other end. Mail messaging provides "anytime delivery"; MQI messaging provides "immediate delivery"

---

* There will always be some degree of healthy debate and lively controversy (as in all software development and evolution) about the objectives, functions, suitability, and applicability of software such as mail messaging for online messaging. At this writing, most mail messaging systems are not suitable for online messaging, primarily because of their original, more casual performance objectives and design points. In fact, the 1984 and 1988 versions of ISO's X.400 mail messaging do not mention online messaging while the 1992 version acknowledges, but does not address, online messaging. At this writing, most X.400 implementations are based on the 1984 specification. Similarly, TCP/IP-based mail messaging and other simple mail messaging also does not yet address online messaging between programs.

(even if programs are not ready to accept the messages). The complete spectrum of messaging runs from fascimile transmissions through mail messaging to online/real-time MQI messaging, as shown in Fig. D.1.

Fascimile (FAX) transmissions are between persons (indirectly through fax devices), mail distributions are between persons (indirectly through workstation or PC devices and programs), and online messaging is directly between programs without necessarily involving persons or devices. Thus, distinctions can be quickly recognized based upon which combinations of persons, devices, and programs are involved. Generally speaking, fax involves persons/devices, mail involves persons/devices/programs, and online messaging involves programs.

D.2   Communication Between Persons (Mail Messaging)

Communication between persons may or may not involve programs and computers, but when it does it is called mail messaging and is a queue-based communication. Communication between programs is discussed in Chap. 8.

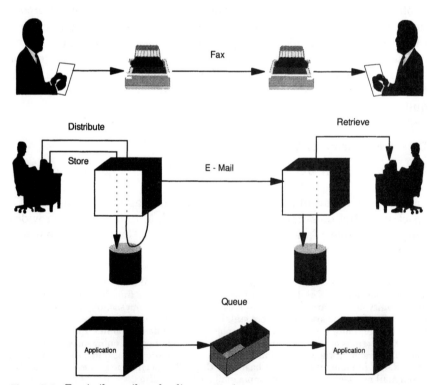

Figure D.1   Facsimile, mail, and online messaging.

D.2.1 Pairs and groups of persons

Persons can communicate pairwise (two at a time) or groupwise (three or more at a time). In addition, for both pairs and groups, persons can exchange information (in a two-way manner) or simply distribute information (in a one-way manner). Programs communicate in a way very similar to persons, using all sorts of information flow patterns, which are described in Chap. 5.

D.2.1.1 Pairs of persons. One person can communicate concurrently and pairwise with multiple other persons, as shown in Fig. D.2. In this example, person 1 is communicating separately with persons 2, 3, and 4. Persons 2, 3, and 4 are not necessarily communicating with one another. There are three pairs of communicating persons, whereby person 1 is either exchanging information with the other three persons or merely distributing information to them. The three persons could also be merely distributing information to person 1.

Reversing the discussion, persons 2, 3, and 4 could be viewed as clients to person 1 as a server, who concurrently serves all of the other persons. This is similar to a medical doctor who (concurrently or serially) serves several patients in his or her office.

D.2.1.2 Groups of persons. Persons can communicate and relay communication among groups, as shown in Fig. D.3. In this example, person 1 is communicating separately with persons 2, 3, and 4. Persons 2, 3, and 4 are not necessarily communicating with one another. Person 2 communicates further with persons 6 and 7 after communicating with person 1. Similarly, person 4 communicates further with person 8. All of this communication can be either in an exchange manner or in a distribution manner (either direction) or in a combination of both manners.

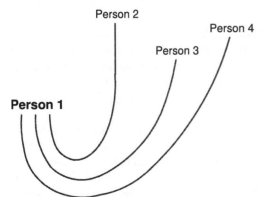

Person 2

Person 4

Person 3

Person 1

Figure D.2 Persons can communicate in a pairwise manner.

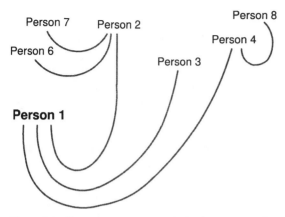

Figure D.3   Persons can communicate in a groupwise manner.

The message flow in this diagram is that of a tree structure, as described in Chap. 5. It can also be viewed in reverse as a nested client-server structure where person 2 and person 4 are playing both a client role to person 1 and a server role to persons 6, 7, and 8.

Many other more complex patterns occur constantly and naturally among people everywhere communicating for all sorts of purposes. These person-to-person flows happen, however, in relatively slow motion since all message flows are gauged by the human actions that occur between message flows.

D.2.2   Connections between persons

Persons can communicate directly using connections, as shown in Fig. D.4. In this case, the connection is private between the persons. A face-to-face conversation uses the air space as a connection between the persons. Each person is dedicated or bonded to the other for some period of time.

D.2.2.1   Devices using connections.   Persons can communicate through devices (for example, telephones or computers) that use connections, as shown in Fig. D.5. Communication over connections is *not* messaging, it is just conversing. The persons are communicating indirectly but are dedicated or bonded to one another for a particular period of time.

D.2.2.2   Telephone connection.   Persons can communicate through telephone devices, which are based on connections, as shown in Fig. D.6. Both persons, as well as the telephone connection between them, must be constantly available or communication stops. In addition, the persons must share a common language or communication cannot start. Again, a telephone conversation is *not* messaging.

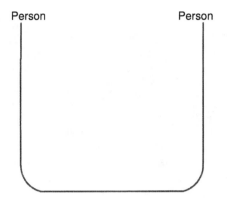

Person   Person

Figure D.4   Persons can communicate using connections.

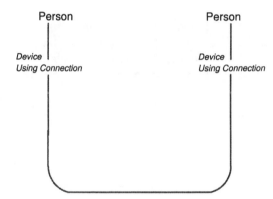

Person   Person

Device   *Device*
Using Connection   *Using Connection*

Figure D.5   Persons can communicate through devices using connections.

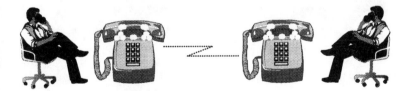

Successful call depends upon:

Constant communication line availability

Constant people availability

Common, defined language

Figure D.6   Telephones are based on connections.

D.2.3   Queues between persons

Persons can communicate indirectly using queues, as shown in Fig. D.7. A very simple example of communicating through queues is the stack of notes often found on an office desk or the collection of notes left for children or a spouse on a refrigerator door or elsewhere.

D.2.3.1   Devices using queues. Persons can communicate through devices that use queues, as shown in Fig. D.8. The devices provide the queues, and the persons are not communicating directly; they are messaging. A simple example of a device that uses queues is an ordinary tape cassette player from which one can retrieve music "queued" earlier by a recording artist. You and the musician are messaging (in a very broad sense).

D.2.3.2   Answering machine queue. Telephone answering machines are based on queues, as shown in Fig. D.9. The called party may be busy or absent when the calling party attempts to communicate. The telephone answering machine records and plays back the audio (message) queues.

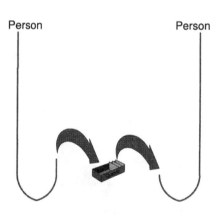

Figure D.7 Persons can communicate using queues.

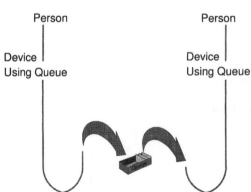

Figure D.8 Persons can communicate through devices using queues.

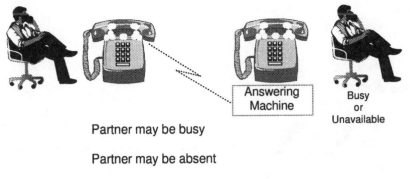

Partner may be busy

Partner may be absent

Messages can be picked up at later time

Figure D.9 Telephone answering machines are based on queues.

D.2.4 Mail messaging

Persons can communicate through both devices and programs, as shown in Fig. D.10. The devices intercommunicate directly through programs; the persons intercommunicate directly through the devices and indirectly through the programs.

D.2.4.1 E-mail operations. Persons can send and receive electronic mail (electronically stored mail or e-mail), as shown in Fig. D.11. During a particular time period, one person can be sending mail that will be received at some unspecified later time by another person. E-mail messages are growing exponentially by the day across businesses, industries, governments, academia, and everywhere. E-mail messages are being served by mail messaging. An e-mail message is, obviously, a mail message (less the leading "e").

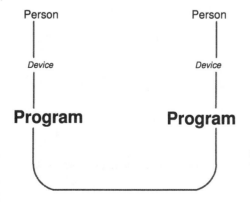

Figure D.10 Persons can communicate through devices and programs.

Person   Person

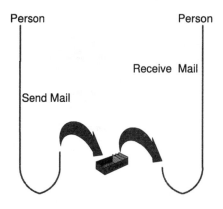

Receive Mail

Send Mail

Figure D.11 Persons send mail and receive mail.

D.2.4.2   Send mail.   The first step is sending mail, at the convenience of a person, as shown in Fig. D.12. There is usually no immediate (or perhaps not any) response period specified or demanded. The recipient (or recipients) may respond quickly, slowly, or not at all.

D.2.4.3   Receive mail.   The second step is receiving mail, at the convenience of a person, as shown in Fig. D.13. This second step may not occur if the mail is lost or if the recipient neglects to receive the mail. There is usually no immediate (or perhaps not any) response that needs to be taken with any urgency, especially if the mail is part of a distribution list.

D.2.5   Mail messaging machinery

Mail messaging involves input, storage, and output devices, as shown in Fig. D.14. Mail is first created, then stored, then distributed, then stored again, then retrieved, and finally acted upon as necessary at one or more destinations.

Person

Send Mail

Figure D.12   First, mail is sent (at human convenience).

Person

Receive Mail

Figure D.13   Later, mail is received (at human convenience).

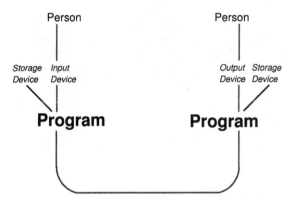

Figure D.14   Mail involves input, storage, and output devices.

D.2.5.1 Input devices. Input devices are necessary for humans to create mail, request distribution of mail, and request retrieval of mail.

D.2.5.2 Storage devices. Storage devices are necessary to store and retrieve mail at the origin and destination(s) and at stopover points between the origin and destination(s).

D.2.5.3 Output devices. Output devices are necessary for humans to see inbound and outbound mail to act upon it.

D.2.5.4 Mail message transport. Mail messages are distributed (not exchanged) between storage devices, as shown at the extreme left and right edges in Fig. D.15. The sender person requests the mail messaging program to retrieve a previously stored mail message from a storage device and distribute it to one or more persons (each specified by name and location or by alias name). Mail directory files are often vital to the mail messaging program in order to accomplish this step.

The transport network selected for mail message distribution (or online message moving) is a separate consideration from choosing between mail and online messaging. Any particular type of transport network might be considered for selection for either or both types of messaging traffic. See Chap. 7 for a discussion of application (messaging) and communication (transport) environments.

Figure D.15 shows two program-to-program connections, one for mail traffic and one for nonmail traffic, just to emphasize the choice that must be made about whether to provide separate or shared transport networks for mail and nonmail messaging traffic. Mail messaging traffic can certainly interfere with nonmail messaging traffic if overloaded network facilities are shared. However, careful network analysis can lead to monetary and other savings if the mail and nonmail messaging traffic can share network facilities. This is a hard choice to make without research, analysis, or just experimentation. Messaging traffic patterns and growth expectations are important. What size are the messages? Where are they going? When are they moving? What are the information flow patterns of today and tomorrow? Are traffic volumes, peak periods, directions, pairings, message sizes, and rates predictable?

Messaging of both the mail and online types happens in an application environment but conditionally requires one or more communication environments to interoperate the separate messaging environments.

D.2.5.5 Store-and-forward (S/F) operations. Mail messages are distributed (not exchanged) between storage devices and one or more store-and-forward operations can be involved, as shown in Fig. D.16. At each stopover point between an origin and any of the destination

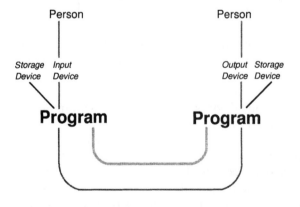

Figure D.15 Mail is distributed across transport networks between storage devices.

points, the mail messages can be replicated and the extra copies redistributed. At these stopover points, the entire mail message (regardless of its size) must be assembled before it can be forwarded (or replicated).

D.2.5.6 Drifting delivery responsibility. The responsibility for mail delivery is transferred from program to program as the mail moves, as shown in Fig. D.17. Delivery responsibility drifts (and shifts) with the movement of the mail message. A mail message can be from a few bytes in length to tens of thousands of bytes in length. When the messages are longer, they must be segmented and reassembled by the mail messaging programs before they can be forwarded.

This shifting of responsibility for mail delivery from one program to another leaves the person at the origin without any responsibility for the ultimate mail message delivery or deliveries. That is, there is no end-to-end (person-to-person or program-to-program) protocol necessary and the person sending the mail message can simply forget the message after requesting that it be distributed. Similarly, the mail program at the origin can also forget the mail message after it is success-

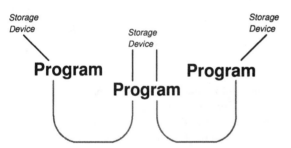

Figure D.16 Mail can involve store-and-forward operations.

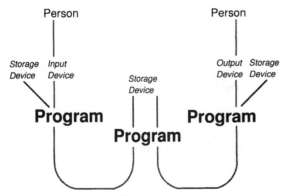

Figure D.17 Responsibility for mail delivery is repeatedly transferred.

fully transferred to the next program, whether that program is at the ultimate destination or at a stopover (intermediate) point between the origin and one of the destinations.

In short, mail messages move (are distributed) in a "pass-the-baton" manner similar to that used in a relay race. The mail message sender can, of course, request an independently flowing mail distribution message from the destination (or destinations) back to the origin to acknowledge the message's arrival at a destination program location, but this mail-ack message is just another piece of mail for a human to receive and read.

Mail message delivery is much like the shipment of goods by freight in that you create the goods (a message), ship them by an agent (the mail messaging program), and then wave them goodbye, trusting that they will arrive at their proper destination sometime.

D.2.5.7   CCITT X.400 interpersonal mail.   X.400 interpersonal mail operates at human speeds.

CCITT X.400, although referred to as messaging, is mostly oriented to the distribution of administrative data such as electronic mail. It is not designed for online/real-time program-to-program communication and provides no transactional support. Instead, it provides a comprehensive structuring of information (notes, files, and documents) required for person-to-person communication.

D.2.5.8   Electronic mail and electronic data interchange.   Electronic mail takes many forms and is provided by many vendors (for example, LotusNotes-generated notes and simple TCP/IP-based notes). Because of the wide variety of e-mail software packages, many application-level gateways are already installed to effect interoperation between such mail messaging systems as IBM's PROFS office system and the many TCP/IP-based systems.

Electronic data interchange (EDI) is a specification for message formats to be used when exchanging messages between different companies and industries (for example, between suppliers, manufacturers, and distributors). EDI consists mostly of format specification and is only a lighter-duty protocol specification. E-mail, by contrast, consists mostly of protocol specification and has very little format specification.

D.2.5.9   Vendor-independent messaging (VIM).   The Vendor-Independent Messaging (VIM)* consortium is one of seven or eight emerging mail-messaging consortia. VIM is an X.400-like office mail interface ori-

---

* VIM replaced the now-obsolete Open Messaging Interface (OMI) term.

ented toward people (who are, of course, intercommunicating at human speeds). VIM is sponsored by a computer vendor consortium including Apple Computer, Inc., Borland International, Inc., IBM Corporation, Lotus Development Corporation, Novell, Inc., and others.

The objective of the VIM API is

> to provide the services necessary to empower Application developers to provide *mail-aware* and *mail-enabled* applications across a wide range of messaging and operating system platforms
>
> to enable inter-application collaboration in a *non-real time* manner.*

It is significant to note in this statement of objective that VIM is an interface for a *non*-real-time, *mail* system. VIM is intended for inter-personal messages, files, documents, and notes and is based heavily on the Lotus Notes software product.[†]

In contrast, the MQI is intended for general message traffic between any programs located in diverse application environments (operating systems, networks, and vendor hardware) where the programs are inter-communicating primarily in an online, real-time manner in a transactional (or nontransactional) environment with subsecond (or second) end-to-end response times. The asynchronous nature of the MQI allows partner programs to be busy or unavailable, in which case messages are queued at the target location, and the message-generating program can continue executing.

Of special importance for business transactions, the MQI offers assured message delivery, message recovery, and interaction with syncpoint management for safe movement of online messages between Message Queue Managers.

D.3 Communication Between Programs (Online Messaging)

Communication between programs is very similar to communication between persons as far as the message flow patterns are concerned. Pairs and groups, exchange, and distribution are involved. A major difference is that communication between persons usually involves more steps (create, store, send, replicate, reassemble, redistribute, store, retrieve, analyze) than does communication between programs (PUT

---

* Page 1, paragraph 3, *Vendor-Independent Messaging Interface Functional Specification, Version 1.00,* dated March 23, 1992.

[†] Other mail messaging APIs related to VIM are OCE (from Apple), MAPI (from Microsoft), SMF (from Novell), and XAPIA (from a consortium of X-400 e-mail vendors). Yet other consortia are likely to emerge before a unification of mail messaging software and interfaces occurs.

and GET, with all other steps provided automatically beneath the online messaging interface).

Mail messaging consists of store/send followed by store/retrieve-receive (that is, two-step operations at both origin and destination), while online messaging consists of just PUT and GET (that is, one-step operations at both origin and destination). Mail messaging speed is gauged by human actions between mail message distributions (not exchanges) and is, therefore, a relatively slow-motion process. Pure online message speed is gauged by program actions between online message exchanges (not distributions) and is, therefore, a relatively high-speed process.

D.3.1 Programs, devices, and persons

Programs can communicate with or without persons and devices being involved as shown in Fig. D.18. Communication between programs is controlled by programs, not by persons or devices. Programs initiate communication with persons and with devices and with other programs (through queues). Messaging programs are expecting immediate delivery of a message to a queue or from a queue. Messaging programs are not, however, expecting any communication with another *program* (except indirectly through the presence of messages in one of the queues that the program is reading).

D.3.2 Communication and application environments

Application environments (and applications and programs within them) communicate across one or more communication environments, as shown in Fig. D.19 and discussed in Chap. 7.

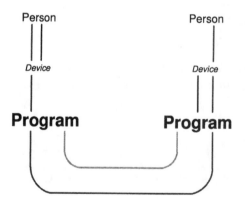

Figure D.18 Program-to-program communication, with and without persons.

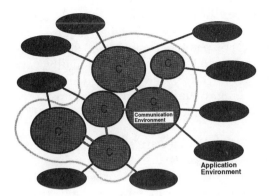

Figure D.19 Application environments communicating across communication environments.

D.3.3 Online messaging

This entire book is dedicated to online messaging, so only some highlights of online messaging are included in this appendix.

D.3.3.1 Exchange and distribute. Program pairs and groups can exchange and/or distribute messages, as shown in Fig. D.20. Whereas mail messaging deals primarily with one-way distribution from one origin to multiple destinations, online messaging deals primarily with two-way exchange between one origin and just one destination or a very limited number of destinations. Broadcast distributes a message to all available destinations; multicast distributes a message to a subset of all available destinations identified by destination address values or otherwise (for example, by group IDs). Online messaging uses predominantly exchange patterns between programs unlike the distribution patterns between persons for mail messaging.

D.3.3.2 Programs and queues versus persons and mailboxes. Programs and persons are similar; queues and mailboxes are similar, as shown in

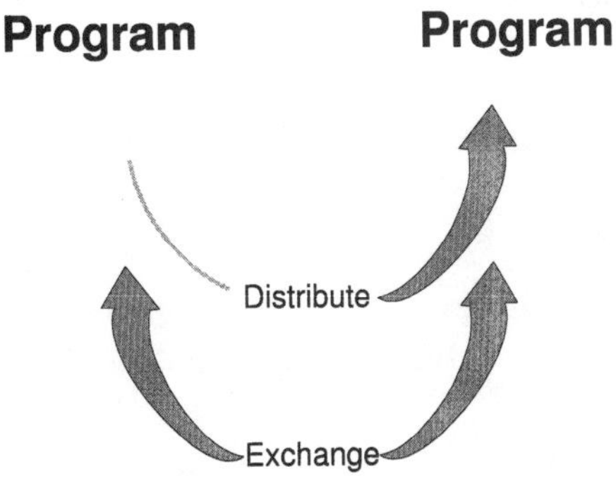

Program   Program

Distribute

Exchange

Figure D.20 Programs can exchange and/or distribute messages.

Fig. D.21. Mail messaging requires the specification of both the target person name and mailbox location. Online messaging requires the specification of only the queue (corresponding to a mailbox) and not the program (corresponding to the person). Hence, person and mailbox are tied together with mail messaging, while program and queue are not tied together with online messaging. Usually only a single person can access a particular mailbox, while multiple programs can (if permitted) access a single queue.

Unlike mail messaging where target values are specified as two values (mailbox and person), online messaging calls for target values that are specified as single queue name value without reference to the target program. The single queue name value can, however, be qualified with additional indicators for association (for example, queue_manager.queue).

D.3.3.3   Message queuing.   One program PUTs a message and another program GETs the same message, as shown in Fig. D.22. Online messaging automatically moves messages to remote queues, as shown in Fig. D.23. Thus, local programs can transparently PUT a message to a queue without concern for whether the queue is in a local environment or in a remote environment.

Unlike mail messaging, online messaging can optionally use safe-moving protocols for shuffling messages between local and remote queues.

D.3.3.4   Message moving.   Online messaging automatically moves messages to remote queues, as shown in Fig. D.24. Messaging and queuing (that is online messaging) automatically moves messages from where they *are* to where they *should be,* if such is necessary (for example,

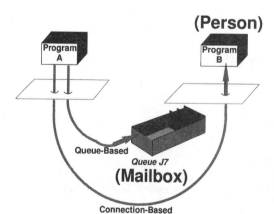

(Person)

Program A

Program B

Queue-Based

Queue J7

(Mailbox)

Connection-Based

Figure D.21   Programs and queues versus persons and mailboxes.

Program   Program

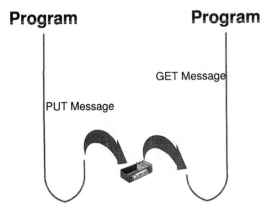

GET Message

PUT Message

Figure D.22   One program PUTs a message; another program GETs the same message.

Program   Program

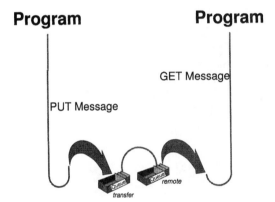

GET Message

PUT Message

Figure D.23   Local programs can transparently PUT to remote queues.

between different application environments). Online messaging also uses primarily only memory queues with disk queues available as backup for all or some of the messages within all or some of the memory queues.

D.3.3.5 Message moving example.   An example of online messaging message moving is shown in Fig. D.25. In this example, program CO (top left of diagram) is communicating with program CB (top right of diagram). The messages that program CO PUTs go into remote queue Queue1 while the messages that program CB PUTs go into remote queue Queue2. Both programs are reading from local queues and writing to remote queues without being aware that the online messaging software is shuffling messages between the queues.

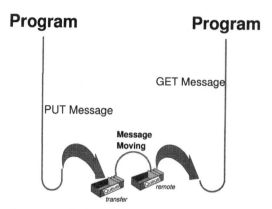

Figure D.24 Online messaging automatically moves messages to remote queues.

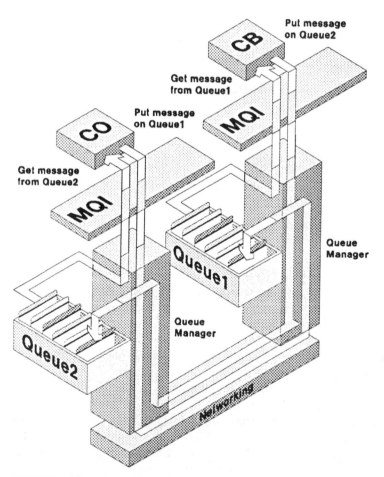

Figure D.25 Messaging and queuing moves messages for applications.

This example is unlike mail messaging in that the queues are being filled and emptied at computer and network speeds instead of at human, computer, and network speeds. There are no humans involved in this example.

D.3.3.6 Mixed mail and online messaging. It is inevitable that questions arise about mixing mail and online messaging within the same application or even within the same application program. More and more the administrative operations (well suited to mail messaging) are being integrated with business-critical operations (better suited to online messaging) to get better productivity out of people. It can be a great convenience to have simultaneous access, personwise or programwise, to both types of messaging. As of this writing, there are no products that offer easy simultaneous access to both types of messaging, and (unfortunately) either mail messaging or online messaging software must be chosen for the mixed traffic or the mail and online messages must flow over distinctly separate messaging software.

Mail messages over MQI messaging. Some categories of mail messages could be intermixed with normal program-generated messages. A big consideration in doing this is the degree to which these lesser-valued mail messages may adversely affect the speed and performance of the higher-valued online messages. High-value, large-volume online message traffic needs to be delivered immediately. Lower-valued, dispensable mail messages perhaps do not deserve immediate delivery.

Comparing...	Mail Messaging	Online Messaging
Communicating Parties	Persons	Programs
Communicating Speed	Human Speed	Computer Speed
Urgency of Delivery	Anytime	Immediate
Usual Response Demand	Minutes, Hours	Subseconds or Seconds
Primary Mode of Communication	One-way	Two-way
	Distribution	Exchange
Primary Message Content	Notes, Files, Documents	Signals, Transaction Messages
Traffic Value	Lower Value and Expendable	Higher Value and Precious
Performance Required	Any	Very High
Target Specification	Mailbox Location and Person Name	Queue Name, No Program Name
Associated Systems	None	CICS and Other Transaction Systems

Figure D.26 Comparison of mail and online messaging.

Online messaging over mail message-handling systems. Some categories of online messages could be intermixed with normal mail-generated messages. A big consideration in doing this is the degree to which these higher-valued online messages may be adversely affected by the intermixed presence of the lesser-valued mail messages. High-value, large-volume, online message traffic needs to be delivered immediately and mail transport networks are not usually designed for immediate (or even assured) delivery.

D.4   Comparing Mail and Online Messaging

A comparison of mail and online messaging is shown in Fig. D.26. The leftmost column identifies characteristics; the remaining two columns compare mail and online messaging with respect to these characteristics. We suggest that you read down the rows of the table one by one, scanning the middle column and then the rightmost column to quickly recognize each comparison and contrast.

This comparison table is simplistic in that it gives only a rough approximation of the comparison of messaging systems at large. Individual mail messaging systems (for example, LotusNotes) or individual online messaging systems (for example, MQSeries Family) may not compare exactly as this table indicates.

Glossary

This small glossary contains only those terms that are particularly meaningful to, or peculiar to, online messaging (messaging and queuing) or to the MQI (Message Queue Interface). The primary purpose of this glossary is to make reading this book a little bit easier if questions arise about our usage of some terms. If a word does not appear in the small selection within this glossary, then the normal usage of the word should be assumed.

ACID (properties) A—Atomicity; C—Consistency; I—Isolation; D—Durability. Atomicity refers to the results of a logical unit of work; parts of a unit of work are either all committed or all rolled back. Consistency indicates that a completed logical unit of work transforms a shared resource (for example, a database or file) from one valid state to another. Isolation means that changes to shared resources that a logical unit of work affects do not become visible outside the logical unit of work until it is committed. Durability means that changes that result from logical unit of work commitment survive subsequent system or media failure. These are four properties of a logical unit of work in a transaction processing system such as CICS or IMS. The phrase "ACID test" or "ACID properties" is often used in application requirement descriptions.

address In general, a label used to specify the location (logically or physically) of an entity such as a program, a device, a queue, or some other computer or network entity. Computer addresses are often binary numbers or alphanumeric characters that are not very mnemonic (easy to remember). They may even be completely or partially unprintable, undisplayable, and not human readable. Addresses stay constant and seldom have substitute values. (See Hauzeur [1986] in the Bibliography for an excellent discussion of address and name labels.)

address space 1. A collection of identifiers for labeling logical or physical locations; for example, a collection of printer addresses or network node addresses. 2. Any part of a computer memory reserved for exclusive use by a system or subsystem; for example, an MVS operating system address space reserved for the execution of CICS.

agent (code) "Helper code" providing indirect access to some set of services; for example, a database manager or application subsystem or networking service such as messaging and queuing can be considered, in a broad sense, agent code. The term "agent code" is often shortened to "agent."

alias queue (name) A queue name that is an alias or substitute for another queue name. More properly referred to as **queue (name) alias.** When an application or a queue manager uses an alias queue, the alias name is resolved and the requested operation is performed on the queue with the resolved name. The term "alias queue name" is often shortened to "alias queue."

alias queue definition   A queue definition used to specify an alias or substitute queue name. See **alias queue.**

AOR   CICS Application-Owning Region component. See also **TOR.** (See Wipfler [1989] in the Bibliography.)

API   see **application programming interface.**

APPC   Advanced Program-to-Program Communication supported by SNA logical unit of type 6.2. APPC has varying syntax (format) across implementations but consistent semantics (meaning). See also **CPI-C,** which is an interface with both consistent syntax and semantics.

application   A collection of application programs that collectively serve some useful purpose or purposes (for example, a business purpose, personal purpose, scientific purpose, entertainment purpose, or some other purpose). Each individual program is an application program. A collection of application programs that are designed to work together constitute an application.

application (code)   A program (or set of programs) written for business, scientific, or other purposes.

application data   Within a message, data that belongs exclusively to an application program (or to a collection of application programs).

application environment   A collection of computing software facilities that are each accessible by an application program executing within a single computer (in the nondistributed case) or across computers (in the distributed case). In OSI terms, an application environment is a collection of application services, where services include operating system, database, networking, and other services necessary for the application program to complete its execution and fulfill its purpose. See **application service.**

application programming interface (API)   The set of calls through which an application accesses an MQSeries queue manager's services and gets results back. Generally, an API is any such interface to any set of services.

application service   A set of functions (usually related) accessible by an application program executing within a single computer. In general terms, an application service is a set of functions, and an application environment is a collection of services within which are functions. See **application environment.**

asynchronous messaging   A method for communication between programs in which the requesting program can proceed with its own processing, without having to wait for a reply to its request.

asynchronous operation   An operation whereby the requestor is not blocked (or suspended) while the request is being processed. A number of requests can be issued and results of the processing can be later accessed in any order. (See Wipfler [1989] in the Bibliography; Chapter 8 of her book provides a discussion of the asynchronous operation provided by CICS.)

asynchronous processing   See **asynchronous operation.**

attributes   Characteristics and control information associated with queues and messages.

back out   To remove changes (including deletes and adds) that previously have been made to recoverable resources (for example, one or more databases). Synonymous with **roll back** or **undo**.

basic queue   A queue in which messages can be stored.

basic queue name   The simplest form of a queue name.

batch   A group (batch) of commands or messages brought together for processing or transmission. Pertaining to activity involving little or no user action. Contrast with **interactive.**

batch processing   Processing that occurs serially with large collections (batches) of input.

batch window   A "window of time" (for example, from 8:00 P.M. until 6 A.M. or between 4 P.M. and 5 P.M.) during which batch operations must be started and completed, usually in some particular sequential manner. Quite often, this window of time can be insufficient to execute all of the required programs.

browse   In message queuing, to look at a message without removing it from the queue. See also **GET.**

browse cursor   An indicator used when browsing a queue to identify the message that is next in sequence.

bulk data transfer   Delivery of data to and from end users and between applications. This can include warehousing of data and mailbox delivery of message responses to end users. In the simplest terms, the transfer of large amounts of data.

call return (model)   See **RPC.**

chain (flow pattern)   A series of relay flow patterns. See **relay.**

CICS   See **Customer Information Control System.**

client (program)   A requestor of services. (See Orfali [1994] in the Bibliography for a very thorough discussion of client programs and matching server programs.)

client-server model   A particular information-flow model involving just two programs, whereby one program continuously plays the role of client, by only requesting information, and the other program continuously plays the role of server, by only responding to requests from the client program. Frequently, programs within a large number of stand-alone workstations play the role of client programs, and programs within a smaller number of host computers of varying sizes and capabilities play the role of server programs. This allows consolidation of such servers as file servers, print servers, network servers, and intercompany communication servers. (See Orfali [1994] in the Bibliography for a very thorough and easy-to-understand discussion of the client-server model.)

C/S   See **client-server model.**

command list (CLIST)   In MVS-TSO, a data set, or a member of a partitioned data set, that contains TSO commands to be performed sequentially in response to the EXEC command.

commit   To make coordinated changes to recoverable resources; for example, to update multiple databases to a consistent state such as deleting a record from one database and simultaneously adding it to another to accomplish a transfer.

commit scope   The range of time, data, and programs covered by a commit operation.

common programming interface (CPI)   A consistent set of APIs across application platforms.

communication environment   The networking middleware, networking software, and networking hardware that supports the distributed application environments.

completion code   A data value returned to an application program at the completion of an MQI call.

Complex Transaction Enabling (CTE)   The application helper code that makes complex information flow patterns easier to accomplish.

connection handle   The identifier, or token, by which a program knows the instance of queue manager to which it is connected.

connection-based communication   Intercommunication through logical connections.

connectionless (CL)   Characterized by the absence of logical connections between communicating parties (programs). This provides the connectivity of M programs to N programs (or M persons to N programs) without the overhead of managing $M \times N$ sessions.

connectionless communication   An OSI term that describes communication occurring between two programs of any type, whereby there is not a logical connection between them. This term was originally used to characterize communication between components of the lower network model layers but has also been "misused" for communication between programs within the application layer.

connection-oriented (CO)   Characterized by a dependence on logical connections between communicating parties (programs).

connection-oriented network   A network consisting of communication accomplished solely by the use of logical connections between communicating parties.

context   Information about the originator of a message.

control data   Data used by systems and networking software for control purposes.

coordinated update   An update operation coordinated by a third-party program.

correlation identifier   An identifier used to associated two or more messages. These messages may be all outbound from a given point, both inbound to and outbound from a given point, or inbound and outbound through many points.

CPI   See **common programming interface.**

CPI-C   See **Common Programming Interface (for) Communications.**

Common Programming Interface (for) Communications (CPI-C)   An API for connection-based communication programming having both consistent syntax (format) and consistent semantics (meaning) and being supported by both SNA LU 6.2 and TCP/IP networking.

CPI-MQ   Replaced by MQI. Was originally Common Programming Interface for Message Queuing.

CTE   See **Complex Transaction Enabling.**

CTS   Common Transport Semantics, as identified in the Networking Blueprint and, later, in the Open Blueprint. See **MPTN.**

Customer Information Control System (CICS)   An IBM licensed program that enables transactions entered at remote terminals to be processed concurrently by user-written application programs. It includes facilities for building, using, and maintaining databases.

DAE   See **Distributed Automation Edition.** Also known as Distributed Application Environment.

DASD   See **direct access storage device.**

DataTrade   A messaging/queuing application system designed to support securities trading and other types of distributed applications.

DB2   A relational database system.

DBCS   See **double-byte character set.**

DB/DC   Database/data communications.

DCE   See **Distributed Computing Environment.**

dead-letter queue   A queue to which a queue manager or application should send messages that cannot be routed to their identified destination.

dequeue   To remove an element (message) from a queue.

direct access storage device (DASD)   A storage device in which access time is effectively independent of the location of the data. Usually implies a physical disk device.

DISOSS   Distributed Office Support System.

distributed application   A set of application programs that are running in many different computers but which, collectively, constitute the application.

Distributed Automation Edition (DAE)   A messaging/queuing system that is part of the IBM Industrial Sector's PLANT FLOOR SERIES of products. Later renamed Distributed Application Environment.

Distributed Computing Environment (DCE)   An environment defined by OSF (the Open Software Foundation consortium) providing a number of support elements for the creation of distributed (cooperative or client-server) applications.

Distributed Management Environment (DME) Management component defined by OSF as a companion to DCE (Distributed Computing Environment).

Distributed Program Link (DPL) An RPC-like mechanism used by CICS. (See Wipfler [1989] in the Bibliography.)

distributed processing Computer processing that occurs in a distributed manner within many different computer processors that are interconnected with some form of computer networking. See also **distributed application.** (See Wipfler [1989] in the Bibliography.)

distributed services Software services that are distributed across a network.

Distributed Transaction Processing (DTP) A type of CICS application programming. (See Wipfler [1989] in the Bibliography.)

DLC Data link control (SNA).

DME See **Distributed Management Environment.**

double-byte character set (DBCS) For support of languages comprising more than 255 characters and therefore requiring more than 1 byte to represent a single character.

DPL See **Distributed Program Link.**

DTP See **Distributed Transaction Processing.**

dynamic queue A message queue not defined in a static manner.

ECI See **External Call Interface.**

EDI See **electronic data interchange.**

Electronic Data Interchange (EDI) A set of defined formats used for interchanging information between companies.

enqueue To add an element (message) to a queue.

environment See **application environment.**

ESA Enterprise Systems Architecture.

exclusive access A type of access whereby only one program is allowed access to a message queue at any one time.

External Call Interface (ECI) The CICS ECI can be used by applications not written in the CICS style to invoke CICS transactions.

FAA See **Financial Application Architecture.**

FAP 1. Financial Application Program. 2. Formats and Protocols (SNA).

FEPI Front-end programming interface.

FIFO First-in first-out. A technique for processing a queue.

Financial Application Architecture (FAA) Developed by the Finance Industry as an architectural base for future application solutions.

flow pattern See **information flow pattern.**

formats   Patterns of data recognizable by programs.

fully qualified name   The most unambiguous form of a name accomplished by adding qualifiers (additional values) either preceding or following the original name.

function shipping   A feature of CICS that accomplishes transparent access to remotely located functions. (See Lamb [1993] or Wipfler [1989] in the Bibliography for a detailed discussion.)

GET   In message queuing, to either browse a message or remove a message from a queue. See also **browse** and **PUT.**

GUI   Graphical user interface.

handle   The identifier, or token, by which a program knows an MQSeries object or an instance of an MQSeries queue manager. See also **connection handle** and **object handle.** Also, the identifier, or token, by which the queue manager knows the program using the MQI.

helper code   Software that is usually called middleware.

heterogeneous   Environments that include unlike systems, perhaps including both IBM and non-IBM systems.

HLPI   High-level programming interface.

IAA   See **Insurance Application Architecture.**

IDL   See **NIDL.**

IMS   See **Information Management System.**

information flow pattern   The pattern (collection of paths and directions) by which information flows among and between programs (and people and devices).

Information Management System (IMS)   IBM's hierarchical database software product that manages information on databases for large numbers of terminals and workstations.

Information Technology (IT)   This usually includes all hardware, software, procedures, and practices to accomplish data processing using computers.

information unit   A discrete collection of data.

initiation queue   A special type of queue used by a queue manager to indicate that a trigger event has occurred.

instance   A single copy of a given program.

instantiation   The creation of a new object in a given class. The creation of multiple copies of a given program, when such becomes necessary or desirable.

Insurance Application Architecture (IAA)   Developed by the insurance industry as an architectural base for future application solutions.

interactive   Characterized by immediate interchanges between programs, persons, devices, and mixtures therefrom. For example, an interactive program implies continuous interchange between a person and the program. Contrast with **batch.**

interenterprise systems Systems that connect a company's data processing system(s) with those of one or more other enterprises. For example, a manufacturing company can benefit by interconnections with both suppliers and customers, as well as distributors.

interface A well-defined specification for identifying the relationship between two programs, where one program is requesting services of the other. The interface is usually defined by the program providing the services, whereby the types of services are identified, the method of requesting each service is described, and the possible responses are explained. For example, the MQI specification identifies the services that are available from the MQSeries products, how to request each service, and what possible responses to expect. That is, MQI specifies the relationship between the application program and the MQSeries product.

IP Internetworking Protocol. See **TCP/IP**.

ISC Intersystem Communication facilities (CICS). Communication between separate systems by means of SNA networking facilities or by means of the application-to-application facilities of VTAM. ISC links CICS systems and other systems, and may be used for communication between user applications, or for transparently executing CICS functions on a remote CICS system.

IT See **Information Technology**.

IWS Intelligent workstation. See also **PWS**.

JCL See **job control language.**

job control language (JCL) A control language used to identify a job to an MVS (or other) operating system and to describe the job's requirements.

join (flow pattern) The information flow pattern where one point collects information from multiple other points. See **split.**

lattice (flow pattern) A split operation followed by a join operation in an information flow pattern.

local Belonging to the queue manager to which an application program has connected using the MQCONN call; contrast with **remote.**

local environment The collection of programs and services within a single computer.

local queue A queue that belongs to the queue manager to which the application is connected. Contrast with **remote queue.**

local queue manager The queue manager in a local processor.

local queue name The name by which any queue is known in the local environment. Some of these queues are local; others are remote.

local resource name The name by which a resource is known in the local environment.

logical connection A connection between two programs, where the programs can be of any type and serve any purpose.

logical route A list of logical addresses (for example, SNA LUs).

logical unit (LU)   One instance of layers 6, 5, and 4 of the OSI or SNA model.

logical unit of recovery (LUR)   The term for updates performed by a transaction between two points of consistency. Synonymous with **unit of recovery.**

logical unit of work (LUW)   The CICS term for updates performed by a transaction between two points of consistency. It can span more than one LUR. Also, a sequence of processing actions (for example, database changes) that must be completed together. The collection of processing actions has ACID properties.

LU   Logical unit (SNA protocol between units): 1—display terminal, coaxial type, single display, 2—cluster controller/3270 data stream, 3—printer, 4—peer-to-peer (precursor to SNA), 5—host, 6.2—program-to-program communication (for example, APPC).

LU 6.2   IBM Systems Network Architecture's logical unit of type 6.2.

LUR   Logical unit of recovery.

LUW   Logical unit of work (or LUOW).

mail messaging   Messaging between human beings.

mail-enabled application   An application program that has access to mail-messaging services.

mainframe interactive (MFI)   A type of communication between nonprogrammable display terminals (for example, an IBM 3278 display station) and a mainframe computer, whereby the display terminal is controlled by a program (or set of programs) in the mainframe computer.

mapping   A substitution process through which one value is substituted for another (for example, one name for another or one address for another) or through which one value is associated with another value (for example, a name with an address or an address with a route). (See Hauzeur [1986] in the Bibliography for an excellent discussion of mapping.)

MCA   See **message channel agent.**

message-driven processing (MDP)   A software technology that provides support for distributed applications. It employs messaging as a fundamental means of communication. Applications can be created from independent asynchronous processes, with messages between these processes staged by means of queues.

message   In message-queuing applications, a string of bytes sent from one program to another program. In system programming, information intended for the terminal operator. A unit of information including a string of both control and data bytes.

message channel   A logical connection between queue managers.

message channel agent (MCA)   The part of a queue manager that controls and uses a message channel to adjacent queue managers. See **message mover.**

message channel program (MCP)   See **message channel agent.**

message descriptor   Data within a message that records the properties of a message.

message monitor A special program that monitors the flow of messages between programs such that it executes alternately with application programs (first the application program, then the message monitor, then another application program, and so forth).

message mover A program for "shuffling" queue elements (messages) among queue manager programs, usually across networks.

message priority An indicator that measures the relative importance of individual messages.

message properties Parameters that determine how the MQSeries queue managers process messages.

message queue Synonym for **queue.**

Message Queue Interface (MQI) IBM's queue-based communication programming interface that is part of the Open Blueprint. The MQI is supported by all of the IBM MQSeries products. (See IBM MQSeries [SC33-0850] in the Bibliography for complete details of the MQI syntax.)

Message Queue Manager (MQM) Synonym for **queue manager.**

message queuing A programming technique in which each program within an application communicates with the other programs by putting messages on queues and getting message from queues.

message routing Transferring a message across or between networks.

message traffic Collections of messages moving between queues and programs.

message unit One message.

messaging Intercommunication without connections. A method for one program in a network to communicate information with others without a logical connection between the programs. The communication can take place synchronously (the requester waits for function execution to be completed, then resumes) or asynchronously (the requester can proceed immediately without waiting). Asynchronous messaging is the default; synchronous messaging must be accomplished by application logic within the communicating programs and through the use of application level protocols. (See IBM MQSeries [GC33-0805] in the Bibliography.)

messaging and queuing Asynchronous communication between application programs that allows each program to operate independently of every other program by filling and empting queues in lieu of communicating across logical connections to other programs. (See IBM MQSeries [GC33-0805] in the Bibliography.)

MFI See **mainframe interactive.**

middleware That variety of software that is "helper code" between an application program and something more complex, such as operating system services, networking, file access, or device support. Messaging and queuing is communication middleware accessed by the MQI and provided by the MQSeries product family. (See King [1992] in the Bibliography.)

model queue A template that can be used to generate a dynamic queue.

mover   The component of an MQSeries queue manager that moves (sends and receives) messages between itself and another queue manager whenever messages are to be stored in a remote queue. See **message mover.**

MPTN   See **Multi-Protocol Transport Network.**

MQ   A shortened form of "messaging and queuing," which is the purpose of the MQSeries product family and the MQI.

MQI   See **Message Queue Interface.**

MQM   See **Message Queue Manager.**

MQSeries   The family name for the products that implement (support) the Message Queue Interface (MQI).

MRO   See **Multi-Region Operation.**

MSC   See **Multi-System Communication.**

Multiple Virtual Storage (MVS)   A multiprocessing operating system existing with IBM S/390 and S/370 hardware.

Multi-Protocol Transport Network (MPTN)   A networking architecture that allows one network to concurrently support the protocols of many other networks. (See IBM [GG24-4338] in the Bibliography.)

Multi-Region Operation (MRO)   (CICS interface for transaction routing.) CICS provides facilities such as SVCs (operating system supervisor calls) or the use of MVS cross-memory facilities to accomplish intercommunication. (See Wipfler [1989] in the Bibliography for a complete discussion of MRO.)

Multi-System Communication (MSC)   IMS support for transaction routing. (See Wipfler [1989] in the Bibliography.)

MVS   See **Multiple Virtual Storage.**

name   In general, a label used to identify an entity, such as a program, a device, a queue, or some other computer entity. Computer names are usually alphanumeric in representation and somewhat mnemonic (easy to remember), especially when they are specified in programming languages. They are usually thought of as being printable, displayable, and human readable. Names and addresses are closely related; a name implies an entity, while an address implies the location of that entity. (See Hauzeur [1986] in the Bibliography for an excellent discussion of name and address labels.) NOTE: *Program* names are not specified in the MQI; *queue* names are. Names can move and frequently can have one or more substitute (equivalent) values.

name resolution   The process by which one value of a name is substituted for another, less useful value.

name scope   The set of computers across which a particular name value is valid.

namespace   A registry of labels (name values) used for identifying computer entities, such as programs, queues, devices, and such. The actual collection is usually defined by some specification (for example, two to six alphanumeric characters with no imbedded blanks, allowing leading blanks but prohibiting trailing blanks).

network   A data communications network. New technologies are combining video, voice, and data traffic into a single, multimedia communication network.

Network Interface Description Language (NIDL)   Defined for OSF/DCE to provide a centralized definition of the resource manager interface. Used in creating RPC-based distributed programs. Usually shortened to **IDL.**

networking   The use of a communications network (or collection of networks) for communication between computers.

Networking Blueprint   An IBM-created diagram (first distributed in March 1992) that suggests how local computing environments are (and will be) related to networking software. The purpose of the diagram is to illustrate relationships of networking and application programming. Details are available in *Networking Blueprint Executive Overview* (IBM order number GC31-7057). "Messaging and Queuing" and "MQI" are components of the Networking Blueprint. See **Open Blueprint.**

NIDL   See **Network Interface Description Language.**

node   In a network, a point at which one or more functional units connect channels or data circuits.

NPT   Nonprogrammable terminal (formerly **MFI** or mainframe interactive).

NV/AS   NetView/Access Services.

object   An instance of a class. For example, a queue, a process, or a queue manager. (See Coad [1990] and Yourdon [1989] in the Bibliography.)

object code only (OCO)   Software shipped without source code.

object descriptor   A data structure that identifies a particular object. Included in the descriptor are the name of the object and the object type.

object handle   The identifier, or token, by which a program knows the MQSeries object with which it is working.

OCO   See **object code only.**

OLTP   See **Online Transaction Processing.**

one-way message   A type of message that requires no reply message. It flows in only one direction and exists by itself.

Online Transaction Processing (OLTP)   Transaction and time/event-based delivery functions (for example, routing, store and forward, decompose/recompose, diary). Online implies relatively immediate action; transaction processing implies processing a sequence of messages that constitute a transaction.

OOPS   Object-oriented programming system. (See Coad [1990] and Yourdon [1989] in the Bibliography.)

Open Blueprint   An IBM-created diagram (first distributed in April 1994) that depicts integrated application and communication environments where components can be selected from multiple vendors and multiple platforms using mixed protocols, formats, languages, and such. The Open Blueprint is a superset of the earlier Networking Blueprint. (See IBM [G326-0395] in the Bibliography for a complete discussion of the Blueprint and its value.)

Open Software Foundation (OSF)   A widely supported, industry-based consortium of which IBM is a founding member.

Open System Interconnection (OSI)   The computing/networking model defined and controlled by the International Standards Organization (ISO). (See Halsall [1988] in the Bibliography.)

OSF   See **Open Software Foundation.**

OSI   See **Open System Interconnection.**

parallel processing   Processing that overlaps within a time period or occurs concurrently.

peer-to-peer model   A communication model in which every communication program can have the same privileges, responsibilities, and roles as every other program. There are not necessarily any server-only or client-only programs. See **client-server model.**

persistence   The characteristic of a message that assures that it will be safe-stored in transit from one queue to another queue or safe-stored within a single queue.

persistent messages   Messages that are permanent and survive restarts of the queue manager.

physical route   A list of physical addresses (for example, link stations).

platform   A very general term that is used to reflect one or more aspects of an application environment, including operating system, database, networking, computer hardware, programming languages, and other application services.

priority message   A message identified with the highest priority value.

priority queue   A queue containing messages with the highest priority value.

process   An instance of the execution of a program.

process monitor   See **trigger monitor.** An application program that monitors the contents of one or more queues and creates or dispatches processes to service the messages on those queues. A trigger monitor may also cause processes to terminate when they are no longer required, change processes from serving one queue to another, and generate alerts in response to queue-depth thresholds or other critical criteria.

processor   The functional computer unit (computer hardware) that interprets and executes instructions.

program   A stored sequence of computer instructions that can be executed (by computer hardware or software) to accomplish a purpose.

protocol   A well-defined specification, usually recorded through a detailed written publication, for identifying the operating relationship between two programs where the two programs are similar and are providing the same set of services to other programs. Protocols are "rules" for exchanging information (control and otherwise) between the two similar programs, usually across a communications network between the programs. For example, MQSeries products interoperate with one another through protocols. That is, MQSeries protocols specify the relationship between two MQSeries products. Protocols are often shown as sequences of lines with arrow heads on both ends between programs arranged side by side. Protocols exist between like types of programs. See also **interface.**

PUT In message queuing, to place a message on a queue (local or remote). See **GET.**

PWS Programmable workstation.

queue Named object that applications can PUT messages on and GET messages from. It is owned and maintained by a queue manager. A software mechanism for saving messages. A repository for messages. A queue definition plus storage space for messages. (See IBM MQSeries [GC33-0805] in the Bibliography.)

queue alias See **alias queue.**

queue attribute A data value that indicates one of the properties of a queue.

queue-based communication Intercommunication through queues, in lieu of logical connections.

queue definition A queue that does not have storage space for messages but which identifies the characteristics of the storage space and messages therein.

queue discipline The procedure by which queue managers manage and control message queues.

queue element A single message within a message queue.

queue manager An MQSeries system program that provides queuing services to applications. It provides the MQI so that programs can access messages on the queues that the queue manager owns. Software that controls queue space usage. The MQSeries program that supports the MQI and executes PUT, GET, and other requests received from the application program. (See IBM MQSeries [GC33-0850] in the Bibliography.)

queue manager instance A single queue manager with a name that is unique within a network of queue managers. A particular copy of an MQSeries queue manager when multiple identical copies are allowed by the local operating system to be operational.

queue manager name The 48-character name of an MQSeries queue manager.

queue monitor A special application program that continuously monitors the state of one or more queues (for example, to check the number of messages in a queue or to verify that there is an active program associated with the queue).

queue name The label by which a message queue is known and accessed.

queue object A message queue.

queuing A technique for deferring delivery of a message when the destination or the route to the destination is busy or not available. It is this capability that allows asynchronous messaging to take place.

reason code A return code that describes the reason for the failure or partial success of an MQSeries call.

relay (flow pattern)   The information flow pattern that consists of receiving and then passing information.

remote   Not belonging to the queue manager instance to which an application program is connected. Contrast with **local.**

remote environment   The application environment on a remotely located computer.

Remote Procedure Call (RPC)   Defined within the OSF/DCE environment as a means of simplifying the development and maintenance of distributed systems. A call-return mechanism based on a two-element, closed chain information flow pattern.

remote queue   A queue that belongs to a queue manager that is not the queue manager to which the application is connected. Contrast with **local queue.** A message queue located in a remote processor.

remote queue manager   Queue manager in remote processor.

reply message   A type of message that is a reply to an earlier request message.

reply queue   The queue to which a reply message should be sent.

reply-to queue   The name of a queue to which the application that issued a PUT request wishes a reply message or a report message to be sent.

reply-to queue manager   The name of the queue manager that controls a queue to which the application that issued a PUT request wishes a reply message or a report message to be sent.

report message   A type of message that gives information about another message. A report message usually indicates that the original message cannot be processed correctly for some reason.

request message   A type of message that contains a request.

requesting program   The application program that sends a request message.

resolution   A substitution of one value for another, less useful value.

resolved queue name   The queue manager name plus the local queue name.

resource   A program, database, or other computer entity that is managed and accessed.

Resource Manager Interface (RMI)   An interface used to access application services (for example, DB2 and MQI) from a CICS application environment.

responding program   An application program that sends a reply message to a requesting program.

reversal transaction   A transaction designed to completely undo the results of a previous transaction.

RMI   See **Resource Manager Interface.**

roll back   Synonym for **back out.**

route   A list of addresses, logical or physical. (See Hauzeur [1986] in the Bibliography for an excellent discussion of routes, addresses, and names.)

routing   Transferring an information unit across or between networks. Delivery of a message to a target destination, often through a heterogeneous network.

routing queue   Queue containing elements (messages) that need to be routed elsewhere (or that have already been routed from elsewhere). See **transmission queue.**

RPC   See **Remote Procedure Call.**

RTR   A common abbreviation for *router*.

serial processing   A series of processing steps that must occur in a particular sequence without overlap.

server (program)   The companion to a client program that provides some service (for example, file access or printing) to a client program. (See Orfali [1994] in the Bibliography for a very thorough discussion of server programs and matching client programs.)

shared access   A type of access to a queue that permits programs to concurrently access a single queue.

single-phase commit   Method used within the MVS operating system by batch and TSO when an action that is in progress can be completed and all changes that are part of that action can be made.

SLU   Secondary logical unit.

sockets API   A connection-based communication API directly using the TCP/IP networks without benefit of communication middleware.

source queue   The queue name specified on the GET request.

split (flow pattern)   The information flow pattern where one point distributes information to multiple other points. See **join.**

SPM   See **Syncpoint Manager.**

SQL   See **Structured Query Language.**

stateful message/stateless process   Method by which all state table information is stored within a message and none is stored within any program. That is, programs process messages and then intentionally fail to store any information about that processing since the information is stored in the message instead. Most processes employ stateless message/stateful process, the converse of this method.

static attributes   Attributes that do not change or cannot be defined dynamically.

store-and-forward (S/F)   A transfer mechanism consisting of two stages whereby data is first stored for an indeterminate period of time and then later forwarded upon the stimulus of a command or the passage of time.

structures   See **information flow pattern.**

Structured Query Language (SQL)   A language for accessing data.

stub   See **stub code.**

stub code   Ancillary code that is linked into the caller and callee programs when using the Remote Procedure Call mechanism for program-to-program communication.

subsystem   In MVS, a secondary or subordinate system, usually capable of operating independently of, or asynchronously with, a controlling system.

synchronous   Having the quality of absolute dependence upon time, usually serially among parties two at a time.

synchronous messaging   A method for communication between programs in which the requestor waits for a reply before resuming its own processing.

synchronous operation   A simple mechanism to issue a request for a function to be executed. The application then resumes execution after the request has been completed.

synchronous processing   Serialized processing between programs.

syncpoint   An intermediate or end-point during processing of a transaction at which an update or modification to one or more of the transaction's protected resources is logically complete and error free.

Syncpoint Manager (CICS or IMS)   A third-party program that helps accomplish coordinated updating of databases among application programs.

syncpoint processing   A type of processing that accomplishes coordinated updating of databases among application programs.

target queue   The queue identified by the queue name specified on the MQPUT (or MQPUT1) request. The queue into which a message should be placed.

TCP/IP   Transmission Control Protocol/Internet Protocol. Two networking protocols used by the Internet network of networks. (See Comer [1988] in the Bibliography.)

time independence   That characteristic of online messaging and queuing that permits applications to operate without regard for the time during which other programs are executing.

time-independent processing   Asynchronous processing that does not depend upon time or time periods or concurrent availability of all processes.

time-sharing option (TSO)   An option on the MVS operating system; for System/370, the option provides interactive time sharing from remote terminals.

TOR   CICS Terminal-Owning Region component. See **AOR.** (See Wipfler [1989] in the Bibliography.)

TPF   See **Transaction Processing Facility.**

traffic patterns   Patterns by which messages flow among programs (and devices and people).

transaction (Tx or Trx or Trans or Xaction)   A bounded collection of messages with a definite start and end. The collection accomplishes some useful purpose such as updating a database, dispensing cash, or inquiring about the availability of seats on an airplane flight.

Transaction Processing Facility (TPF)   IBM operating system product developed for airline and other reservation/scheduling applications. It was previously known as the ALC, or airline control program.

Transaction Routing (TR)   Transferring a collection of messages from one point to another (or, alternatively, among many points). From an end user standpoint, a means of providing the user with access to several independent systems concurrently. The user sees one single point from which service is delivered and has only one logical session with the network. The transaction routing support takes care of the physical connections through the network, selecting alternate paths as required.

transfer (flow pattern)   The information flow pattern that moves information from one point to another.

transmission queue   A special queue used for automatically exchanging messages among MQSeries products (that is, among queue managers). A queue in which messages destined for a remote queue manager are placed. Queue containing elements that need to be routed elsewhere (or that have already been routed from elsewhere).

tree (flow pattern)   An information flow pattern composed of nested split operations.

trigger event   An event that causes a queue manager to put a trigger message on an initiation queue.

trigger message   A message that indicates that a trigger event has occurred.

trigger monitor   An application program that processes an initiation queue.

triggering   The process by which programs can be alerted that a message or a specific number of messages has arrived in a queue.

TSO   Time-sharing option subsystem facility within the MVS operating system that allows many terminal users to access the MVS operating system simultaneously.

two-phase commit   A coordinated update mechanism whereby one program plays the role of "coordinator" (a sort of orchestra leader) and one or more other programs play the role of "updater" (a sort of orchestra instrument player) using a two-phase (two-step) operation. In phase one, the coordinator program requests a positive indication from each of the updater programs that they are prepared for a coordinated update operation soon to follow. In phase two, after receiving a positive indication from all of the updater programs, the coordinator sends a signal to all of the updater programs to actually perform their portion of the aggregate of updating operations and send an indication of whether their updating was successful or not. If the coordinator detects that one or more updaters were unsuccessful in their updating operations, it sends a signal to all updaters to back out the just-performed updating (that is, revert to the previous version just before the updating). The procedure that allows a set of autonomous processes or agents eventually all to commit or roll back together. This helps to assure the correct execution of a logical unit of work.

UDP messaging   The sending of low-value, expendable datagrams across an Internet Protocol (IP) network through the use of the User Datagram Protocol (UDP). (See Comer [1991] in the Bibliography.)

undo   Synonym for **back out.**

unit of recovery (UOR)   Synonym for **logical unit of recovery.**

unit of work (UOW)   Synonym for **logical unit of work.**

UOW   See **unit of work.**

verb   That part of a programming interface specification that serves to identify a request and to specify parameters and parameter values. An abstract
term that is often implemented by a single macro (or programming CALL
structure) in a programming language.

VIM   Vendor-Independent Messaging. A mail-enabled interface that is being
sponsored by IBM and other vendors in a consortium for mail messaging.

virtual circuit (VC)   A logical connection across a network.

workflow computing   (As defined by Forrester): "Users, computers and networks adding and extracting value from information as it 'flows' through the
extended organization. Example: an insurance claim that is routed automatically between agent, adjuster, managers, accountants, and payables clerk.
Think of it as a 'Value River' with many 'tributaries'." With message-driven
processing, it is a form of distributed computing that networks several transactions together into a single 'workflow' or complex transaction. This allows
both the application programmer and end user to view these related transactions as a single entity, potentially improving productivity and reducing business cycle times.

workflow manager (WFM)   A control program that regulates and manages
information flow patterns among programs, persons, and devices.

Xaction   See **transaction.**

XRF   Extended recovery facility (continuous system availability).

XTI   X/Open Transport Interface. An interface defined by the X/Open consortium and used for accessing a transport network. See also **Multi-Protocol
Transport Network.**

X.25 messaging   Packet switching performed in an X.25 data network.

X.400 messaging   Mail-oriented messaging between persons at human
speeds, as defined by the OSI network model.

X.500 directory   A directory used by X.400 mail messaging.

Bibliography

This section contains three separate lists of publications related to this book. Each list is arranged in sequence by author name and title.

The first list contains references to computer and business trade press articles about messaging and queuing that appeared between the time IBM first announced the MQSeries product family in 1992 and the publication of this book in 1995. The growing interest in messaging and queuing and the wider discovery of its usefulness will ensure that articles such as these will continue to appear regularly in the computer and business trade press.

The second list contains references to textbooks and journal articles that are in some way relevant to messaging and to message queuing. Most of these have been explicitly referenced once or more within the book; the remaining references contain very relevant material.

The final list contains references to the IBM publications that cover the MQSeries product family and related topics. Though this list was correct at the time this book went to press, the introduction of new products inevitably means that it will become increasingly dated.

Trade Press Articles

Ambrosio, Johanna, 1993, "Middleware brings host of data to PC apps," *Computerworld,* December 13, p. 24.

"Apertus Sells Its Middleware to IBM's Networking Software Lab," 1994, *Computergram International,* September 20.

"The Application Is Dead: Long Live the Business Object," 1994, *Software Futures,* February 1, pp. 208–226.

Barker, Paul, 1994, "IBM gives CICS software new lease on life," *Computing Canada,* vol. 20, no. 12 (June 8), pp. 1, 8.

Barney, Doug, 1994, "IBM outlines WorkGroup blueprint," *Infoworld,* vol. 16, no. 48 (November 28), p. 55.

———, 1994, "IBM to showcase its workgroup strategy based on UltiMail app," *Infoworld,* vol. 16, no. 44, October 31, p. 5.

Baron, Talila, 1994, "IBM Intros Messaging Server, Apps Suite," *Communications Week,* Nov. 21, p. 4.

Brett, Charles C., 1992, "Messaging vs. RPC vs. Conversational; the 1990s debate," *OTM Spectrum Reports,* vol. 6, no. 4, pp. 34–41.

Brown, Bob, and Wayne Eckerson, 1993, "IBM at work on object-oriented work flow management system," *Network World,* vol. 10, no. 34 (August 23), p. 1.

Business Wire, 1993, "IBM Announces Enhanced Message Queuing Products," *World News Today,* September 16.

Business Wire, 1994, "Apertus Technologies Announces Database Integration and Unix Platform Support for the IBM MQSeries," *World News Today,* June 14.

Business Wire, 1994, "Early, Cloud, and Company Introduces CallFlow," *World News Today,* May 10.

Business Wire, 1994, "Early, Cloud, and Company Offers Service Enhancements," *World News Today,* June 6.

Center, Kathy, 1994, "A new way to think about business processes," *Tapping the Network Journal,* vol. 5, no. 1 (Spring), pp. 22–24.

Cole, Barb, 1995, "IBM licenses message oriented middleware technology to Hitachi," *Network World,* vol. 12, no. 4 (January 23), p. 52.

"Control with Message Queuing: Some users may prefer an alternative to TP monitors for managing transactions," 1994, *Information Week,* July 4, p. 66.

Cooney, Michael, 1994, "IBM plans big role for CICS, MQI," *Network World,* vol. 11, no. 20 (May 16), pp. 1, 66.

Cox, John, 1993, "New Tools Tighten DEC-IBM Links," *Communications Week,* August 16, p. 13.

———, 1994, "Middleware Tool Has Workflow Manager," *Communications Week,* January 5, p. 85.

Dawson, Mike, 1994, "Making distributed databases a reality," *Systems Management 3x1400,* vol. 22, no. 9 (September), pp. 60–66.

Dickman, Alan, 1994, "The RPC-vs.-Messaging Debate: Under the Covers," *Open Systems Today,* August 15, p. 58.

Dix, John, 1994, "Leaving a mark on DCE," *Network World,* April 4, pp. 33–38.

Dolgicer, Max, 1993, "The middleware muddle: Getting the message," *Data Communications,* vol. 22, no. 14 (October), pp. 33–34.

Dowding, P., 1994, "Who is using workflow software and for what?" *Information Management Technology (UK),* vol. 27, no. 2 (March), pp. 81–83.

Eckerson, Wayne, 1993, "Middleware company broadens its appeal," *Network World,* vol. 10, no. 2 (January 11), pp. 91–92.

Fogarty, Kevin, 1994, "IBM to buy up key piece of broad middleware plan," *Network World,* vol. 11, no. 36 (September 5), p. 6.

———, 1994, "IBM middleware beefs up workflow, replication," *Network World,* vol. 11, no. 30 (July 25), p. 40.

———, 1994, "IBM positions middleware at the heart of its groupware strategy," *Network World,* vol. 11, no. 40 (October 3), p. 48.

———, 1994, "IBM puts message-oriented middleware in the spotlight," *Network World,* vol. 11, no. 37 (September 12), p. 62.

Frye, Colleen, 1994, "Move to Workflow Provokes Business Process Scrutiny," *Software Magazine,* vol. 14, no. 4 (April), pp. 77–80.

Gaffin, Adam, 1993, "Vendor ships message-based middleware," *Network World,* vol. 10, no. 49 (December 6), p. 34.

———, 1994, "IBM, others join to promote key client/server technology," *Network World,* vol. 11, no. 30 (July 25), p. 11.

"Getting the Message," 1994, *Unix News,* June, p. 64.

Girishankar, Saroja, 1993, "IBM Adds DCE to Development Ware," *Communications Week,* August 9, p. 5.

———, 1993, "Masking Net Differences: Whether it's based on RPCs or messaging, middleware is evolving to give users uniform access to applications on multiplatform nets," *Communications Week,* November 22, p. 41.

———, 1994, "E-Mail Provides Transport for Workgroup Applications," *Communications Week,* June 13, p. 15.

———, 1994, "Enhanced MQSeries Aids in App Development," *Communications Week,* June 13, p. 15.

———, 1994, "IBM to Integrate MQ Middleware with DCE," *Communications Week,* May 9, p. 3.

———, 1994, "PeerLogic and IBM to Merge Middleware," *Communications Week,* September 12, p. 5.

Goldberg, Arthur P., 1992, "Message Queuing or RPC: Why Not Both?" *Communications Week,* April 20.

Guruge, Anura, 1993, "IBM's Networking Blueprint and AnyNet," *Business Communications Review: Internetworking Supplement,* August, pp. 43–47.

Higgins, Kelly Jackson, 1994, "Transaction Processing Moves to Users," *Open Systems Today,* September 12, p. 48.

"Hitachi, Getting Even Closer to IBM, Adopts MQSeries Messaging Technology," 1995, *Computergram International,* January 16.

"Hitachi to adopt IBM on-line communications software," 1995, *World News Today,* January 13, p. 1.

"Hitachi Will Develop Applications For IBM: MQS Network Messaging Software," 1995, *World News Today,* January 13.

Horwitt, Elisabeth, 1994, "IBM MQ series plays PeerLogic Pipes," *Computerworld,* vol. 28, no. 38 (September 19), p. 49.

———, 1994, "Middleware gaps closing," *Computerworld,* vol. 28, no. 28 (July 11), pp. 1, 14.

"How the Chicago Mercantile Exchange developed a realtime clearing system with message oriented middleware," 1994, *I / S Analyzer Case Studies,* vol. 33, no. 7 (July 1994), pp. 12–16.

"IBM and Hitachi agree on common messaging for computer communications," 1995, Business Wire, January 12.

"IBM & PeerLogic Products Can Combine," 1994, *Computerworld,* September 19, p. 49.

"IBM & PeerLogic To Integrate Middleware," 1994, *Communications Week,* September 12, p. 5.

"IBM & Systems Strategies Ink Devel & Mkt Pact," 1992, *Network World,* December 7, pp. 2, 74.

"IBM Enhances MQSeries Middleware," 1994, *Communications Week,* June 13, pp. 15, 21.

"IBM gets the team spirit with apps tools," 1994, *Network World,* September 26.

"IBM Has Groupware Strategy," 1994, *Computerworld,* October 31, p. 14.

"IBM Heralds CICS with Everything," 1993, *Network Week,* April 16.

"IBM Hursley Extends Capabilities of Newly-Acquired MQSeries," 1994, *Computergram International,* November 17.

"IBM Launches Its IBM WorkGroup Concept for Groupware," 1994, *Computergram International,* November 17.

"IBM MQSeries Tech To Be Used By Hitachi," 1995, *Software Industry Report,* January 23, p. 8, 9.

"IBM Outlines Workgroup Strategy," 1994, *Newsbytes News Network,* November 14.

"IBM positions middleware at the heart of its groupware strategy," 1994, *Network World,* Oct. 3, p. 48.

"IBM Rolls Out Networking Products," 1993, *Communications Week,* September 20, p. 87.

"IBM To Acquire Apertus Middleware Tech," 1994, *Network World,* September 9, p. 6.

"IBM To Enhance Middleware Tools," 1994, *Network World,* November 7, p. 2.

"IBM To Unveil Client/Server Product Line," 1993, *Network World,* March 29, pp. 1, 8.

"IBM Updates the AS/400," 1994, *Network Week,* July 8.

"Internetworking product guides: Network software," 1993, *Data Communications,* vol. 22, no. 15 (October 21), pp. 91–103.

King, Steven S., 1992, "Message-Delivery APIs: The Message Is the Medium," *Data Communications,* April, pp. 35–43.

———, 1992, "Middleware! Making the network safe for application software," *Data Communications,* March, pp. 58–66.

Korzeniowski, Paul, 1992, "Choosing a Path to Client-Server Apps," *Communications Week,* March 23, pp. 1, 53.

———, 1992, "Message Queuing Standards: So Far, Only Talk," *Communications Week,* May 18, pp. 1, 76.

Krill, Paul, 1994, "FlowMark to get DCE," *Open Systems Today,* March 21, p. 30

Levitt, Jason, 1994, "Choosing Cross-Platform Network APIs," *Open Systems Today,* February 21, p. 62.

"Making the Client-Server Connection," 1995, *Data Communications,* vol. 24, no. 2 (February), p. 17.

Mendler, Camille, "IBM Adds 'Messaging Middleware'," *Communications Week International,* May 30, p. 31.

"Middleware Providers," 1994, *Communications Week,* June 20, p. 49.

"Middleware To Go Mobile," 1995, *Computerworld,* January 23, p. 1.

"Migration or Bust," 1994, *Unix News,* June, p. 21.

"More IN Store from IBM," 1993, *Communications Week,* December 13, p. 79.

Moser, Karen D., 1992, "API Expands Cross-Platform Links," *PC Week,* March 9, p. 13.

"Moving Beyond RPC," 1992, *The SAA and Open Software Spectrum,* May, pp. 51–59.

Musthaler, Linda, 1994, "Start making sense: Getting different software elements to talk . . . ," *Communications Week,* February 7, p. 44.

Nelson-Rowe, Laurel, 1993, "MVS Linked to Client-Server Systems: Messaging Software Connects PC Users to Mainframe Apps," *Open Systems Today,* December 6, p. 24.

"Open OLTP in the 1990s," 1994, *Software Futures,* June 1.

"Openware Takes on IBM's MQ with Space, Shadow/Direct from Italy," 1994, *Computergram International,* Dec. 15.

"PeerLogic And IBM To Link Middleware Solutions," 1994, *World News Today,* September 7.

Skjellum, Anthony, Nathan E. Doss, and Kishore Viswanathan, 1994, "Inter-communicator Extensions to MPI in the MPIX (MPI eXtension) Library," Department of Computer Science & NSF Engineering Research Center for Computational Field Simulation, Mississippi State University, Miss. 39762, July 22. [Available through the Internet WWW at http:\\www.cs.msstate.edu/dist_computing/mpi-fqq.html.]

Stahl, Stephanie, and John Soat with Joseph E. Panettieri, 1993, "Mid dle [sic] Ware Saves Programmers and Costs," *Information Week,* November 1, pp. 62–68.

"Systems Strategies Bundle IBM Software," 1992, *Digital News & Review,* December 21, p. 3.

"Systems Strategists Extend Middleware Tool," 1993, *Network World,* January 11, pp. 91–92.

Vitaliano, Franco, 1993, "Get the message," *Computerworld,* vol. 27, no. 37 (September 13), p. 54.

Weissman, R. F. E., 1994, "Unleasing the power of client/server computing," *Object Magazine,* vol. 4, no. 1 (March/April), pp. 38–39.

Wilson, Tim, 1994, "IBM to Launch CICS Software," *Communications Week,* May 16, p. 7.

Books and Journal Articles

Bohl, Marilyn, 1984, *Information Processing,* 4th ed., Science Research Associates, Chicago.

Booch, Grady, 1994, *Object-oriented Analysis and Design with Applications,* Benjamin/Cummings, Redwood City, Calif.

Coad, Peter, and Edward Yourdon, 1991a, *Object-Oriented Analysis,* Prentice Hall, Englewood Cliffs, N.J.

———, 1991b, *Object-Oriented Design,* Prentice Hall, Englewood Cliffs, N.J.

Comer, Douglas E., 1991, *Internetworking with TCP/IP, 2d ed., Vol 1: Principles, protocols and architecture,* Prentice Hall, Englewood Cliffs, N.J.

Cypser, Rudy J., 1978, *Communications Architecture for Distributed Systems,* The Systems Programming Series, Addison-Wesley, Reading, Mass.

———, 1991, *Communications for Cooperating Systems OSI, SNA, and TCP/IP,* The Systems Programming Series, Addison-Wesley, Reading, Mass.

Deitel, Harvey M., 1984, *An Introduction to Operating Sytems,* Addison-Wesley, Reading, Mass.

Derfler, Jr., Frank J., and Les Freed, 1993, *How Networks Work,* Ziff-Davis Press, Emeryville, Calif.

Gane, C., and P. Sarson, 1979, *Structured Systems Analysis,* Prentice Hall, Englewood Cliffs, N.J.

Gray, Jim, and Andreas Reuter, 1993, *Transaction Processing: Concepts and Techniques,* Morgan Kaufmann, San Mateo, Calif.

Halsall, Fred, 1988, *Data Communications, Computer Networks and OSI,* Addison-Wesley, Avon, England.

Hauzeur, Bernard M., 1986, "A Model for Naming, Addressing, and Routing," *ACM Communications on Office Information Systems,* vol. 4, no. 4, pp. 293–311.

Horowitz, Ellis, and Sartaj Sahni, 1976, *Fundamentals of Data Structures,* Computer Science Press, Inc., Rockville, Md.

Jacobson, Ivar, Magnus Christerson, Patrik Jonsson, and Gunnar Overgaard, 1992, *Object-Oriented Software Engineering,* Addison-Wesley, Wokingham, England.
Kleinrock, Leonard, 1976, *Queueing Systems Volume 2: Computer Applications,* John Wiley, New York.
Lamb, Rob, 1993, *Cooperative Processing Using CICS,* 1993, McGraw-Hill, New York.
Lockhart, Harold W., 1994, *OSF DCE: Guide to Developing Distributed Applications,* J. Ranade Workstation Series, McGraw-Hill, New York.
Martin, James, and Kathleen Kavanagh Chapman, 1987, *SNA: IBM's Networking Solution,* Prentice Hall, Englewood Cliffs, N.J.
McMenamin, Stephen M., and John F. Palmer, 1984, *Essential Systems Analysis,* Yourdon, New York.
Meijer, Anton, 1987, *Systems Network Architecture: A Tutorial,* Pitman, London.
Meyer, Bertrand, 1988, *Object-oriented Software Construction,* Prentice Hall, Englewood Cliffs, N.J.
Orfali, Robert, and Dan Harkey, 1993, *Client/Server Survival Guide with OS/2,* Van Nostrand Reinhold, New York.
Orfali, Robert, Dan Harkey, and Jeri Edwards, 1994, *Essential Client/Server Survival Guide,* Van Nostrand Reinhold, New York.
Rosenberry, Ward, David Kenney, and Gerry Fisher, 1993, *Understanding DCE,* O'Reilly & Associates, Sebastopol, Calif.
Rumbaugh, James, Michael Blaha, William Premerlani, Frederick Eddy, and William Lorensen, 1991, *Object-Oriented Modeling and Design,* Prentice Hall, Englewood Cliffs, N.J.
Schwartz, Mischa, 1977, *Computer-Communication Network Design and Analysis,* Prentice Hall, Englewood Cliffs, N.J.
Shirley, John, 1992, *Guide to Writing DCE Applications,* O'Reilly & Associates, Sebastopol, Calif.
Sommerville, Ian, 1989, *Software Engineering,* 3d ed., Addison-Wesley, Wokingham, England.
Stevens, Wayne P., 1991, *Software Design: Concepts and Methods,* Prentice Hall, Englewood Cliffs, N.J.
Tanenbaum, Andrew, 1988, *Computer Networks,* Prentice Hall, Englewood Cliffs, N.J.
Walker, John Q., II, and Peter J. Schwaller, *CPI-C Programming in C: An Application Developer's Guide to APPC,* J. Ranade Workstation Series, McGraw-Hill, New York, 1994.
Wipfler, Arlene J., 1989, *Distributed Processing in the CICS Environment: A Guide to MRO/ISC,* McGraw-Hill, New York.
Yourdon, Edward, 1989, *Modern Structured Analysis,* Prentice Hall, Englewood Cliffs, N.J.
Ziegler, Kurt, 1991, *Distributed Computing and the Mainframe: Leveraging Your Investments,* John Wiley, New York.

IBM Publications

MQSeries: *AIX Version 2 Application Programming Reference,* SC33-1374.
MQSeries: *AIX Version 2 System Management Guide,* SC33-1373.
MQSeries: *Application Programming Guide,* SC33-0807.
MQSeries: *AT&T GIS UNIX Version 1 Release 1 User's Guide,* SC33-1437.
MQSeries: *Command Reference,* SC33-1369.
MQSeries: *Concepts and Architecture,* GC33-1141.
MQSeries: *Digital VMS VAX Version 1 Release 3.1 User's Guide,* SC33-1144.
MQSeries: *Distributed Queue Management Guide,* SC33-1139.
MQSeries: *An Early Look at Application Considerations Involved with MQSeries,* GG24-4469.
MQSeries: *Example of Using MQSeries,* GG24-4326.
MQSeries: *HP-UX Version 1 Release 3.1 User's Guide,* SC33-1376.
MQSeries: *An Introduction to Messaging and Queuing,* GC33-0805.
MQSeries: *Message Queue Interface, Technical Reference,* SC33-0850.
MQSeries: *MVS/ESA Version 1 Release 1.2 Application Programming Reference,* SC33-1212.

MQSeries: *MVS / ESA Version 1 Release 1.2 System Management Guide*, SC33-0806.
MQSeries: *OS / 2 Version 2 Installation and System Management Guide*, SC33-1371.
MQSeries: *OS / 2 Version 2 Release 3 Application Programming Reference*, SC33-1370.
MQSeries: *OS / 400 Version 1 Release 3.1 User's Guide*, SC33-1145.
MQSeries: *OS / 400 Version 2 Release 3 Administration Guide*, SC33-1352.
MQSeries: *OS / 400 Version 3 Release 1 Administration Guide*, SC33-1361.
MQSeries: *OS / 400 Version 2 Release 3 Application Programming Reference (C and COBOL)*, SC33-1334.
MQSeries: *OS / 400 Version 2 Release 3 Application Programming Reference (RPG)*, SC33-1199.
MQSeries: *OS / 400 Version 2 Release 3 Application Programming Reference (C and COBOL)*, SC33-1363.
MQSeries: *OS / 400 Version 3 Release 1 Application Programming Reference (RPG)*, SC33-1362.
MQSeries: *Planning Guide*, GC33-1349.
MQSeries: *Programmable Command Formats*, SC33-1228.
MQSeries: *Programmable System Management*, SC33-1482.
MQSeries: *SCO UNIX Version 1 Release 3.1 User's Guide*, SC33-1378.
MQSeries: *Sun Solaris Version 1 Release 1 User's Guide*, SC33-1439.
MQSeries: *SunOS Version 1 Release 3.1 User's Guide*, SC33-1377.
MQSeries: *Tandem Guardian Version 1 Release 3.1 User's Guide*, SC33-1146.
MQSeries: *UnixWare Version 1 Release 3.1 User's Guide*, SC33-1379.
MQSeries: *VSE / ESA Version 1 Release 3.1 User's Guide*, SC33-1142.

MQM MVS/ESA: *Application Programming Reference*, SC33-1212.
MQM MVS/ESA: *Licensed Program Specifications*, GC33-1350.
MQM MVS/ESA: *Messages & Codes*, SC33-0819.
MQM MVS/ESA: *Problem Determination Guide*, SC33-0808.
MQM MVS/ESA: *Program Directory*.
MQM MVS/ESA: *System Management Guide*, SC33-0806.

MQM/400: *Administration Guide*, SC33-1361.
MQM/400: *Application Programming Reference (C and COBOL)*, SC33-1363.
MQM/400: *Application Programming Reference (RPG)*, SC33-1362.
MQM/400: *Licensed Program Specifications*, GC33-1360.

ACF / TCAM Version 3 General Information, 1994, GC30-3235.
ACF / TCAM Version 3 Planning Guide, SC30-3240.
Advanced Communications Functions for TCAM Version 3, SC30-3240..
Distributed Application Environment, General Information, GG28-8049.
Introduction to Networking Technologies, GG24-4338.
Networking Blueprint: Executive Overview, SC31-7057.
Open Blueprint Introduction (White Paper), G326-0395.
Open Blueprint Technical Overview, GC23-3808.
Open Networking Reference: A Business and Technical Perspective, GG24-4110.
OSF / DCE Application Development Reference, 1993, SR28-4995.
OSF / DCE User's Guide and Reference, 1993, SR28-4992.
Transaction Processing: Concepts and Products, GC33-0754.

Index

1-way (open) flow type, 60, 179, 360–361, 384, 436
2-way (closed) flow type, 60–61, 180
20-questions examples, 27–31, 39–43

abort, presumed, 215
absorption point, 53, 76
abstract syntax notation version one, 209
access:
exclusive, 381, 430
information, 83, 87
program 47–48, 74
queue, 155, 381–384
shared, 48–49, 129, 440
access control list, 203
access security, 4
account management, 226
accounting management, 226
ACID properties, 7–8, 425
ACL (see access control list)
adaptation:
availability, 27, 142, 364–365
time, 39, 142, 364–365
adaptive bridge, 194
adaptor (or adapter) (see adaptation)
added value, MQSeries products', 163–167
address, 425
Internet, 187–188, 196, 198
logical, 198
network, 198–199
physical, 198
qualified, 70
queue manager, 438
resolved, 172–173, 199
unqualified, 70, 431
unresolved, 172–173, 191, 439
address resolution, 172–173, 199
Address Resolution Protocol, 199, 210

address space, 8, 77, 425
address-to-address mapping, 433
Advanced Interactive Executive (AIX), 388–389
Advanced Program-to-Program Communications, 137, 160–161, 426
advantages of queue, 44–45
aged-out message, 313–314
(See also expiry)
agent code, 425
AIX (see Advanced Interactive Executive)
AIX 3.2, 395
alias mapping, 192, 293, 425
alias name, 192, 293, 425
alias queue, 192, 293, 438
alias queue definition, 192, 293, 426
alias queue name, 192, 293, 425
answering machine, 18–19, 46, 404–405
AOR (see application-owning region)
Apertus Technologies, Inc., 386
API (see application programming interface)
APPC (see Advanced Program-to-Program Communications)
application, 54, 83–91, 111–113, 426
code, 426
data, 10, 32, 55, 86, 426
definition of, 54, 94–95, 111
development tools, 87–88, 90, 135
distributed, 94–97, 112–113, 185, 429
element, 186
enabler, 83–84, 91–92
enabling services, 135–137
formats, 10, 32, 55
function, 170
heritage, 277
incorporating existing, 276–278
information flow, 96–97, 111–113
legacy, 277

ABOUT THE AUTHORS

BURNIE BLAKELEY was the chief architect of the MQI specification and assisted in the development of the IBM MQSeries family of messaging and queuing products. Earlier, he helped develop the Advanced Program-to-Program Communication (APPC) interface for SNA LU 6.2 and edited the *APPC Transaction Programmer's Reference Manual*. He now works with multimedia, high-speed networking architectures, and interfaces.

HARRY HARRIS was the development manager responsible for implementing the MQI on the MVS/ESA operating system, the first product of the IBM MQSeries product family. He now provides consulting services to MQSeries customers in the United Kingdom and around the world.

RHYS LEWIS was the senior developer in the team that implemented the IBM MQSeries product family on the AIX/UNIX and OS/2 operating systems. Earlier, he was the author of *Practical Digital Image Processing*. He now is involved in extending the range of MQSeries products to many more operating systems and networking protocols.